Construction Engineering in Underground Coal Mines

Scott G. Britton
*Practicing Mining Engineer
Collegeville, PA*

Published by
Society of Mining Engineers
of the
American Institute of Mining, Metallurgical, and Petroleum Engineers, Inc.
New York, New York • 1983

The Institution of Mining and Metallurgy

Copyright © 1983 by the
American Institute of Mining, Metallurgical, and Petroleum Engineers, Inc.

Printed in the United States of America
by Guinn Printing, Inc., Hoboken, New Jersey

All rights reserved. This book, or parts thereof, may not be reproduced in any form without permission of the publisher.

Library of Congress Catalog Card Number 82-071994
ISBN 0-89520-403-7

This book is dedicated to my grandfather, Kenneth E. Holley (1903–1978)—a practical ceramic engineer and his daughter, Carolyn Holley Britton, my mother, who allowed me to build treehouses and dig holes unabated.

Preface

This book was written to provide a guide to underground construction projects commonly used in coal mining. No underground mine can operate without transportation, ventilation, drainage, and haulage systems or the management to direct their construction. Yet, there exists no single source to which an engineer or supervisor can turn to gain the basic steps for completing the required task.

This book, therefore, discusses all the construction engineering aspects for *necessary* coal mine projects. It looks at materials, labor, design alternatives, site characteristics, installation time, costs, and management of each project. I gathered much of the information while working as an engineer and underground construction foreman and the rest has been graciously donated by equipment manufacturers.

This book is intended for both practicing and student engineers of all disciplines, however, the mining engineer will find it most useful. It is a basic text, which supplements the information found in *Elements of Practical Coal Mining* and the *SME Mining Engineering Handbook*.

I would like to thank my wife, Nancy Britton, for her support and coordination of the manuscript preparation. I would also like to thank those manufacturers who freely gave much of the technical data to make this text very useful.

Scott G. Britton

Collegeville, PA

Table of Contents

Preface v
Introduction xi

Section 1. Practical Construction Engineering Techniques 1

Chapter

1 General Construction Parameters 3
  Geologic Parameters 3
    Tectogenic Parameters 3
    Lithology Parameters 9
  Physical Parameters 11
    Mining Height 11
    Gas 12
    Water 13
    Humidity and Weathering 13
    Seam Pitch 14
  Regulatory Parameters 15
    Ventilation Projects 16
    Drainage Projects 19
    Transportation Projects 19
    Haulage Projects 20
  Mining Operational Parameters 21
    Safety Parameters 21
    Cost Parameters 22
    Supply Parameters 22
    Labor Parameters 22
    Project Material Parameters 22
    Planning and Management Parameters 23
  Conclusions 23
  References and Bibliography 23

2 Transportation—Railroad Construction 25
  Railroad Design Factors 25
    Maximum Transported Loads 25
    Clearance 25
    Railcar and Traffic Parameters 26
    Mining Entry Dimensions 26
    Roadbed 26
    Locomotion 27
    Maintenance Scheduling 27
    Rail Weight and Tie Selection 28
    Precautions for Transporting Equipment 31
  Laying Track 32
    Preparation 32

Assembly	34
Maintenance	39
Installing Turnouts	41
Turnout Preparation	46
Turnout Installation	48
Installing Trolley Wire	54
Preparation	54
Assembly	56
Cutting a Trolley Switch	56
Maintenance	58
References and Bibliography	59

3 Ventilation Construction — 63

Building Overcasts	63
Alternative Designs	64
Installing the Overcast	72
Maintaining and Inspecting the Overcast	81
Constructing A Regulator	82
Designing the Regulator	83
Building the Regulator	84
Maintaining the Regulator	85
Constructing A Stopping	86
Permanent Stoppings	86
Temporary Stoppings	89
Construction of a Door	91
Man Doors	91
Full-entry Doors	94
References and Bibliography	96

4 Pump Stations, Sumps, and Drainage Systems — 99

Analyzing and Designing Effective Underground Drainage Systems	99
Inputs to the Mine Environment	99
Outputs from the Mine Environment	101
Engineering the System	102
Construction of the Main Pumping Station	113
Preparing the Site	114
Installing the Main Pump	118
Operation and Maintenance	121
Construction of a Portable Pumping Station	122
Designing for Inputs	123
Operation and Maintenance	123
References and Bibliography	124

5 Belt Drives, Takeups, and Transfer Points — 131

Designing for Efficient Systems	133
Analyzing Belt Spillage	134
Alternative Installation Aspects for Drives, Takeups, and Transfer Points	135
Installing Belt Drives, Head Rollers, and Takeups	144
Preparation	150
Installing the Belt Drive and Takeup	153

Installing the Suspended Head Roller	157
Operation and Maintenance	158
Installing the Transfer Points	158
Design Parameters Affecting Transfer Point Performance	159
Building the Transfer Point	160
Operation and Maintenance	161
References and Bibliography	162

6 Belt Line Construction 163

Analyzing the Parameters of Effective Belt Line Construction	163
Rope, Structure, and Belt Integrity	163
Belt Line Alignment	164
Loading and Dumping Interaction	164
Materials, Labor, and Equipment for Construction	165
Installing a Belt Line Extension	171
Preparation	172
Belt Line Assembly	174
Tailpiece Assembly	177
Operation and Maintenance	179
References and Bibliography	182

7 Surge Bins, Shops, and Special Projects 197

Analysis for Effective Performance	198
Parameters to Consider	198
Materials, Labor, and Equipment	199
Installing a Surge Bin	199
Preparation	201
Assembly	204
Operation and Maintenance	204
Installing an Underground Repair Shop	205
Preparation	205
Assembly	208
Maintenance	210
Special Projects	210
Installing a Rotary Dump	210
Installing the Underground Crusher Station	216
Building an Underground Foreman/Field Engineering Office	218
References and Bibliography	220

Section 2. Principles of Construction Engineering Management 223

8 Construction Project Management 225

Organization, Communications, and Responsibilities of Management	225
Construction Management Organization	225
Communications for Project Management	227
Responsibilities of Management	228
Analysis of Worker Motivation	234
Supervisory Leadership	237
Leadership and Coordination	237

	The Directing Function of Management	237
	References and Bibliography	239

9 Design Tools for Construction Engineering — 241

Computer Aids	241
Traditional Mining Software	241
CPM Uses for Mine Construction Scheduling	243
Minicomputer Software for Mine Construction	252
References and Bibliography	253

10 Cost Analysis of Underground Support Construction — 259

Developing Cost Accounting Techniques	259
The Nature of Project Costs	260
Allocating Direct and Indirect Project Costs	261
Capital Investment Decisions for Support Projects	263
The Cost Control Process	268
Cost Estimating for Support Construction	270
General Approach to Cost Estimating	270
Transportation	272
Ventilation	274
Drainage	275
Haulage	276
Mine Support Projects as a System	279
Summary	279
References and Bibliography	280

Appendix — 283
Index — 309

Introduction

Every underground coal mine requires support systems to make coal production possible. These support systems use construction engineering principles and practices for successfully completing the project.

Traditional approaches to construction projects rely heavily on the coal mine supervisor having knowledge of the required steps. This knowledge, in past years, was gained from on-the-job training. This training system was sufficient only because much of the everyday construction work was simply designed and completed. The traditional approaches also employed the neighborhood principle: i.e., rely on the past work of other companies in similar mining conditions for the right approach. In other words, what worked for them should work for you. Unfortunately, this is not always the case.

Construction projects in today's modern coal mines are complex. Projects such as laying track, building overcasts, and installing belt lines are now being engineered to increase the projects' useful life, lower the operating and maintenance costs, and maintain an efficient and effective mining complex.

The justification for this text is two-fold. First, the information on current construction engineering techniques as applied to post 1969 coal mines are scattered and generally unavailable to the average mine engineer. Because of this, there exists no widely recognized source specifically discussing construction engineering techniques for everyday underground construction projects.

Secondly, as mining operators become more cost conscious, there will be an effort to reduce unnecessary waste and miscellaneous costs. By designing and installing the most efficient support projects, mine management will maximize the usefulness of the project and minimize the cost of the labor, material, and time needed for construction.

The objective, therefore, is to provide a basic applications handbook for practical construction projects in underground coal mines. The handbook would serve as a guide to completing necessary construction projects by illustrating effective techniques now used in the coal industry.

In meeting this objective, the text considers the basic requirements of five typical construction areas, namely: (1) track, (2) ventilation, (3) pumps and drainage systems, (4) belt lines, belt drives, takeups and transfer points, and (5) surge bins, storage areas, shops, and special projects.

Each chapter focuses on one aspect of these project areas from the viewpoint of the construction supervisor or mining engineer. By discussing each project in this manner, the reader will better understand the essential elements necessary to complete the task.

The book is divided into two main sections. The first section, Chapters 1 through 7, thoroughly discusses the techniques of completing each type of construction project. Included in the discussion are design alternatives, estimated material needs and associate costs, the required time for completion, the necessary manpower, and a step-by-step guide to finishing the task from beginning to end. The second section of the book, Chapters 8 through 10, discusses such items as construction management, engineering design tools, computer techniques, and cost analysis decisions for underground construction engineering.

Section 1
PRACTICAL CONSTRUCTION ENGINEERING TECHNIQUES

The first section of this text delves into construction engineering techniques for successfully completing necessary underground production support projects. I will discuss alternative design concepts, installation practices, materials and material costs, completion time, and necessary labor for individual projects. The emphasis of this section is on describing each project thoroughly and providing a practical guide for the supervisor or engineer.

There are four main construction categories for typical underground mining projects. These are:

1) *Transportation* of men and supply materials. This refers to track and wire installation and roadbed grading and ballasting.

2) *Ventilation,* composed of overcasts, man doors, regulators, and stoppings.

3) *Drainage,* including sumps, pumping stations, drainage networks, pumps, and face drainage.

4) *Haulage* of coal and rock from the face to the outside. The focus here is in the installation of belt lines, belt drives, transfer points, and vibrating bins/feeders.

The first section shall fully discuss these four categories in regard to the construction engineer or supervisor designing, building, or improving a system. In addition, special projects such as surge bins, rotary dumps, and underground offices not classified under these categories will be discussed.

For manageable purposes, I have limited the installation discussions to one or two approaches I feel have merit. The variation of installations in coal mining is endless, each with its own advantages and disadvantages. Therefore, the reader should not dismiss or restrict his thinking to only the approaches in the text. It serves as a *guide* only.

1

General Construction Parameters

This chapter discusses the basic underground mine environment parameters one must consider when analyzing the site of potential construction projects. By studying the mine site parameters influencing the design, engineering, and building of any underground structure, the goal of maintaining effective and efficient mine systems is possible.

There are a number of parameters influencing underground construction. These factors are grouped into four major categories, namely: geologic, physical, regulatory, and design parameters.

Each category plays a decisive part in determining the type or design of the construction project. What follows is a discussion of each category and its impact on any underground project.

GEOLOGIC PARAMETERS

To begin with, coalbeds are formed from the general geologic process illustrated in Fig. 1. Therefore, the geologic parameters influencing construction engineering in underground coal mining are a combination of *tectogenic stresses* and *inherent strata characteristics*. Tectogenic processes refer to the formation of faults, folds, joints, and rock cleavage. Strata characteristics refer to a rock's lithology and engineering properties. Both parameters affect the underground environment chosen for a construction project. Weak or brittle roof rock, jointing or other planes of weakness, or faulting within a coal seam horizon will directly influence an engineer's decision as to project design.

Tectogenic Parameters

When an underground coal mining operation encounters a warp in the seam horizon, it is due in part to the process of tectogenesis. Because coal mining is expanding into deeper, less accessible seams, as well as seams with greater strata deformation, an understanding of the stresses and forces present is necessary to take advantage of, or design against, local conditions.

There are six types of strata deforming features commonly encountered in coal mining. These are folds, joints, faults, unconformities, intrusions, and partings or splits.

Folds: Folds are bends in the strata of the seam horizon (Fig. 2). The major features of folds are the *hinge line* (axis) and the *limbs* of the fold. The hinge line is the line of maximum angle of the folded strata. The limbs of the fold refer to either side of the fold, and they extend in either direction away from the hinge line. A third term, *plunge,* refers to the angle at which the fold is leaning away from the horizontal.

In relatively flat-bedded deposits typical of bituminous coal strata, three types of folds are common: monocline, structural terrace, and drag folds.

Monocline folding is characterized in relatively flat strata by the sudden change of dip in the strata, ranging from a few degrees to even 90° (Fig. 3). Often these monocline folds are termed rolls in coal mining terminology. Monocline folds can be troublesome during roadbed grading or overcast installation.

Structural terrace folding (Fig. 4) occurs in gently dipping seams at places where the seam assumes a horizontal position. The transition to a terrace fold in coal seams goes relatively unnoticed by mine operators because of their unobtrusive nature, but terrace folding does indicate areas of minor strata disturbance.

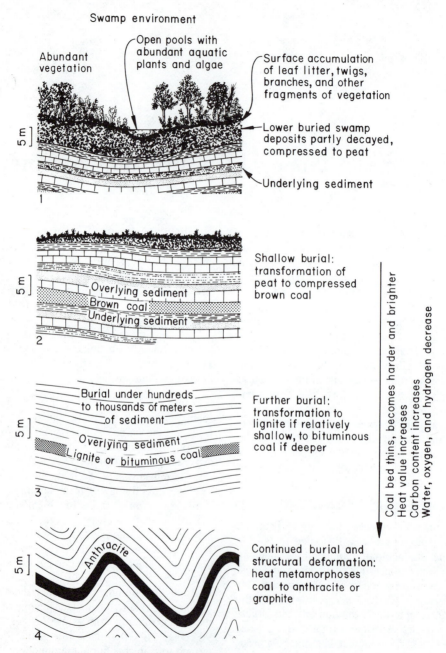

FIG. 1. The process by which coal beds form begins with the deposition of vegetation. Protected from complete decay and oxidation in a swamp environment, the deposit is later buried and subjected to mild metamorphism, which transforms it into lignite, bituminous coal, or anthracite, depending on depth of burial, temperature, and amount of structural deformation (Press and Siever, 1978).

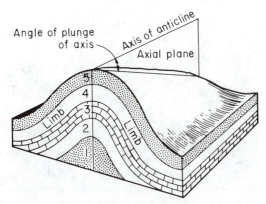

FIG. 2. Diagrammatic illustration of the parts of a fold (Press and Siever, 1978).

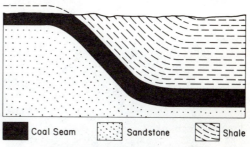

FIG. 3. Typical monocline folding.

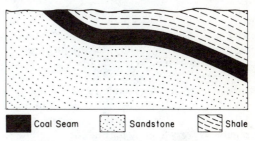

FIG. 4. Structural terrace folding.

Drag folds are formed in the coal seam or roof/floor strata when a strong bed slides past a relatively weak bed. This sliding results in shear forces which buckle the strata in the weak bed (Fig. 5). In coal mining operations, this is usually seen when competent sandstone or limestone roof has moved in relation to the coal seam, causing drag folding to appear in the coal seam.

In the anthracite coal fields of eastern Pennsylvania, folding is a more prominent feature. In these coal fields, the typical fold is termed disharmonic. It is exemplified when the limbs of the fold are not uniform throughout the stratigraphic column (Fig. 6). Construction engineering techniques in these areas can be unusually difficult due to these distorting fold structures.

The construction engineer designing an underground structure near evidence of folding must remember several points. One, folding represents the condition of "premining" ground stresses present in the strata. Two, anchoring into prestressed competent strata in the postmining state may result in subsequent failure from induced shear forces. Finally, folding also creates further strata fracturing to some degree, and can be a sign of additional weakness planes within the coal horizon.

Joints: Joints are fractures weakening the competence of rock strata. Joints are commonly found in all three mining horizons—roof, floor, and seam. In coal mining, joints in the coal itself are referred to as cleat (Fig. 7). "Face" cleats are joints running parallel to the maximum compressive stresses while "butt" cleats are joints found normal to the maximum compressive stress.

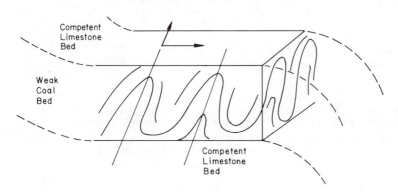

FIG. 5. Drag folds forming in weak bed. In this case, it is coal acting as the weak bed.

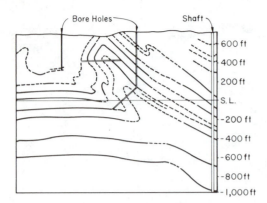

FIG. 6. Cross-sectional view of major folding in coal strata (Darton, 1940). Metric equivalent: ft × 0.3048 = m.

FIG. 7. Prominent face cleat and butt cleat in coal (Crickmer and Zegeer, 1981).

Joints are the result of strata rupturing due to a stress-strain force exerted on it during folding. Jointing causing collapse of roof strata during mining is very common (Fig. 8). Therefore, engineering personnel should chart all of the joint occurrences known about the coal property during mining.

It would be presumptuous to assume all cleats are a sign of major faults or folding, even though joints occur around and near such deformation. Mine operators encountering face cleats during production may have a good, but not certain, indication of strata disturbance in the seam horizon. Construction engineers designing projects such as sumps, overcasts, or surge bins in heavily jointed areas should carefully analyze the patterns and joint occurrence before deciding on one design approach or project orientation.

Faults: Faulting in coal seams is defined as strata fracturing where movement has occurred. During coal mine production or construction work, faults tend to be a nuisance and generally unsafe, due to the fracturing of the strata on either side of the fault.

There are several terms describing a fault, namely, *hanging wall, footwall, dip,* and *strike* of a fault. The hanging and footwall terms refer to the individual strata on either side of the fault, depending upon the relative position of the strata (Fig. 9). The strike is the direction at which the fault runs in the horizontal plane, while the dip is the angle between the horizontal plane and the plane of the fault.

Faulted strata usually makes mining difficult. Construction projects such as belt lines, railroads, or extensive ventilation networks in faulted

FIG. 8. Shale roof has fallen because of fractures (Crickmer and Zegeer, 1981).

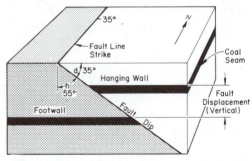

FIG. 9. Normal fault showing the hanging wall, footwall, dip, and strike features.

ground must be designed with additional reinforcement. Greater safety measures are necessary to protect men and equipment during construction and subsequent mine use. Further considerations for designing in faulted ground depends upon the lithology of the rock. For example, clastic sandstone rock behaves differently than nonclastic mudstone. In addition,

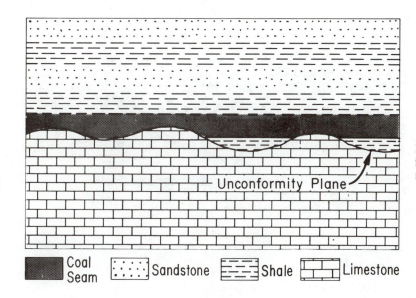

FIG. 10. Typical unconformity found in sedimentary strata.

weathering complicates faulting problems by increasing strata creeping in the fractured rock. Further discussion on faulting consideration for construction projects follows in later sections.

Unconformities: An unconformity in sedimentary strata is the eroded surface or nondepositional plane separating older and younger rock formations (Fig. 10). Although there are a great number of types of unconformities, depending upon rock types and tectonic history, coal mining encounters unconformities in only one instance. This case is best illustrated when a coal seam gradually thins into the roof or floor at areas where no faulting of strata is evident.

The concern for unconformities in mining construction centers around the presence of the new rock strata for construction purposes. Engineers and supervisors must take into account a greater portion of roof strata above the seam or floor strata below the seam when anchoring into these strata for support. It helps if unconformities are thought of as planes of weakness in the coal seam horizon, and their presence may create additional problem areas during development and construction.

Intrusions: Intrusions are interruptions of the coal seam horizon due to events after the formation of the coal deposit. The diapiric nature of these interruptions occurs in two major forms, *depositional* and *injective*.

Depositional intrusions include coal seam interruptions from stream and glacial erosion (Fig. 11). Such formations are commonly referred to as *washouts* or *horsebacks*. During mining, these areas are characterized by weak ground and variable strata changes, making construction projects such as overcasts, belt lines, or transfer points difficult to design without some installation flexibility built in.

Injective intrusions include clastic dikes and veins. These structures (Fig. 12) tear into and through the coal horizon causing fractured areas and producing premining stresses of a local nature. Construction projects in areas of dikes or clay veins are often support projects with a long life and may need reinforcement on a large scale, depending upon the size of the intrusion. Extensive cribbing or roof truss supports may be necessary for safe mining.

Partings and Splits: Partings and splits in the coal seam are due to the depositional variations of the strata formation. Intermittent flooding of ancient rivers deposited the parting material, usually shale, in the coal-forming basin. This causes the phenomenon of splitting (Fig. 13). If the split becomes thick enough, this parting is referred to as a middleman.

Although partings are a nuisance to mining production and construction, mining engineers are mainly concerned with partings in highly fractured ground and in areas where partings form part of the roof or floor, since ground problems may intensify during mining. Anchorage in parting material can also be a problem area.

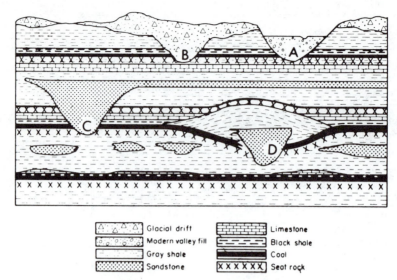

FIG. 11. Some features affecting continuity of coals. Coal removed by modern stream erosion at A; preglacial erosion at B; by a stream after coal deposition at C; and at D, the stream was present throughout the time of peat accumulation (Crickmer and Zegeer, 1981).

This concludes tectogenic parameters. Several factors need to be stressed, and they are:

1) The six tectogenic features affect coal mine construction in direct proportion to their occurrence within the coal seam horizon.

2) Folding, jointing, and faulting features tend to affect the project design more than unconformities, intrusions, or partings.

3) Disturbed ground requires a better, more flexible project design to ensure safety and effective performance.

It should be remembered that full consideration of tectogenic parameters usually occurs during initial design and premining stages. Only special support projects (surge bins, rotary dumps, and so on) may require a second local analysis of the conditions to better estimate the construction requirements.

The second major category of geologic parameters deals with the lithology of the coal seam horizon, including immediate roof and floor strata.

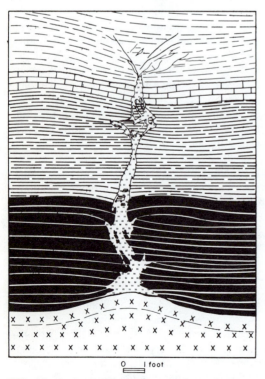

FIG. 12. A clastic dike interrupts the coal and overlying strata (Crickmer and Zegeer, 1981).

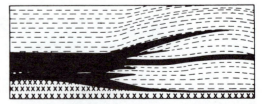

FIG. 13. Splitting of a coal seam (Crickmer and Zegeer, 1981).

Lithology Parameters

It is a well recognized fact of geologic engineering that rock lithology influences the behavior of the rock to mining-induced stresses. When a construction supervisor or mining engineer recognizes the strata characteristics he is working with, he can take a better approach to safe effective design and construction. In recognizing the appropriate rock strata, the engineer or supervisor should understand some of the basic features of the four major rock types found in coal mining. They are: (1) sandstones; (2) siltstones, shales, and argillites; (3) limestones and dolomites; and (4) conglomerates and breccias.

Other less common sedimentary types are sediments and evaporites. Since sediments and evaporites are not usually encountered, they will not be discussed. Each major rock type and its general behavior under mining-induced stresses will be discussed.

Sandstones: In the majority of coal mining circles, a sandstone roof or floor rock is considered to be a competent construction material. When mixed with shales or as a cementing base for a conglomerate type of rock, sandstone can be a treacherous material making it difficult to design any competent anchorage system for mine structures.

There are several common minerals making up most sandstones. Quartz and chert, micas, feldspars, and other fine rock particles are present in different percentages with the cementing material. A variety of sandstones are possible in coal horizon strata, prompting a brief review of its characteristics. This review takes the form of an engineering analysis of sandstone as a construction material, based on US Bureau of Mines (USBM) investigations. Table 1 summarizes the characteristics of sandstone. The extreme variance between these engineering characteristics illustrate the wide range of com-

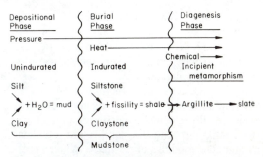

FIG. 14. The forming of shales, siltstones, and argillites (modified from Jackson, 1970). Metric equivalent: ft × 0.3048 = m.

petency a sandstone may exhibit at any given mining situation. Sandstone has a relatively high shear strength for sedimentary rock, and this coupled with its low tangent of the angle of internal friction testifies to the fact it is usually more competent than other common rock types.

The construction supervisor or engineer encountering a sandstone roof or floor may face a variety of rock characteristics which can aid or hinder a mine project. Anchorage into the sandstone strata should be investigated before long-life mine projects are installed, or when structures with high vibration (i.e., belt lines, feeders, crushers, etc.) are installed. However, the rule of thumb states sandstone is a better-than-average rock to work with.

Siltstones, Shales, and Argillites: Siltstone is the unindurated product of river silt and mud. Shale is the indurated product of siltstone and mudstone, while argillite is the partially metamorphosed state of shale. The process, first illustrated by Twenhofel (1937), is shown in Fig. 14. Argillites will not be covered because of their relatively infrequent occurrence in sedimentary rock proper.

Shale is a laminated rock that easily splits along bedding planes. USBM compressive strength tests show shale varies between 230

TABLE 1. Engineering Characteristics of Sandstones

Type	Compressive Strength, psi		Dynamic Modulus of Elasticity, psi		Modulus of Rupture, psi		Shear Strength, psi	Tangent of Internal Friction Angle
	Max	Min.	Max	Min.	Max	Min.		
Sandstone	34,000	5,000	8.0 × 10⁶	0.8 × 10⁶	3,600 (13,000 avg.)	600	2,450	7°

Metric equivalent: psi × 6.894 757 = kPa.

MPa (33,500 psi) and 71 MPa (10,300 psi) with an average of 75 MPa (11,000 psi). Siltstone averages 27 MPa (4000 psi) for compressive strength. Table 2 summarizes the static properties of siltstone and shale.

As an engineer or construction supervisor reviews design plans for projects constructed in shale or siltstone rock, the two important parameters controlling the stability are the shear strength and modulus of elasticity. When a project exceeds these limits, there will be an increased likelihood of failure. A successful design will keep within the proven limits of stress and strain of the rock. Projects such as belt drives, head rollers (suspended), or other anchor-point structures are particularly susceptible. Since shale is quite a variable strata, local rock conditions in the project site area should be known before determining maximum point loading of such structures. With this inspection of the local site, the engineer or supervisor will gain more insight into the parameters influencing the project design.

One rock type not included in this discussion is a variety of siltstone called fire clay. Fire clay is chiefly composed of the kaolin clay mineral group. When exposed to air and water, this rock can rapidly disintegrate to something a little better than mud. When fire clay is present near the roof, or makes up the immediate roof, it should be taken down before any project is started, since it will surely decay and crumble with time.

Limestone and Dolomite: This type of rock is characterized by the carbonate fraction of the material exceeding the noncarbonate fraction. Calcium carbonate is the main fraction material in limestone, while the mineral dolomite dominates the material fraction of dolomite.

The static properties of limestone show this rock to be quite variable in testing. Based on USBM tests of mechanical rock properties, limestone exhibits the properties shown in Table 3.

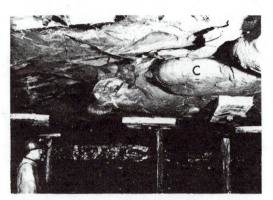

FIG. 15. Large concretion in roof (C); note smooth polished surface (Crickmer and Zegeer, 1981).

In general, limestone makes a stable construction material for point-anchoring or bearing great weights with its relatively high shear strength. The midwestern coal seams of Illinois and western Kentucky generally find consistent limestone roof rock over the underground mining operation. When considering excavations for construction projects such as overcasts in limestone, the question of whether to drill and blast the limestone top or cut the top down may be raised. The engineering properties of limestone in the excavation site will help determine the better approach for the operation to take.

Conglomerate: Conglomerate rock in coal mining is relatively scarce as either roof or floor rock. Strictly speaking, there is little conglomerate per se, but it is commonly manifested as a shale base with concretions or a sandstone base with clastic material larger than 9.5 mm ($3/8$ in.) diam (Fig. 15). This type of roof deteriorates over time when exposed to air and water, so this roof rock should be taken down or supported before any extensive project is started.

In concluding the lithology discussion of typical coal horizon strata, several points need to be reviewed:

1) The lithology of the roof and floor rock

TABLE 2. Engineering Characteristics of Shales

Type	Compressive Strength, psi		Elasticity, psi		Modulus of Rupture, psi		Shear Strength, psi	Tangent of Internal Friction
	Max	Min.	Max	Min.	Max	Min.		
Shale	33,500	10,300	9.9×10^6	1.5×10^6	4.2	0.3	1,160	19.7
Siltstone	45,800	4,000	9.3×10^6	1.0×10^6	5.0	1.1	720	7.7

Metric equivalent: psi × 6.894 757 = kPa.

Note: Since argillites are not common in coal mining strata, this rock type will not be fully discussed.

TABLE 3. Engineering Characteristics of Limestone and Dolomite

Type	Compressive Strength, psi		Shear Strength, psi	Modulus of Rupture, psi		Tangent of Internal Friction	Dynamic Modulus of Elasticity, psi	
	Max	Min.		Max	Min.		Max	Min.
Limestone/Dolomite	37,600	5,300	2,150	5,200	400	12.6	14.1×10^6	1.2×10^6

Metric equivalent: psi $\times$ 6.894 757 = kPa.

influences the integrity of any construction project over time.

2) The static properties of the rock may force several design alternatives to be considered (i.e., drilling and blasting vs. cutting, etc.).

3) All of the rock types exhibit variable static properties, illustrating how the rock strata can sometimes be strong and competent or weak and treacherous in the same mine.

The final point stresses the valuable information to be gained with an on-site inspection of the local strata conditions.

In recapping geologic parameters, the construction supervisor or mining engineer should be aware of both the tectogenic and lithologic processes of the local construction site. All the information pertinent to the design of the project will aid in safer construction and more efficient use during the mine life.

PHYSICAL PARAMETERS

Physical parameters are the second major category of parameters influencing construction design and development. These parameters describe the influence of the physical mining environment on the construction design. The main parameters I shall discuss are mining height, gas, water, humidity and weathering, and seam pitch. These five parameters alone or in combination with each other can reduce the best designed projects to ineffective and unsafe structures. An engineer designing any of the ventilation, drainage, transportation, and haulage facilities for a mine must consider these parameters for possible advantages and disadvantages. The first parameter for discussion is mining height.

Mining Height

Mining height is a physical mining parameter affecting nearly all necessary construction projects. Each facet of support project construction needs analyzing when the engineer or supervisor has doubts about the best design for the specific mining height of the operation. I shall briefly discuss mining height with respect to ventilation, transportation, drainage, and haulage projects.

Ventilation projects such as overcasts, booster fans, regulators, and undercasts all vary greatly with the mining height of the operation. As a general rule of thumb, a low coal mine ventilation project will require more flexibility in design than a moderate or high coal mine. This flexibility is necessary because of the cramped working area and the need to keep costs to manageable levels. Questions of drilling and blasting or cutting the roof, floor, or ribs are raised, as well as questions of labor, material delivery, and safety. Low coal designs should include the compensating ventilating area requirements necessary because of the restricted height.

Mining height on transportation projects is important in both track and rubber-tire mines. Minimum and maximum mining height between roof and floor is necessary for maintaining clearances of the transportation equipment in lower coal. Too little clearance reduces the safety and effective design of the system and increases the transportation time of vehicle movement. In low coal, the grading of the floor becomes a large parameter in obtaining the correct clearance.

To illustrate the importance of mining height influence, let's look at the example of a room-and-pillar mine in southwestern Virginia which is working the Splashdam seam. It has a mining height of 813 mm (32 in.) and a poorly designed and installed track and trolley system. One producing unit of the mine was particularly difficult to reach because the construction supervisor failed to take into account a slight dip in the mining height while laying the track, which effectively reduced the clearance of the entry. Both track and trolley were laid right past the low spot for a great distance. The first time through the dip with track equipment revealed the height to be only 660 mm (26 in.). The vehicle was unable to pass. The men refused to walk to the section, and a strike ensued. If more attention had been paid to the installation, such a costly incident could have been avoided.

When installing a large, mine-wide drainage system, there usually is a central collection point. At this point a main mine pump, designed to handle the actual and future projected waste water generated in the mining operation, is installed. This pump installation, if not mounted on the surface (as with a drift mine), must have a minimum height for operation and installation. In addition, the water collection area (sump) usually requires a minimum amount of excavation to meet projection requirements. The mining height, therefore, impacts the cost of drainage pumping when excavation work is needed. Low coal mines have a greater problem with any drainage installation because of constricted mining heights. A simple check of necessary clearance when designing the installation will reduce future problems and give an estimate of the excavation requirements.

Mining height in haulage systems plays an important design role. Mining height helps determine whether belt lines are hung from the roof or supported by stands on the floor (Fig. 16). Mining height also affects the decision of either hanging the terminal head pulley or supporting it by the drive system, as shown in Fig. 17. Design decisions on such items as concrete belt drive or transformer pads may be altered by the mining height of the operation.

Gas

The presence of methane gas in the mine affects any ventilation projects more than the drainage, transportation, or haulage systems. Ventilation projects comprise the foundation of ventilating systems designed to do the following two things: (1) provide enough pure air to the miners and assure an adequate supply of oxygen, and (2) dilute, render harmless, and remove the noxious gases of mining operations and the physical mining environment.

The second point applies directly to the design of ventilation projects because of gas emissions and dust generation due to the mining operations. Overcast or undercast designs should account for the minimum acceptable airway diameter in all cases. Regulators must be designed and adjusted to changing gas dilution requirements. Drainage, transportation, and haulage projects are impacted by gassy conditions with the explosion- and fire-proof designs of pumps, motors, and conveyor systems. In

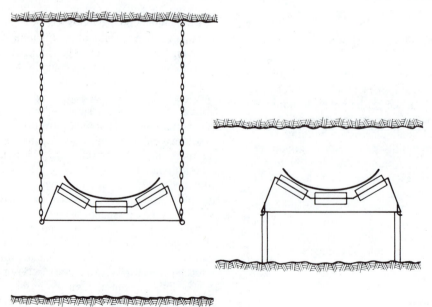

Vertical View

FIG. 16. Seam height affects the belt line support system design.

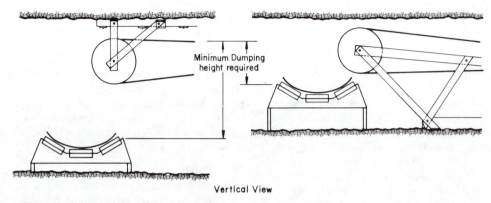

FIG. 17. Seam height affects terminal support choices by defining the available minimum dumping height.

addition, gassy conditions also limit the type of tools that can be used for construction. Diesel or gasoline powered picks, loaders, mixers, or drills are not easy to obtain a permit for, so electric replacements are used. However, gassy conditions per se should not stop a construction project because alternatives do exist.

Water

The presence of water in the underground environment can cause tremendous problems when not dealt with effectively.

Ventilation projects such as undercasts and overcasts are particularly susceptible to water buildup in critical areas. Undercasts accumulate water in the subparts of the structure. Overcasts, on the other hand, may collect water on either side of the structure walls and wing walls (Fig. 18). Regulators, doors, or stoppings are not unduly affected by water in moderate amounts. Flooding water will certainly affect any ventilation structure.

Drainage projects are constructed in response to the water in the mine environment. The flow of water into the production sections should be determined and designed for, as well as the water in outby areas. Hopefully, main drainage projects will be constructed before their need, since it is very difficult to work with water up to one's waist.

Transportation and haulage projects are adversely affected by water. Corrosion and acidic water will quickly damage the integrity of either rails and ties or belt stands and point anchors. In addition, water streaming from the roof can lower visibility, reduce the safety of the environment, and the moisture can cause material buildup along belt lines and transfer points by sufficiently wetting the material.

Humidity and Weathering

During the summer months, hot, moist, outside air enters the underground mine environment. Because the mine stays at a nearly constant 18.3°C (65°F) year-round, the outside air cools and moisture condenses on the roof strata. In the winter months, conditions in the mine reverse. The cold outside air travels through the

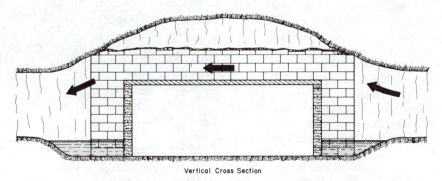

FIG. 18. Water collecting behind the wing walls of an overcast.

mine environment and warms up. The warming air absorbs all the available moisture, drying the mine out. This yearly moisture cycle, called weathering, can cause terrific problems for the construction engineer.

Ventilation projects are susceptible to this type of weathering because of the relative isolation of completed projects and the dependency of effective sealing techniques. Mine conditions can change quickly, catching mine management off guard. Overcasts on return air splits are of particular note. Because of relatively infrequent travel in this area (once-a-week inspection trips) sloughing and roof falls due to weathering can occur and accumulate. This is especially true in older mines.

Humidity affects drainage projects by having a varying effect on drainage patterns during winter and summer months. Engineers designing systems for a specific minimum and maximum flow rate at mines with humidity and water problems should analyze these fluctuations to avoid damaging pumping equipment from corrosion and dry pumping.

Humidity makes rails slick, fire clay bottoms soft, and rots the ties on transportation systems. However, for construction purposes, humidity and weathering affects the installation of new railway systems very little. Maintenance of the rail system, unfortunately, is another matter.

Haulage project construction is also affected very little by humidity. The largest source of problems comes from the sloughing roof and subsequent corrosion of anchor ties of roof-suspended equipment (head drives, belt-line chains and plates, etc.). This leads to system breakdown and possibly failure at critical points in the system. Humidity can certainly cause increased maintenance of the haulage system. Engineers and designers should observe the immediate roof and gage its response to continued weathering. This is especially important for main-line haulage, or mine-life terminal projects. For studying humidity effects, the psychrometric chart is useful. This chart lets the engineer calculate the amount of moisture he can expect to enter the mine environment in the summer, or how much moisture will be picked up by the air in the winter. By calculating these limits, the engineer can gage the relative effects of humidity in his particular mine, and determine the extent, if any, of the weathering problem on his projects.

Weathering is also a physical parameter which can cause a great deal of concern over time. Weathering encompasses the degradation of the mining environment during the mine life. This includes sloughing, roof falls, and floor heave. All these features can damage construction projects unless the proper design analysis is undertaken. Floor heave, for example, can block an airway or shut down a transportation route, as shown in Fig. 19.

Seam Pitch

Seam pitch can be classified as either a geologic or physical parameter. For construction engineering purposes, it is better to regard seam pitch as a physical parameter of the mining environment, mainly due to the degree of control the mine management has over pitch with grading and other techniques.

All construction projects are only slightly affected by seam pitch. Only in the anthracite coal fields and some far western fields may seam pitch cause a problem. Otherwise, most US coal seams, as mentioned before, are quite flat, or pitch slightly (3° or less). On pitches greater than 10°, however, there may be a definite design change in all projects.

Ventilation projects in steeply pitched seams should be designed with some flexibility as to wing and structure wall variations (Fig. 20). This pitch also affects water buildup and drainage requirements by holding back some mine water.

Drainage projects are affected by seam pitch when the pitch is due to folding or just tilting. Folding produces a rolling effect on the seam

FIG. 19. Floor heaving in mine entry (Crickmer and Zegeer, 1981).

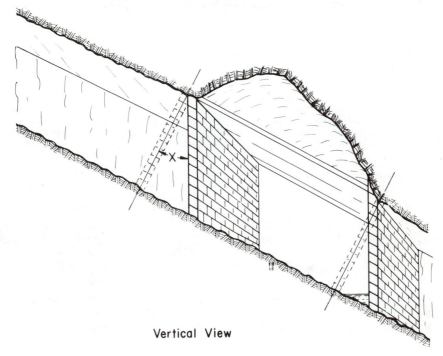

Vertical View

FIG. 20. Wing walls should be flexible in design for steeply pitching seams.

strata forcing the engineer to analyze the flow patterns of the mine waste water to optimize the mine sump and pumping system.

Transportation projects can be virtually redesigned in steep pitching seam. In some cases, an ordinary transportation system such as a railway must be replaced by specialized systems. Construction engineers working with these systems (monorails, etc.) are faced with a unique set of parameters to analyze. Currently, such systems are beyond the scope of this text.

Haulage systems are handled with much of the same parameter analysis techniques as the transportation system. Steep pitching seams may require additional driving power on the belt lines, or utilize antislip devices. In addition, specially designed haulage systems may find an application in some extreme cases.

In concluding the discussion of physical parameters, some points need emphasizing. These are:

1) The analysis of physical parameters influencing construction projects is usually accomplished in the initial premining design stages. Follow up checks of local conditions at the time of construction may be necessary and desirable.

2) Some physical parameters (water, mining height, etc.) affect the design of projects more than others (pitch, gas, etc.).

3) The failure to examine the physical parameters influencing the design projects could compromise the safety of the mining operation and lower the overall performance of the system.

Table 4 illustrates the relative importance of each physical parameter for the main construction categories.

REGULATORY PARAMETERS

The need to examine the regulatory parameters surrounding the construction of underground projects stems from the 1969 Federal Coal Mine Health and Safety Act (CMHSA). This set of statutes first defined and categorized the proper materials, design approach, and minimum safety standards for underground construction. Since that time, the task of effective design has become increasingly complex. I shall discuss federal requirements of materials, designs, and safety for each of the typical underground construction projects. In addition to the federal laws, some state laws will be cited where applicable for illustration.

TABLE 4. Physical Parameter Impacts on Project Construction

Project Parameter	Ventilation	Drainage	Transportation	Haulage
Mining Height	MA	MI	MA	MA
Gas	MA	NC	NC	MI
Water	MI	MA	MI	NC
Humidity and Weathering	MI	NC	NC	MI
Seam Pitch	MI	MA	MA	MA

MA = major impact; MI = minor impact; NC = No change.

Ventilation Projects

For the purpose of the regulations, there are three types of ventilation devices requiring discussion: overcasts, doors, and stoppings.

Overcasts: Overcasts are a staple of any underground ventilation system. Designing these structures to take advantage of local conditions, plus geologic, physical, and mining parameters hinges upon the required materials, design approaches, and safety parameters allowed in both federal and state regulations.

The federal requirements about materials for overcasts are quite straightforward. Essentially, all permanent overcast materials must be of substantial and incombustible material. The suggested list includes concrete, concrete blocks, cinder blocks, brick, or tile. Metal stoppings are approved if designed and built for ventilating purposes. This definition is found in Section 75, Part 316-2 under criteria for approval of ventilation system and methane and dust control plan.

Some state plans have a slightly more concise definition of materials for overcasts. The Commonwealth of Pennsylvania, for example, requires the following:

"Overcasts and undercasts shall be constructed tightly of incombustible material . . . of sufficient strength to withstand possible falls from the roof." (Section 244)

The State of West Virginia in Section 23-2-4 has a broader requirement for overcasts and undercasts by defining these structures to "be constructed of incombustible material and maintained in good condition." Other coal-producing states have similar requirements.

One further requirement of the Federal CMHSA regulations involves premining submission of the overcast construction plans to the district offices. This required item is incorporated in the ventilation system and methane and dust control plan of the mining operation. The operator must submit a complete list of materials to be used in the construction of the overcast (Section 75.316-2). Changes in building materials are just a matter of resubmitting a list for approval.

There are two general design approaches to overcast excavation; cutting the top or drilling and blasting the top. Each approach has particular requirements to fulfill under the Federal CMHSA and various state requirements.

Although cutting overcasts, or drilling and blasting overcasts are not directly mentioned in either federal or state regulations, the regulations apply to this construction work as equally as the cited face areas or travelways.

When cutting the top, the roof bolts in place initially are cut down along with the rock. Replacing this roof support falls under Section 75, Part 200-7 of the federal statute. This section gives a complete description of the criteria for approval of roof control plans, their installation, and materials needed.

The Commonwealth of Virginia emphasizes temporary support methods when cutting the top and providing for rebolting. Sections 45.1-40, 45.1-41, and 45.1-42 clearly set forth the approach for resupporting top when it is cut. The West Virginia and Pennsylvania regulations are similar to Virginia's and are cited in Sections 22-2-26, 22-2-27, and 253, respectively.

Drilling and blasting the roof rock requires using permissible explosives. Because of this, there are additional statutes on the use, handling, and storage of explosives. These regulations are outlined in Subpart N, Section 75, Parts 1301, 1303, 1304, 1305, 1307, and 1308 of the federal law. Usually, such regulations do

GENERAL CONSTRUCTION PARAMETERS

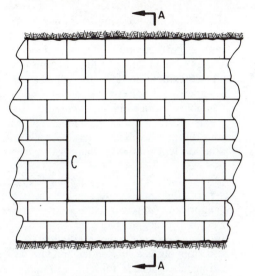 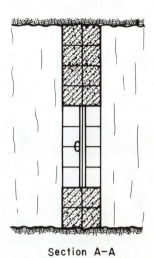

FIG. 21. Typical man door installation (side opening type) in a permanent stopping.

not pose undue restraints on efficient working practices and designs.

The State of Utah has a very detailed approach to using explosives for underground work, including Sections 49 (Underground Transportation), 50 (Underground Storage), 51 (Blasting Rules), and 52 (Misfires). Likewise, the Commonwealth of Kentucky concisely defines the use of explosives in all phases of industrial application. The Kentucky regulations are cited in the 805th volume of the Kentucky Annual Report, Sections 4:075 through 4:140, or for convenience, the Kentucky explosive and blasting regulations of 900 through 911.

In addition to explosive regulations when drilling and blasting the rock, the regulations for the approved roof control plan must be followed.

Individual safety practices for overcast construction are fairly straightforward. For the most part, such references are covered in the roof control plans in both federal and state statutes.

The thrust of these regulations is to maximize the safety of the working place and minimize the exposure of the worker to hazards. Since this is also a key concern for the construction supervisor, being familiar with the regulations is a must.

Doors: There are two types of doors under this category, man doors and full-entry doors. Man doors are used to provide access between entries separated by a stopping line (Fig. 21). Full-entry doors allow men, materials, and equipment to pass through a stopping line with no difficulty (Fig. 22).

By law, today's coal mines must have a man

 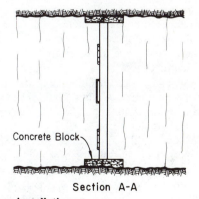

FIG. 22. Typical full-entry door installation.

door in stopping lines no more than 152 m (500 ft) apart. Some states vary the exact wording (every five entries, etc.), but the intent is still the same. The federal statutes do not directly address the construction or required material necessary for either man door or full-entry door construction. However, such plans need to be submitted to the district manager under Section 75.316-1, as part of the information to be submitted by the operator. Some states, however, do give fairly detailed guidelines on door construction and placement.

The Commonwealth of Virginia, for example, has the following passage on stopping man doors:

"45.1-59(g) To provide easy access between the return belt, and intake escapeway entries, substantially constructed mandoors properly marked so as to be readily detected shall be installed at least every fifth crosscut in the stopping lines separating such entries."

Virginia also goes on to describe full-entry doors and their operation. In Section 45.1-60, the following regulations are cited:

"(a) in gassy mines, the ventilation shall be so arranged by means of air locks, overcasts, or undercasts that the passage of haulage trips or persons along the entries will not cause interruption of the air current; provided, however, that in mines or in developing sections where air locks are not practical single doors shall be used to course the air, and unless operating mechanically, shall be attended constantly while the mine is in operation. Air locks shall be ventilated enough to prevent accumulations of methane therein.
(b) Doors shall be kept closed except when men or equipment is passing through the doorways. Motor crews and other persons who open doors shall see that the doors are closed before leaving them."

Again, the regulations are general enough to allow a wide range of materials to be used for door construction.

Some states also add regulatory descriptions to either or both door types. Utah, for example, uses the following regulation to describe the man door requirement:

"Section 43(h) A fireproof door large enough to permit the passage of a person shall be installed in permanent stoppings at intervals not to exceed 500 feet. The doors must be designed and installed to be self-closing."

Here, Utah has set a minimum number of conditions for materials and design. Pennsylvania has similar regulations regarding doors in Section 244-b and 244-c of their mining laws.

In constructing doors, the engineer or mine designer has a variety of materials and designs available to him to help cut costs, increase the useful life of a door, and reduce construction time. Further detailed discussions of door construction will follow in Chap. 3.

Stoppings: A discussion of regulatory parameters of stoppings should consider the two main categories of stoppings, temporary and permanent. Temporary stoppings are designed for short life, low cost, and easy installation. A permanent stopping is a substantial structure designed and installed usually for a life of three years or more.

The Federal CMHSA regulations discuss permanent stopping construction under Section 75.316-2(b), saying these structures shall be substantial and built of incombustible material, including concrete, concrete blocks, cinder blocks, brick, or tile.

State regulations defining permanent stoppings closely follow the federal criteria. Pennsylvania also requires the stoppings be reasonably airtight (Section 243-f). West Virginia statutes allow fire-resistive material as well as incombustible material for permanent stopping construction. One of the states directly addressing the question of temporary stoppings is Illinois. In this state, the distinction between main and submain (cross-entry) entries and panel entries allows temporary stoppings to be erected for periods of up to three years. The passage of Illinois law allowing this condition is cited below.

"Section 17.05. In crosscuts connecting main and cross entry inlet and outlet air courses, the permanent stoppings shall be erected of masonry, concrete, or other incombustible material, and shall be erected within 600 feet of the face of main and cross entries at all times. Temporary stoppings of wood or other equally effective material shall be maintained, as nearly air tight as possible, between the last permanent stopping and the crosscut nearest the face in main and cross entries. In room and stub entries the stoppings shall be built of wood or other equally effective material. All stoppings shall be kept in good condition, so as to keep the air up to the working faces."

This concludes the discussion of regulatory parameters of ventilation projects. By familiarizing one's self with the cited passages of federal and state laws on these ventilation projects, an engineer or construction supervisor can minimize any adverse impacts of these parameters on all ventilation devices, including those not discussed.

Drainage Projects

When discussing drainage projects such as sumps, pumping stations, pipe networks, and face drainage, there are few direct federal regulations pertaining to electrical pumps, monitoring devices, and power cables other than permissibility in gassy environments. The general guidelines fall under Subpart F, Section 75.500. A direct listing of materials for pumping stations and listing of pump types is given (Appendix A, Section 75.500). Therefore, the general requirement of permissibility of pump motors, cables, housing, and accessories is a design boundary an engineer should be aware of at all phases of installation.

State regulations follow closely the federal statutes on electrical equipment such as pumps, the incoming cables, and the safety of the installation. This is especially true for pump networks needed in the production face areas for drainage. A review of state regulations before a final design is approved may aid in reducing any future problems with regulatory parameters.

Transportation Projects

As mentioned in the beginning of this section, there are two common modes of transportation for men, materials, and supplies. These are rail and rubber-tired methods. For the purposes of reviewing the regulatory parameters affecting transportation construction, we will discuss the rail mode of transportation only. Construction projects for rubber-tired travel are not extensive (i.e., battery-storage barns, repair shops, etc.), and will be discussed in Chap. 7, under Special Projects. The rail construction discussion will be divided into two discussions, the rail system and the trolley wire system.

The rail system regulations in the Federal Code deal mainly with the design and construction of the roadbed. These statutes are cited in Section 75.1403-8 and -10. Another requirement (Section 75.508-1) is mapping the rail system on the mine map for proper identification. As for other items such as workmanship, materials, or construction safety, the Federal CMHSA leaves these to state jurisdiction.

West Virginia mining law devotes an extensive section to haulage roads and equipment. The Section, 22-2-37, discusses requirements for the track, switches, joints, frogs, and other elements of rail construction. The main points are covered in Parts a, b, and c, and are cited here for convenience:

"(a) The roadbed, rails, joints, switches, frogs and other elements of all haulage roads shall be constructed, installed and maintained in a manner consistent with speed and type of haulage operations being conducted to insure safe operation. Where transportation of personnel is exclusively by rail, track shall be maintained to within five hundred feet of the nearest working face.

(b) Track switches, except room and entry development switches, shall be provided with properly installed throws, bridle bars, and guard rails; switch throws and stands, and where possible, shall be placed on the clearance side.

(c) Haulage roads on entries developed after the effective date of this article shall have a continuous, unobstructed clearance of at least twenty-four inches from the farthest projection of any moving equipment on the clearance side."

The State of Utah provides for similar regulations and also requires that, "Rails shall be secured at all joints by means of plates or weld" (Section 59-k). The Commonwealths of Virginia, Kentucky, and other coal mining states have similar regulations for the construction of railroad beds and tracks.

When a rail system is electrified, the Federal Code contains specific subparts on trolley wire and direct current systems. Subparts F and K either directly or generally pertain to d-c system installation. Table 5 summarizes the applicable sections. Construction foremen and engineers designing the installation methods should review the applicable requirements and also check the work during and after installation for compliance.

TABLE 5. Applicable Federal Sections on d-c and Trolley Wire Systems

Subpart	Section-Part	Description
F	75.509	Electric power circuit; deenergization
F	75.510, -1	Energized trolley wires, repair
F	75.516	Power wires; support
K	75.1000	Cut out switches
K	75.1001, -1	Overcurrent protection
K	75.1002, -1	Location of trolley wires
K	75.1003, -1, -2	Other guarding requirements

State regulations on trolley wire systems are quite similar to federal regulation, often emphasizing these requirements by separate sections. To facilitate a review, Table 6 summarizes the applicable state sections on trolley wire installation of some example states.

Even though state codes do not appear to add more regulations to federal requirements, engineers and construction supervisors should be familiar with their applicable state laws.

Haulage Projects

Of the four construction categories defining typical underground construction, haulage projects, outside of ventilation requirements, has the most federal and state regulations concerning minimum materials, design, and safety criteria. In discussing coal haulage projects (i.e., belt or rail), belt haulage and belt installation are of primary interest.

Belt haulage practically touches all aspects of federal law. Construction foremen and engineers must be aware of applicable regulations in Subparts D, F, G, L, O, and R of Section 75 of the Federal Code. Because of the extensive number of requirements, supervisors should be especially concerned with proper installation. To facilitate the review of federal regulations, Table 7 summarizes the applicable sections on belt haulage. These regulations apply mainly during initial design stages of belt haulage systems, although installations require careful review. It should be noted there exists a variety of design approaches, materials, and minimum safety procedures available to engineers maximizing efficiency and least-cost methods for belt haulage.

Rail haulage also comes under several federal statutes. Underground coal hauling by rail, however, is slowly being phased out by more efficient modern haulage designs. Federal subparts contained in Section 75 that are applicable to rail haulage include Subparts D, F, K, L, R, and S (Table 8). Persons interested in minimum rail haulage regulations should consult these subparts.

State requirements for belt and rail haulage underground necessarily follow federal requirements. Each state groups these requirements in various sections of state law, and design engineers or construction supervisors should review their applicable state requirements.

This concludes the section on regulatory parameters. Federal and state requirements on materials, designs, and minimum safety practices should be considered in all stages of design and during practical installation. In many cases, these codes provide flexible guides which the engineer can utilize to his advantage for cost-effective design and implementation.

TABLE 6. Some State Regulations on Trolley Wire Systems

State	Article	Section(s)	Description
Pennsylvania	III-E	322 to 328	Direct current installations
West Virginia	II	22-2 to 22-2-4-1	General electrical provisions/bonding
Kentucky	352	.150, .220	Haulage roads/mine electricity
Illinois	18	18.11, 18.16, 18.18	Electrical regulations
	22	22.02(b), 22.15	Haulage underground
Virginia	6	45.1, -71(c), -74(d, h)	Transportation
	7	45.1-78	Electricity
Utah	VII	61(i)	Transportation of men
	VIII	67(h-n)	Power circuits

TABLE 7. Federal Regulations Concerning Belt Haulage

Subpart	Section	Description
D	75.326	Aircourses and belt haulage entries
F	75.511	Equipment repair of low, medium, and high voltage
	75.516, -1	Power wires; support installed insulators
	Appendix A	Permissible conveyors
G	75.607	Breaking trailing cables and power cable connections
L	75.1100-Z(b)	Quantity and location of fire fighting equipment
	75.1101-1, through 1101-21	Deluge type water sprays, foam generators; main and secondary belt conveyor drives
	75.1102	Slippage and sequence switches
	75.1102, -1 through 1103-10	Automatic fire warning devices
	75.1107-4	Automatic fire sensors and manual actuators
	75.1107-10	High expansion foam devices; minimum capacity
O	75.1403-5	Criteria—belt conveyors
R	75.1707	Escapeways; intake air; separation from belt and trolley haulage
S	75.1805	Examination of electrical equipment

MINING OPERATIONAL PARAMETERS

Mining operational parameters is the category describing those factors directly controlled by the operator. Their influence on any given project is proportional to company policies, operational practices, and supervisory attitudes.

The mining operational parameters under consideration are safety, cost, supplies, labor, project materials, and planning and management.

Although other mining parameters do enter into construction decisions, the main factors mentioned above cover the majority of parameter influence. We shall discuss each one briefly.

Safety Parameters

Safety parameters are those factors present during any construction project which are above and beyond minimum mandated safety requirements. These parameters range from additional company safety rules to specialized machines and equipment.

Engineers and construction supervisors may need to address safety parameters for both the installation materials and personnel. Depending upon the type of project and construction steps required, safety parameters can affect the project greatly. Cutting the roof for an overcast, for example, may require additional supports, or special roof straps, bolts, or plates. Sumps need-

TABLE 8. Federal Regulations Concerning Rail Haulage

Subpart	Section	Description
D	75.327	Aircourses and trolley haulage systems
F	75.508	Map of electrical system
	75.508-1	Mine tracks
	75.510	Energized trolley wires; repair
	75.510-1	Repair of energized trolley wires; training
	Appendix A	Permissible equipment
K	75.1000 through 75.1003-2	
L	75.1100-1	Type and quality of fire fighting equipment
	75.100-2(c) & (d)	Quantity and location of fire fighting equipment
L	75.1707	Escapeways; intake air; separation from belt and trolley haulage entries
S	75.1805	Examination of electrical equipment

ing protection from further weathering may be shotcreted. Workers may require protective clothing or special working periods which may increase the project duration. A final note on safety parameters concerns company emphasis on safety. *Under no circumstances should an engineer or supervisor compromise the safety of the design, materials, or personnel of any project.* Quick decisions, easy solutions, or substandard materials based on reducing the cost or erection time of a construction project may lead to unsatisfactory results.

Cost Parameters

Cost parameters are those factors influencing the material selection, design approach, and project importance based on cost. Depending upon the strategy of the mining management, cost factors can play an important role in overall project construction. Overcasts, for example, can be erected with (1) a preformed metal design, or (2) built from scratch with mine materials. If, after evaluating all other parameters (and finding them equal), the cost of one is decidedly different, then mining management may opt for either based solely on the cost of the project. Often, decisions on the design of drainage networks or haulage systems are greatly influenced by the cost of materials and equipment needed in the design.

When all other parameters are equal, the cost of any project takes on even greater importance. Therefore, engineers should direct their efforts in project design toward optimizing the usefulness and least-cost options of the project. Construction supervisors should follow up these efforts by minimizing waste and using effective time-managing of labor to complete projects on schedule.

Supply Parameters

Supply parameters are those factors involving the scheduling of project materials, the delivery and usage of tools and equipment, and subsequent project construction duration. The basic premise here is to (1) order the right materials (both quantity and type), (2) deliver the right materials on time (at the start of the project), and (3) deliver the right materials on time at the *right place*. In many cases, this is easier said than done.

The delivery of materials for projects such as belt line extensions, track laying, switches, or overcasts may take several shifts to assemble and deliver to the proper construction site. Construction supervisors must be especially aware of what it takes to gather these materials, order them, and deliver everything to the site. Valuable time and money can be wasted going after parts or searching for additional materials. In short, supply parameters can and do have a large impact on the successful completion of a project.

Labor Parameters

Labor parameters deal mainly with management's decisions on construction labor. Basically, the questions boil down to how many and what skills are necessary.

The completion time of a project depends greatly upon the amount of labor provided. The setting and anchoring of a 1067-mm (42-in.) belt drive may require eight to ten manshifts. With two men working, this is only four to five shifts. With one man working, the required time may double, triple, or the job may be even impossible to complete. Mine management pursuing an effective schedule of project completions should consider this parameter closely.

Labor skill is the second component. The completion time of a project is greatly enhanced when thoroughly knowledgeable and highly skilled workers are doing the project. Conversely, inexperienced miners working with unfamiliar tools will lengthen the project duration. Mining management should balance this labor parameter with both skilled and unskilled workers to effectively reduce any negative aspects.

Project Material Parameters

These factors deal with the selection of project materials and its subsequent impact on the construction itself. These materials could influence the design, duration, location, and cost of the project. One example is the decision to buy prehung mine doors for air locks. These metal airlock doors cost much more than doors made of mine materials (i.e., wood). Although prehung doors cost more initially, they will eliminate a significant amount of installation labor and time. In some cases, these prehung doors function better and longer than wooden, mine-built doors, thus reducing the replacement factor. Engineers, purchasing agents, and mine management should consider these material parameters and their subsequent impact on the construction project.

Planning and Management Parameters

Planning and management parameters deal mainly with mine management's approach to construction projects. Planning and management includes the priorities of the mine superintendent down to the construction foreman in regard to scheduling and completing the necessary projects during normal working shifts. This parameter becomes especially important for large mines where construction of ventilating, haulage, drainage, and transportation systems must be efficient and timely for effective performance. Careful planning will reduce support problems with belt moveups, face drainage, stopping construction, and other projects susceptible to scheduling lags. A final note on planning and management concerns the decision balance. Mine management along with the engineering staff must decide the importance of the construction project and make the necessary scheduling plans. This applies to both planning and implementation.

In concluding mining operational parameters, several points stand out:

1) Engineers and construction foremen should be aware of the parameters influencing project design and installation.

2) Mining operational parameters do not influence any given project equally.

3) By examining all parameters involved with project construction, engineers will better design the usefulness and life of a project, while maximizing the project's cost-effectiveness.

CONCLUSIONS

Throughout this chapter I have tried to explain the influence of various geologic, physical, regulatory, and mining operational parameters on construction design and installation. This approach differs from traditional approaches by showing that an analysis of influencing parameters should take place for *every* underground construction project, not just the special ones. By careful consideration of the local conditions, the mine design, and regulations involved, an engineer can and will develop a sound, functional project design. Such formal analysis will, most often, be done during premining planning stages. Spot checks and installation site decisions by construction supervisors will be enhanced when these supervisors are aware of all influencing parameters, and are able to deal with them effectively.

REFERENCES AND BIBLIOGRAPHY

Adler, L., and Sun, M., 1976, *Ground Control in Bedded Formations,* Virginia Polytechnic Institute and State University, Blacksburg, VA.

Anon., 1978. *Mining Laws of Virginia,* Department of Labor and Industry, Charlottesville, VA.

Anon., 1961, *Mining Laws of Pennsylvania,* Department of Environmental Resources, Harrisburg, PA.

Anon., 1977, *Mining Laws of West Virginia,* Central Printing Co., Beckley, WV.

Anon., 1976, *Laws Governing the Mining of Coal and Clay,* Kentucky Department of Mines and Minerals.

Berger Associates, 1975, *Evaluation of Mining Constraints to the Revitalization of Pennsylvania Anthracite,* US Bureau of Mines Contract #50241039.

Billings, M.P., 1972, *Structural Geology,* 3rd ed., Prentice-Hall, Englewood Cliffs, NJ.

Crickmer, D.F., and Zegeer, D.A., 1981, *Elements of Practical Coal Mining,* 2nd ed., AIME, New York.

Darton, N.H., 1940, "Some Structural Features of the Northern Anthracite Coal Basin, Pennsylvania," Professional Paper No. 193, US Geological Survey.

Industrial Commission of Utah, 1976, *General Safety Orders, Utah Coal Mines.*

Jackson, K.C., 1970, *Textbook of Lithology,* McGraw-Hill, New York.

Kelley, V.C., 1960, *Slips and Separations, Bulletin,* Geological Society of America, Vol. 71, pp. 1545–1546.

Parker, J., 1976, *Roof Control in the Coal Mines of Appalachia,* Seminar material given by Parker and Associates, White Pine, MI.

Peng. S., 1980, *Coal Mine Ground Control,* John Wiley & Sons, New York.

Press, F., and Siever, R., 1978, *Earth,* 2nd ed., W.H. Freeman and Co., San Francisco, CA.

US Department of Labor, 1976, *Code of Federal Regulations, Title 30,* Mining Health and Safety Administration.

2

Transportation—Railroad Construction

In this chapter, I shall discuss railroad construction for modern underground coal mines. Because men, materials, and supplies are being moved more often and farther by rail, the efficient and safe construction of the system takes on new importance. Most large coal mines are opting to use track haulage over other support transportation systems because of cost, ease of installation, reliability, and other factors. Engineers tackling the problem of designing such a transportation network must be aware of all the steps required for an effective transportation system design.

This chapter discusses all aspects of efficient trackwork construction. The areas covered include design factors, laying the track, installing turnouts and switches, and installing trolley wire. Each area includes a description of the steps needed to complete the work.

RAILROAD DESIGN FACTORS

At the time of designing any coal mine, decisions about transportation systems are made. These decisions, later drafted into construction prints and installation procedures, reflect the engineer's approach to the most effective system.

As mentioned earlier, rail transportation is taking a greater role in large mining operations. In response to this growth, the designs of these systems should become more sophisticated, and hence cost-effective. The sophistication of rail systems stems mainly from the following design factors:

Estimated maximum transported loads.
Necessary clearance (sides and height).
Railcar and traffic parameters.
Mining entry dimensions.
The mining method.
Type of locomotion (whether trolley, diesel, or battery).
Maintenance schedule.
Rail weight and tie section.
Precautions for transporting mine machinery.

Each one will be discussed briefly.

Maximum Transported Loads

At various times during the life of the mine, heavy equipment such as longwall shields, continuous miners, or belt drives may be transported over the rail system. To avoid the damage and delay of an inadequately supporting track and roadbed, the mining engineer must project the estimated transport loads to use the system. From this projection comes a consideration of rail weight, rail tie type, and required roadbed. If infrequent travel of heavy equipment is planned, (i.e., every five years), then cost considerations should be analyzed. An example of a cost alternative would be the decision to transport heavy equipment such as a continuous miner in two sections and effectively reduce the transport load. By breaking the weight of the load into two lighter sections, lighter less expensive rail, ties, or ballast may be feasible. Engineers facing this design problem should consider the alternatives carefully. A rule of thumb for estimating the design load from the maximum transported load is that the design load should be 15% greater than the maximum transported load to allow for installation irregularities. An added safety factor (approximately 1.5) compensates for any weakness in the materials. In addition, the design load will reflect management's programs in track maintenance, inspection, and repair.

Clearance

During the rail transportation design phase,

the factor of minimum necessary clearance comes under consideration. Both side widths and heights are examined.

Many underground rail systems are placed on an offset centerline from the entry centerline (Fig. 1). This allows more clearance for the walkway side of the track, but still gives sufficient area to the tight side of the track. Minimum recommended side clearances coincide with federal specifications:

Walkway clearance—610 mm (24 in.) from the farthest point of moving traffic.

Tight side clearance—305 mm (12 in.) from the farthest point of moving traffic. Depending upon the projected plans, minimum side clearances may or may not be a problem. Track laid in the same entry as belt haulage must be installed properly to maintain the maximum design clearance for safe and efficient operation (Fig. 2).

Height clearance becomes especially important in low coal operations. According to federal statutes, at least 305 mm (12 in.) of clearance shall be maintained between the highest projection of equipment and the energized trolley wire. When this becomes impossible because of the seam height, additional precautions are necessary. Engineers in low coal operations needing necessary clearances may analyze alternatives such as lower equipment, thinner ties, or less ballasting as possible design choices.

Railcar and Traffic Parameters

The type of traffic a mine transportation system must handle is quite important during design stages. This equipment must be matched to the track weight, gage, and turnout radius to prevent wheel lockup or possible derailing. Locomotives, mine jeeps, personnel carriers, haulage trams, ballast cars, and any other possible rail traffic are included in these considerations. For example, one should not order ballast cars with a wheel gage of 1118 mm (44 in.) for operation on a trackage with a 1067 mm (42 in.) gage. Or, more frequently, a mine locomotive weight will exceed the specified weight per rail of a track system. Since the locomotive is the cornerstone of material and supply movement on track systems, and it is by far the heaviest traffic vehicle on the system, then this parameter between locomotive weight and rail weight capacity becomes the controlling design parameter. Table 1 illustrates the suggested rail weights for

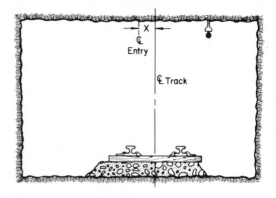

FIG.1. The track centerline is offset from the centerline of the entry.

mine locomotives obtained from the Bethlehem Steel Corp. for aiding in the choice of the minimum rail weights of a system.

Mining Entry Dimensions

Mining entry dimensions also reflect on the size and clearance of the track equipment. Usually, the entry dimensions are firmed up from decisions on rock mechanics, regulations, and management's desires, and becomes an upper boundary for rail design. Flexibility of entry dimensions to accommodate rail transportation factors is infrequent, unless plans specifically call for it (e.g., belt and track in the same entry or double track systems). For the most part, entry dimensions are a fixed parameter during the analysis for track haulage.

Roadbed

The roadbed of the track entry can pose some design problems for engineers. If the stratum is too soft, then heavy equipment traffic can cause

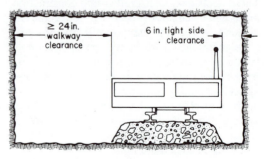

FIG. 2. Minimum recommended clearances for track and transportation safety. Metric equivalent: in. × 25.4 = mm.

TABLE 1. Suggested Rail Weights for Mine Locomotives

Locomotive Weight, Tons	Weight of Rail per Yd*		
	4-Wheel Locomotive	6-Wheel Locomotive	8-Wheel Locomotive
2	20	—	—
4	25	—	—
6	30	—	—
8	30	—	—
10	40	30	—
13	60	40	—
15	60	40	—
20	60	60	—
27	80	70	60
37	100	85	70
50	—	100	90

Source: Bethlehem Catalog 2314.
*Metric equivalents: yd × 0.9144 = m; st × 0.907 184 7 = t.

supplemental hooving and strata-flow away from supporting the track and ballast. When standing water is present, further softening of the roadbed may occur. Of the typical strata types occurring in coal seam horizons, shales and clays are the most susceptible to softening problems, and hence, cause design modifications for track systems. In perspective, however, any design changes based solely on the bottom roadbed are rare.

Unless the mining height of the coal seam prevents proper ballasting, all main line track should be ballasted with material which will: (1) transport and distribute the load of the track and mine traffic to the mine floor, (2) provide a resistance for the track to move in lateral, longitudinal, and vertical directions, (3) provide a drainage base for the track structure, and (4) ease track maintenance of the cross level, surface level, and track alignment. Track ballast used in coal mining applications can consist of broken stone, gravel, cinders, mine tailings, or slag. The standard ballast is usually crushed stone varying in size between 127 to 19.1 mm (5 in. to 3/4 in.). Since mine haulage is classified as number 2 track by standard engineering specifications, the average amount of ballast is 102 to 152 mm (4 to 6 in.) deep with 3:1 slopes beyond the ends of the ties.

Locomotion

The locomotion parameter involves decisions on (1) whether a track system should be battery, diesel, or electrically powered, and (2) the required horsepower of the motor.

In the majority of coal mines, electric locomotion by trolley wire is the accepted method. When steep grades are encountered (e.g., greater than 15°), diesel power may have better possibilities. Battery haulage for areas without electrification, such as a gathering locomotive, is applicable in some mines. These design questions on locomotion are also relatively straightforward and depend mainly on management desires, regulations, and costs.

The required horsepower of the motor is a function of several interacting factors which come together under the term "tractive effort." Since the average speed of mine haulage generally is 16 to 32 km/h (10 to 20 mph), and the weight of the locomotive is known, a tractive effort (TE) quantity can be established from the local tractive resistance factors of the track system. Such design factors are usually chosen during the initial design stages.

Maintenance Scheduling

Like all other mine support systems, trackwork, switches, turnouts, and the roadbeds need attention. Over the mine life of the system, regular maintenance is necessary to maintain the efficiency and reliability of the network. Oiling switches, straightening bent rail, reballasting low spots, and reclipping or spiking the rails down are all examples of regular necessary maintenance. For the design engineer, this means scheduling a portion of time for these activities and developing a milestone chart of track maintenance activities. By completing these projections, the engineer has some flexibility as to the track material and equipment in terms of durability and changeout periods. For example, on mainline transportation track, an engineer can institute a regular maintenance program on a work schedule of six to seven months. Subsequently, his calculations should show a planned useful life of 12 to 15 years for the turnouts and switches with regular maintenance, instead of a useful life of 8 to 10 years without maintenance. In panel development, little maintenance is scheduled during panel life since a milestone chart will show a maintenance period during panel recovery every three years. Similar analysis for submains, butt entries, and siding will reveal a good program of maintenance closely tied to track usage and material.

Rail Weight and Tie Selection

The poundage of rail chosen for mine rail systems is an important parameter. Traditionally, rail weights have been near 18.1 kg/m (40 lb per yd). With modern systems, however, 38.6 kg (85 lb) rail weights are recommended for overall mine use. Table 2 illustrates the current rule of thumb for 18, 27.2, and 38.6 kg (40, 60, and 85 lb) rail applications.

TABLE 2. Typical Mine Rail Use by Weight

Weight per Yd.*	
40 to 60	Light duty, mainly support for belt and rubber tired haulage, panel development
60 to 85	Transportation of personnel, medium-weight equipment; support transportation throughout mine, main and submain development
85 or greater	Heavy duty transportation of men, materials, and supplies; mainline haulage as well as life of mine support system

*Metric equivalent: yd × 0.9144 = m.

To better understand the strength of a rail section, Fig. 3 and Table 3 are presented to illustrate the variety of rail sections available to design engineers. Either the ASCE or ARA-A, B, and C sections of rail are acceptable for mine use in properly designed circumstances (as designated by the American Mining Congress). The length of these sections depends greatly on the intended use and weight of the rail. For example, ASCE 27.2 kg, (60 lb) rail is available in stock lengths of 10.1 m (33 ft) and 11.9 m (39 ft), while ASCE 38.6 kg (85 lb) rail is stocked in 11.9 m (39 ft) lengths only. For mining, the shorter rail lengths offer the advantage of maneuverability and easier transportation, and engineers should specify their requirements when necessary. The amount of rail needed for mining is specified by the number of net tons of track (i.e., a set of rails). Table 4 summarizes the general rail tonnage for selective rail sections.

Each rail weight when properly chosen will provide excellent service for the operation required. Fig. 4 shows the recommended minimum weight of rail as a function of the maximum single wheel load.

The decision to use a particular weight of rail is also governed by economic factors. The cost for ASCE 38.6 kg (85 lb) rail as of January 1980 was over $175 per ton. Lighter rails are naturally less expensive, but require more maintenance while the heavier rails cost more per ton and have a longer life. Premining projections of proper rail selection need to consider both the usage patterns and economic alternatives to the rail weight.

Tie selection is a second big factor. There are two basic types of ties used in underground mining, steel and wood. Steel ties are ties with steel

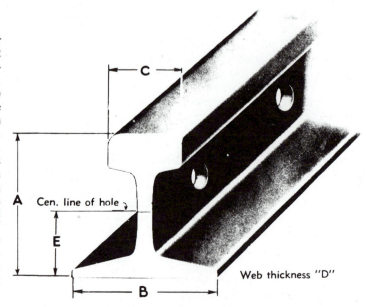

FIG. 3. Rail section (Anon., 1980).

NOTES—Standard Drilling for A.S.C.E. Rails 5.4 to 13.6 kg (12 to 30 lb) inclusive—50.8 × 101.6 mm (2 × 4 in.). 18.1 to 45.4 kg (40 to 100 lb) inclusive—63.5 × 127 mm (2 ½ × 5 in.). Drilling is given from end of Rail to C. L. of first hole, and from C. L. of first hole to C. L. of second hole, etc. Thus a drilling may be 50.8 × 101.6 mm (2 × 4 in.) or 63.5 × 127 mm (2 ½ × 5 in.). If drilling is a three hole drilling the same method is used, thus 63.5 × 127 × 127 mm (2 ½ × 5 × 5 in.). As there is no commonly accepted "standard" drilling for rails other than A.S.C.E. sections, hole spacing for such rails must be given.

TABLE 3. Rail Sections for Mine Application

Weight per Yard	Section Name	Section Number	A Height	B Base	C Head	D Web	E Height of Hole
40	ASCE	4040	3 1/2	3 1/2	1 7/8	25/64	1 9/16
45	ASCE	4540	3 11/16	3 11/16	2	27/64	1 41/64
50	ASCE	5050	3 7/8	3 7/8	2 1/8	7/16	1 23/32
55	ASCE	5540	4 1/16	4 1/16	2 1/4	15/32	1 13/16
60	ASCE	6040	4 1/4	4 1/4	2 3/8	31/64	1 29/32
65	ASCE	6540	4 7/16	4 7/16	2 13/32	1/2	1 31/32
70	ASCE	7040	4 5/8	4 5/8	2 7/16	33/64	2 3/64
75	ASCE	7540	4 13/16	4 13/16	2 15/32	17/32	2 15/128
80	ARA-A	8020	5 1/8	4 5/8	2 1/2	33/64	2 21/64
80	ARA-B	8030	4 15/16	4 7/16	2 7/16	35/64	2 15/64
80	ASCE	8040	5	5	2 1/2	35/64	2 3/16
85	ARA-A	8520	5 3/8	4 7/8	2 1/2	9/16	2 29/64
85	ASCE	8540	5 3/16	5 3/16	2 9/16	9/16	2 17/64
90	ARA-A	9020	5 5/8	5 1/8	2 9/16	9/16	2 37/64
90	ARA-B	9030	5 17/64	4 49/64	2 9/16	9/16	2 11/32
90	ASCE	9040	5 3/8	5 3/8	2 5/8	9/16	2 45/128
100	ARA-A	10020	6	5 1/2	2 3/4	9/16	2 3/4
100	ARA-B	10030	5 41/64	5 9/64	2 21/32	9/16	2 65/128
100	AREA	10025	6	5 3/8	2 11/16	9/16	2 45/64
100	ASCE	10040	5 3/4	5 3/4	2 3/4	9/16	2 65/128
100	PRR	10033	5 1/2	5 1/2	2 13/16	5/8	2 9/32
100	P.S.	10031	5 11/16	5	2 43/64	9/16	2 31/64
105	Dudley	10524	6	5 1/2	3"	5/8	2 43/64
110	AREA	11025	6 1/4	5 1/2	2 25/32	19/32	2 53/64

Source: Midwest Steel Catalogue S-180.
All dimensions shown are in inches.
*Metric equivalents: yd × 0.9144 = m; in. × 25.4 = mm.

ribs and rail clips mounted on wooden planks [1.5 m by 203 mm by 76 mm (5 ft by 8 in. by 3 in.)] to allow easy installation on standard gaged trackwork (Fig. 5). Wood ties for mining are at least 2.4 m long by 203 mm wide by 203 mm thick (8 ft by 8 in. by 8 in.) and cut from selected hardwoods.

Steel ties have the advantages of weight, thickness, and length, as well as durability for long life projects. The cost for steel ties as of January 1980 was $28.50 each, or over $300 per rail length to lay. For underground installations where manual labor is used to move and align the rails, steel ties can save much time and effort during the laying process.

Wood ties need no special preparation for the

TABLE 4. Rail Requirements for Track Length

Weight per Yd*	Section Name	Section Number	Track Footage per Net Ton	Net Tons per Track-Mile
40	ASCE	4040	75.0	70.4
60	ASCE	6040	50.0	105.6
85	ASCE	8540	35.3	149.6
90	ARA-A	9020	33.3	158.4
100	ARA-B	10030	30.0	176.0

*Metric equivalents: yd × 0.9144 = m; ft × 0.3048 = m; st × 0.907 184 7 = t.

rail. Tie plates are sometimes mounted on the ties in a regular fashion to aid in gaging the trackwork. The gage of the track must be constantly checked for tolerance. Two spikes (Fig. 6) are used in a staggered position to give greater strength to the set (Fig. 7). Spiking the rails also requires a substantial amount of room for swinging the hammer, which may be at a premium in some mines. The cost of a wood tie is approximately one-third a steel tie, and is usually obtained from local hardwood sawmills.

In comparing steel and spike ties, the advantages, other than cost, favor the steel tie. The convenience and reduced effort needed for steel ties makes them very popular for underground use. Therefore, through the rest of Chapter 2, I will discuss the laying and assembly of track with emphasis on steel ties.

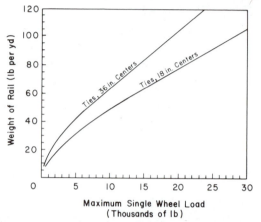

FIG. 4. Weight of rail for tie spacing (modified from Peele, 1943). Metric equivalent: in. × 25.4 = mm; lb × 0.453 592 4 = kg.

Design 8—Outside Stationary Clips

FIG. 5. Steel ties: outside stationary clips (Anon., 1980).

Rail installed after hammering rotary clips to closed position.

TRANSPORTATION—RAILROAD CONSTRUCTION

As a corollary to rail and tie selection, a discussion on rail joints is necessary. This includes using splice bars, bolts, and washers to construct the track, as well as the design of the joint.

There are two joint designs available in mine trackwork with a class 2 rating: *supported* or *suspended*. Supported joints (Fig. 8) have a tie clipped or spiked underneath the rail butts. Suspended joints require at least one tie in either the *X* or *Y* position under the joint splice bars as shown in Fig. 9. Splice bars, or angle bars, are preshaped bars which, when bolted together, form a rigid clamp on the rail (Fig. 10). Special bolts are used in these splice bars which allow easy installation because of the "button head" design (Fig. 11). At the joints where different sized rail are combined in an offset joint, compromise splice or angle bars are needed.

Precautions for Transporting Equipment

Transporting large equipment on rail systems requires several precautions. If the system is electrified, then an insulating material must separate the top of the equipment from the wire. Machinery such as a continuous miner or shuttle car is moved when there is no one inby the

FIG. 6. Track spikes (Anon., 1980).

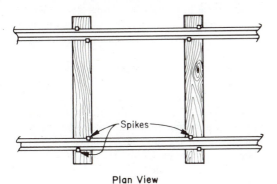

FIG. 7. Staggered spike position when tying down the rails.

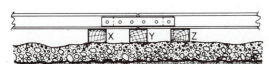

FIG. 8. Supported joint. Note: Wooden blocks represent any materials such as header boards, cap wedges, etc.

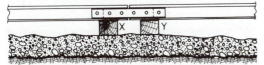

FIG. 9. Suspended joint. Note: Wooden blocks represent any materials such as header boards, cap wedges, etc.

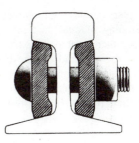

SPLICE BAR TYPE

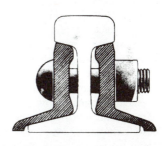

ANGLE JOINT BAR TYPE

FIG. 10. Splice bar and angle joint bar (Anon., 1980).

FIG. 11. Track bolt (Anon., 1980).

moving piece of equipment. Adequate supervision should be present at all times, since careful transportation of large equipment needs direction and control.

Precautions for moving machinery fall under the regulatory guidelines found in 30 CFR 75.1003–2. There is a general split between the guidelines; movement under energized trolley wires with clearance greater than 305 mm (12 in.) and movement under energized trolley wire with clearance less than 305 mm (12 in.). The general precautions for movement with a 305 mm (12 in.) or greater clearance are the following:

1) A certified person should be present at all times during the movement of the equipment.

2) Two inspections should precede the move: (1) the cleaning of coal dust, loose coal, oil, grease, and other combustibles from the equipment to be moved; and (2) the trolley wires, feeder wires, and automatic circuit interruption devices must be examined for proper operation by a qualified person (underground electrical card holder).

3) A record of these inspections should be made and be available for inspection by proper mining authorities.

4) The frames of the equipment being moved must be covered on the top and on the trolley wire side with fire-resistant material such as brattice cloth or conveyor belting.

5) Electrical contact should be maintained between the mine track through the locomotive and lowboy transport to the frame of the mining equipment. An exception to this rule would be for rubber-tired equipment.

If 305 mm (12 in.) of clearance cannot be maintained, such as in many low and high coal mines, then a second set of precautions is necessary. These include:

1) Supplying the trolley power from a source outby the equipment being moved.

2) Setting the automatic circuit interruption device to a point not more than one-half of the maximum current flow if the equipment were to come in contact with the wire.

3) Having a worker stationed with the persons doing the move in direct communication (by trolley phone, telephone, or walkie-talkie) with a responsible person standing at the surface circuit breaker.

4) Permitting no one inby the moving equipment except those assisting with the move.

By following these guidelines, mine management reduces the possibility of a serious accident or mishap while moving any large equipment.

In concluding the design factors to consider for effective track engineering, the important point to remember is the interaction of each design factor in the system as a whole. By developing a complete understanding of all the factors influencing track transportation, a better system will be possible.

LAYING TRACK

The construction foreman or mining engineer approaching the task of laying track must direct or be involved with the three tasks of good trackwork:

Preparation—which includes the subtasks of roadbed grading, material and tool supplying, site orientation, and the operation of stringing rails and ties.

Assembly—which includes the effort to lay the rails and ties, blocking and leveling the track, bonding the rails, and ballasting the roadbed.

Maintenance—which refers to replacement of bad ties, adjustment of tie spacing, replacing or rebending rails.

As shown in the descriptive statements, each task has required subtasks and definite steps necessary for efficient, safe, and cost-effective track laying. Trackwork is usually the first construction task undertaken behind the advancing production units. By law, many mines must keep their track systems within 152.4 m (500 ft) of the working sections, so a regular crew and schedule is used to keep the track close to the units. I shall discuss each task and subtask with regard to the engineer or supervisor's direction of the task.

Preparation

As mentioned earlier, several subtasks are necessary in preparing to lay track, including roadbed grading, material and tool supplying, site orientation, and stringing the ties. To best understand the sequence of steps toward laying the track, a discussion in order of completion of the preparation subtasks is necessary. Let's begin with site orientation.

Site orientation is the process of surveying and establishing the railway position and grade. This task is usually carried out by the mine surveyors. The main points needing to be identified are: (1) the centerline of the trackway (may or may not be entry centerline), and (2) the es-

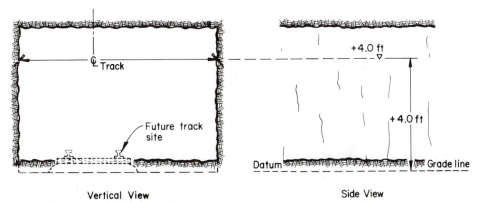

FIG. 12. Elevation offset for grading the trackway. In this case, a + 1.2-m (+ 4.0-ft) offset has been used.

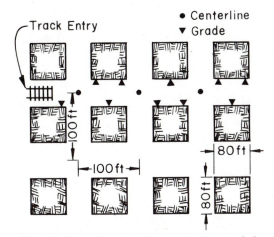

FIG. 13. Plan of typical site orientation needed for efficient track laying. Metric equivalent: ft × 0.3048 = m.

tablished grading height based on the predetermined rail datum line. Both points are extended from established reference spads located in the roof of the entries. The centerline of the trackway is found from entry spads by offsetting its position at right angles at the appropriate distance from the spad, as shown in Fig. 11. The grading height points require the elevation of these spads to be known and used. A profile of the track entry is useful to the engineer when illustrating the concept. The grading spads are spads of equal elevation which identify the basic excavation requirements for grading the track entry. During coal mining, the surveyor will offset this elevation some footage and mark this offset on the rib (Fig. 12) to let the foreman in charge of the grading know where the rail datum line is. By establishing centerline points every 15.2 m (50 ft) and grade points every 12.2 m (40 ft), (Fig. 13), with spads, a reliable reference orientation system will decrease unnecessary guesswork and minimize grading errors. In seams with a pitch, the track entry should be graded with less than 10% change in the grade angle. After these points are established, the roadbed grading task can be done.

Roadbed grading may or may not be necessary, depending upon the datum line, mining height, seam dip, and management's desires. In spots where the bottom is lower than the datum line, ballast is used to build up and support the trackwork. In most coal mines, some amount of roadbed grading is needed. The actual grading is a task best outlined by the following steps.

1) The supervisor or engineer will attach reference string to the grading spads in a crisscross pattern all along the intended grading length (Fig. 14).

2) The cutting of the roadbed at a predeter-

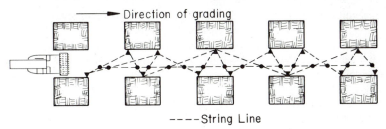

FIG. 14. Typical string line pattern required to control depth of grading operation. Note: String line parallel to entry is the one used by the machine operator.

mined height below the reference string is accomplished by the appropriate continuous miner or roadway driving machine. This cutting operation is a rather difficult task to coordinate and direct at times, since questions of equipment movement, material removal, labor supplies, and time-sharing must all be addressed. Grading the track entry behind the production units is usually done on an idle shift for two reasons: (1) neutral air tends to drift toward the pressure drop at the section, and the amounts of dust generated by the operation forces other activities in the same split of air to stop; and (2) the graded material is transported out of the mine by the mine haulage system.

3) The usual grading operation requires the production unit miner to be backed out from the face area and turned around. Depending upon the relative positions of the power center and miner, the cables may require handling over the conveyor boom and placement on the opposite rib during grading.

4) The shuttle car operator should choose the standard or off-standard car to correspond to the side the miner power and water cables are on. He should also slowly dump the graded material into the section feeder to eliminate material clogging and avoid shearing breaker pins.

5) The miner should slope into the grading at a gentle angle so the shuttle car and other vehicles can negotiate the slope.

6) The miner operator should check the necessary grade height frequently as he is grading the required area. In soft fireclay bottom, the miner has a tendency to dig into the bottom.

7) After completing the grading area [usually 45.7 to 61 m (150 to 200 ft) of 0.9 m (3 ft) grading can be reasonably expected per shift], the miner is trammed back, turned around, and repositioned at the face. All graded material is conveyed out of the mine for disposal.

At the end of the grading task, the roadbed is ready for the delivery of the materials and equipment.

The subtask of *material and equipment delivery* is often overlooked in coal mining, yet it is as important as grading or laying the track. Without the proper and sufficient amount of rails, ties, bolts, nuts, washers, and splice bars (fish plates), valuable time is lost hunting for materials to complete the task. This hunting time can amount to a rather large delay over a working shift and reduce the performance of the track laying crew. The solution is simply to have effective communication among the design engineer, supply supervisor, mine foreman, and construction foreman. The right amount of materials is determined by the engineer, and through effective management, this amount is delivered to the construction site *before* work is to begin. The equipment necessary to do the job should be gathered and passed between the mine foreman and construction foreman on all shifts. By coordinating these activities, the construction task of laying the rail will be smoother.

The final task in preparing to lay track is the *stringing of the rails and ties*. This involves placing the rails and ties in their approximate position for assembly (Fig. 15). By completing this positioning of the rails and ties, the assembly crew can lay the track efficiently and with little trouble. Stringing rails outby a production section can be accomplished by a scoop tractor or shuttle car. The rails are pulled in pairs with a rope attached to each rail and by using a clevis pin (shackle). These ropes are then tied to the scoop or shuttle car and gently pulled into position. Where these are unavailable, a motor and snatch block (*red devil*) will suffice. Stringing rails and ties takes place one to two shifts before the scheduled track laying. A construction foreman or engineer should note that the better method to pull the rails for proper alignment is from the end of the trackwork toward the advancing direction (Fig. 16). This direction allows better alignment of rail ends and minimizes wasted alignment effort during the laying operation.

In conclusion, preparation is important in the management decision process and the subsequent physical effort aimed at efficiently laying track. To illustrate this time progression, Fig. 17 is a flowchart of decisions and followup channels for the typical preparation function. At the end of the stringing operation, the next task is the assembly of the trackwork.

Assembly

At the completion of stringing the rails, ties, and other necessary materials, the assembly mode category begins. This includes laying the track, blocking and leveling the track, bonding the rails, and ballasting the roadbed (if necessary). To accomplish these tasks, a minimum number of men is necessary. Table 5 shows the manpower required for efficient assembly purpose,

TRANSPORTATION—RAILROAD CONSTRUCTION

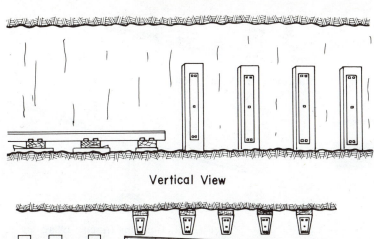

Vertical View

FIG. 15. Stringing of rails and ties before trackwork begins.

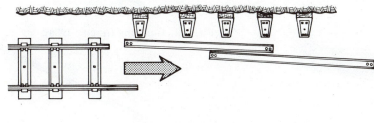

Plan View

FIG. 16. Pulling rails is done by starting from the end of the track and advancing inby.

by subtask. The important point for a construction supervisor to remember is to minimize the physical effort of moving and placing the rails. Too much maneuvering will reduce the effectiveness of the crew by overtiring the workers. The value of proper preparation steps, especially stringing the materials, can be easily seen during track laying. The most effective approach is having the track crew concentrate their efforts in laying the track from the stringing position.

Laying track subtasks begin at the start of the working shift. The crew, consisting of one supervisor and four to six workmen gather the necessary track laying tools. Table 6 outlines the necessary tools for trackwork; the tools are pictured in Figs. 18–26. After collecting the tools, the crew goes to the worksite. The supervisor will string a centerline for the entire length of the work area while the crew sets up the first pair of rails to be laid. To illustrate the

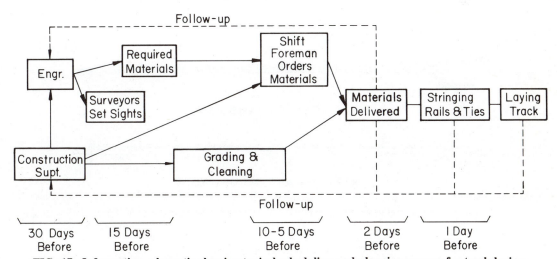

FIG. 17. Information schematic showing typical scheduling and planning process for track laying.

FIG. 18. Railroad pick (Anon., 1980).

FIG. 19. Double face sledge (Anon., 1980).

FIG. 22. Rail tong (Anon., 1980).

FIG. 20. Carbon track chisel (Anon., 1980).

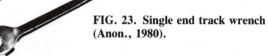

FIG. 23. Single end track wrench (Anon., 1980).

FIG. 21. Pinch point lining bar (Anon., 1980).

FIG. 24. Track jacks (Anon., 1980).

FIG. 25. Track gage (Anon., 1980).

K12 LEVEL AND GAUGE—WOOD

FIG. 26. Level and gage, wood, and step level board (Anon., 1980).

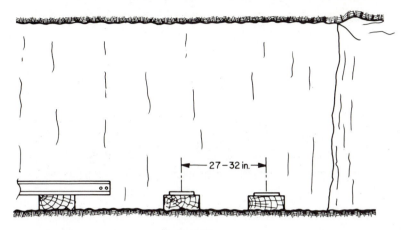

FIG. 27. Positioning of ties before laying rail. Metric equivalent: in. × 25.4 = mm.

procedure, the steps for laying track are:

1) The ties are placed in position approximately 686 to 813 mm (27 to 32 in.) apart along the length of the rails to be laid (Fig. 27).

2) Four workers lift the first rail to be laid and place it end-to-end (butt the ends) with the existing trackwork. Note: The first rail to be laid should be the frontmost rail of the existing trackwork. Track is always laid with staggered joints to avoid excessive weakening. The minimum stagger distance is approximately 0.9 m (3ft).

3) The open end of the rail is jacked up and blocked to level the end to be spliced with the existing trackwork (Fig. 28).

4) A splice or angle bar is placed on either side of the rail and aligned with the bolt holes.

5) Bolts, nuts, and washers are inserted and loosely tightened. The rail is then tapped from the open end to butt the joint together. (Note: For safety purposes, the rail should not be supported solely by the track jack, but should have adequate blocking.)

TABLE 5. Assembly Manpower Requirements

Subtask	Manpower Minimum	Recommended
Laying track (assume proper preparation)	2	4
Blocking and leveling	1	2
Bonding (does not consider welded bonds)	1	1
Ballasting track (with motor and ballast cars)	2	2

TABLE 6. Trackwork Tools

Item	Number Required	Figure Number	Use
Railroad pick	1	18	Trimming clay bottom, arranging ties, and clipping ties to rail
Sledgehammer	1 or 2	19	Arranging ties, adjusting rails, and wedging mechanical bonds
Spike maul	1 or 2	20	Clipping ties to rail, driving rail spikes, adjusting rails, arranging ties
Claw bar (with toe)	1	21	Clipping or spiking ties to rail, aligning rails
Rail tongs	2 or more	22	Moving or aligning rails, and lifting short distances
Track wrench (single end)	1 or 2	23	Tightening track bolts at splices, switches, and track accessories
Track jack [13.6 t (15 st)]	2	24	Lifts and supports rail while clipping ties, helps adjust rails
Track gage	1	25	Maintains gage of rails during work
Track level	1	26	Maintains level of trackwork

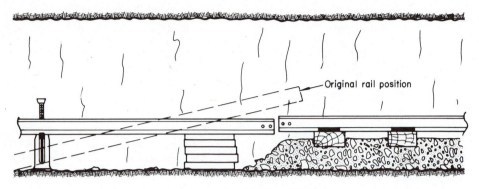

Vertical View

FIG. 28. Jacking inby end of rail to level the rail for splicing into the existing trackwork.

6) The bolts are tightened down to draw the splice bars into the rail and secure the joint. It is advisable to block the joint as soon as possible after making the connection to minimize stress (if a supporting joint is required).

7) The crew begins clipping ties to the rail (if this is the type used) or spikes the first rail to the ties.

8) The joining of the second rail is accomplished in the same manner as steps two through six.

9) The final step of clipping or spiking the ties to the second rail is done with care toward keeping the ties perpendicular to the rails and maintaining proper rail gage. If clips are used, the rail must be "sprung" into position with a claw bar and pick.

At the completion of laying the track, the subtask of blocking and leveling begins. To complete this subtask, a track jack, level, wooden blocks (header boards), wedges, and a sledgehammer are necessary. The steps are few and are repeated over the length of the track to be laid. The idea behind blocking and leveling is: (1) to provide a good support base for the trackwork before the ballast is used, and (2) to re-

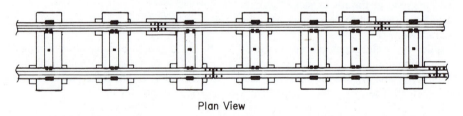

Plan View

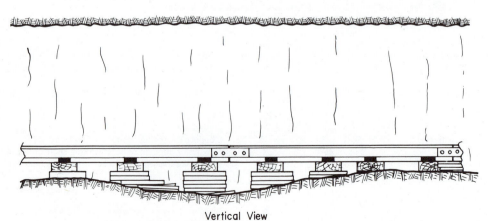

Vertical View

FIG. 29. Blocking and leveling the track across uneven mine floor before ballasting.

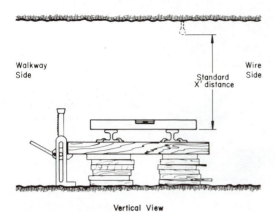

FIG. 30. Leveling the track with a level and jack. First, the wire side is raised to a standard distance from the proposed wire. Then the walkway side is raised to the same height.

move any side-to-side or up-and-down rolling of the trackwork. Effective blocking and leveling will substantially decrease derailing of equipment and maintain consistent speeds.

The steps required in blocking and leveling are as follows:

1) The wire-side rail is raised with the track jack to the appropriate distance from the trolley wire, usually 2 m (6 1/2 ft). If the coal height is less than 2032 mm (80 in.) or wire has not been installed to a level position, a mere blocking of the rail is sufficient. It is important to note the applicable level area influence only extends some 3.0 m (10 ft) on either side of the jack at any one given set up of the jack. Therefore, several repositionings of the jack will be necessary.

2) The rail is firmly blocked and wedged tightly under every other tie over the distance being leveled (Fig. 29).

3) The procedure is repeated for the entire length of the wire-side rail and then the walkway-side rail is blocked up to it by means of a level placed on both rails (Fig. 30).

By following these three steps, the crew eliminates the side-to-side rocking motion. To minimize the rolling motion of up-and-down, the level is placed on a rail longitudinally and checked for near leveling. Unfortunately, the mine roadbed is usually irregular and the amount of leveling can only be estimated. The best approach lies in maintaining consistent and gentle rolls.

After the blocking and leveling subtask, joint bonds (Fig. 31) are hammered on every joint. This ensures good grounding between rails and contact among equipment, rails, and wire. The wedges are usually lead-casted to aid in the grounding. Since only one man is necessary for bonding rail, this subtask is carried out at odd moments during blocking and leveling.

The final subtask is ballasting the trackwork. This is done with tailings, cinders, or stone brought underground in ballast cars. These ballast cars are built to spread the stone through special bottom doors (Fig. 32). Usually, enough ballast to do 30.5 to 61 m (100 to 200 ft) of trackwork is brought underground to spread, requiring two motormen and two motors. Ballasting takes very little time, approximately one-half man-hour per hundred feet. The difficulty lies in coordinating rail movements to allow the ballast cars room and time to work. Usually, weekend work schedules are used for ballasting.

In concluding the assembly task, a natural question of manpower is raised. For effective trackwork a crew of four men can lay between 5 and 10 pairs of rail per shift [45.7 to 91.4 m (150 to 300 ft)]. This rate demands the rail and ties be strung; adequate bolts, nuts, washers, and splice bars be provided; and the crew be knowledgeable in the task. In addition, efficient track laying must be supervised by a foreman familiar with the tasks and confident in the performance of his crew.

Maintenance

The final task of laying track is no less important than the first two. Maintaining an efficient track system improves mine safety and transportation. Also, maintenance activities may not be time-consuming if done on a regular schedule.

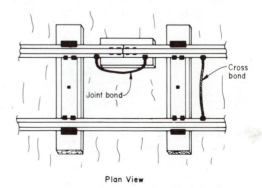

FIG. 31. Bonding the track at every joint. Cross bonds are required every 61 m (200 ft).

FIG. 32. A typical ballast car for coal mine use (courtesy of Difco, Inc.).

The maintenance activities of trackwork are simple. The greatest concern is adequately supporting the trackwork to eliminate bowed, damaged, or unleveled rails. A second inspection includes examining the track joints, switches, and bonding of the trackwork. Loose bolts or bonds should be tightened and secured, and switches should be oiled and adjusted periodically. Proper inspections of the trackways require a committed effort, since damage can occur at any time. Properly maintained trackwork will follow these few inspection guidelines:

1) For mine (class 2) trackwork, every 11.9 m (39 ft) of track should have at least eight non-defective ties with a maximum center to center distance of 1778 mm (70 in.) between them and no more than two bad ties together at any one point. A tie is considered defective when it is: (a) broken through, either crosswise or lengthwise; (b) split to the extent it will not hold spikes, or will allow the ballast to work through; (c) so deteriorated that the tie plate or base of rail can move laterally more than 12.7 mm (0.5 in.) relative to the crosstie; (d) cut by the tie plate through more than half of its thickness; or (e) failing to hold the rail because three of the four clips are damaged beyond proper use from age, wrecks, overuse, etc.

2) Rail should be inspected for any defects or conditions which may be a safety hazard. Tables 7 and 8 summarize those conditions for which to be aware.

3) At the rail joint, each splice bar should be matched with the corresponding rail, and be inspected for the following conditions: (a) If a splice bar is cracked or broken between the middle two bolts, it must be replaced. (b) For mine track, each rail must be bolted with at least two bolts at each joint, one on each side of the joint.

4) A record of inspections and remedial actions should be kept in the mine office along with the regular maintenance records.

Where track is on steep grades, the inspection should also include derails and speed restrictors. Other areas are storage spots for track tools and any other areas needed for the proper operation of the track system.

A suggested schedule of six-month inspections of the entire trackwork will enable early detection and correction of any defects or conditions. Of course, inspections of main-line track are more important than other less traveled rail-

TABLE 7. Trackwork Defects to Inspect for

Defect	Defect Size, in.*	Remedial Action
Bolt hole crack	0- 1/2	Limit speed and inspect within 90 days
	1/2-1 1/2	Limit speed and inspect within 30 days
	1 1/2-larger	Limit speed and schedule replacement
Broken base	0-6	Apply joint bars to defect
	6-larger	Limit speed and replace rail
Rail break		Apply joint bars to break and inspect within 90 days
Damaged rail†		Apply joint base to defect and inspect every 120 days

*Metric equivalent: in. × 25.4 = mm.
†Damaged rail means any rail broken or made unfit for track by abuse or accident such as wrecks, derailments, broken or flat wheels, or similar causes.

TABLE 8. Trackwork Conditions to Inspect for Remedial Action

Condition	If condition requires replacement of rail	If condition does not require replacement
Shelly spots Head checks Engine burn (no fracture)	Limit speed and schedule rail for replacement	Inspect rail for internal defects at intervals of not more than every 12 months
Mill defect Flaking Slivered Corrugated (wavy) Corroded	Limit speed and schedule rail for replacement	Inspect rail at intervals of not more than every six months

ways. Finally, at the detection of problems, mine management should schedule necessary repair or replacement work for weekend or idle shift periods. An experienced supervisor should be present at all work.

INSTALLING TURNOUTS

One of the most critical tasks of railroad construction is installing the turnout. Poor designs and/or installations will cause derails, lower transportation efficiency, and reduce the safety of the system. On the other hand, well designed and/or installed turnouts can enhance the overall reliability of the system.

There are two major tasks for installing turnouts, namely, preparation and assembly. A third task, maintenance, is considered under general track care on a schedule similar to the laying track schedule.

In underground mining, a complete turnout consists of a frog, switch, filler rails, guard rails, and switch stand. It can be either right- or left-handed, and must be ordered as such. I shall discuss each turnout part, beginning with the frog.

A *frog* is the term used to describe the transfer point at a turnout and is derived from its general shape. For underground mines, a frog is either a riveted type (Fig. 33) or the more popular solid type (Fig. 34). Both are approved for mining use by the American Mining Congress (AMC). Different sizes of frogs are distinguished by the *frog number*. The frog number is the ratio of its length (measured on center line of frog) to its width. Fig. 35 illustrates the components and technical information associated with the frog. Frog numbers relate to the frog angle, which is the angle on the turnout radius. For mining, frog numbers rarely exceed 5, but as Table 9 shows, these angles are on the low end of the list. The reason for the low number (high angle) is the tight turning clearances found in room-and-pillar mining. Rib slabbing is often necessary to meet required clearances.

A complete switch for mining purposes consists of two switch points, all clips, bolts, rods, and plates necessary to make the switch work. The switch points are the components needed to make the turnout operate properly. They are located some short distance for the point of bend in the rail (Fig. 36). A typical switch is shown in Fig. 37, designed in accordance with the AMC. Switch plates needed for the switch are a combination of a flat plate and bracing plate for support and guidance. The switch rod and switch clips are usually the side jaw type. This allows for maximum flexibility during installation. Typical designs are shown in Fig. 38.

Filler rails refer to the *curved closure, straight closure,* and *prebent turnout rails* needed in the turnout (Fig. 39). These are made specifically

TABLE 9. Frog Numbers and Angles

Frog No.	Frog Angle
2	28°-04'-21"
2½	22°-37'-12"
3	18°-55'-29"
4	14°-15'-00"
5	11°-25'-16"
6	9°-31'-38"
7	8°-10'-16"
8	7°-09'-10"
9	6°-21'-35"
10	5°-43'-29"
11	5°-12'-18"
12	4°-46'-19"
14	4°-05'-27"

Source: Midwest Steel Catalogue S-180.

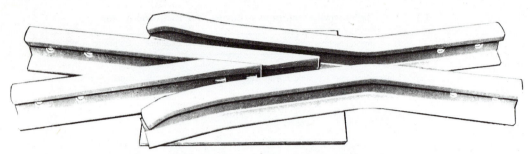

FIG. 33. Riveted frog (Anon., 1980).

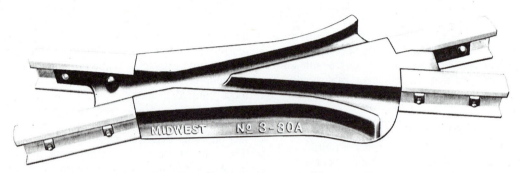

FIG. 34. Solid manganese frog (Anon., 1980).

for the frog angle and are often included by the manufacturer in the complete turnout. When, in some cases, the curved or supplied turnout rail must be bent in-house, the bending will take place during the cutting and laying tasks. Bending rails is accomplished by using a rail bender, or "butterfly" (Fig. 40). The rail bender will bend the rail from the one end of the tool to the fulcrum, or approximately 203 mm (8 in.). Care must be taken not to over or under bend at any one place on the rail. The approach is to make a continuous bend with equivalent pressure at each point. Fig. 40 shows a reversible type of rail bender which allows the hydraulic or mechanical jack to be on the right or left hand side of the fulcrum, permitting more flexibility.

Guard rails are placed in the turnout design to prevent derails. They are bolted to the rail or clipped to the rail like ties. Guard rails are placed on the straight track on the outside rail and the turnout track outside rail opposite the frog. Fig. 41 shows a typical guard rail.

The *switch stand* and connecting rod are the last components of a turnout. The switch stand

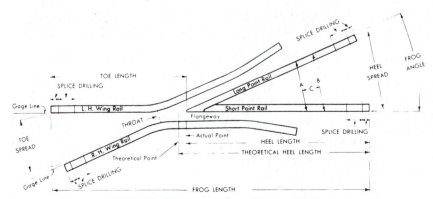

FIG. 35. Diagram of frog (Anon., 1980).

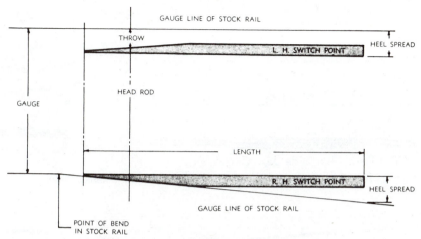

FIG. 36. Switch components (Anon., 1980).

is the mechanical device which allows the switch points to move and the turnout to operate properly. The stand is made up of several components, usually provided by the manufacturer, and is ready for service underground (Fig. 42). The connecting rod is the part between the switch stand and the switch rod. Some are spring loaded to allow movement through the switch, while others are solid for simple design. Fig. 43 illustrates some common types.

A question of switch ties now arises. In most cases, a manufacturer provides several key ties needed to put the turnout together. Their relative position is shown in Fig. 44. Steel ties are particularly useful in this case.

The final topic concerns the overall turnout data for mining installations. To facilitate this discussion, Tables 10 and 11 summarize the data required for turnout design, showing frog numbers, angles, and switch and closure rail details.

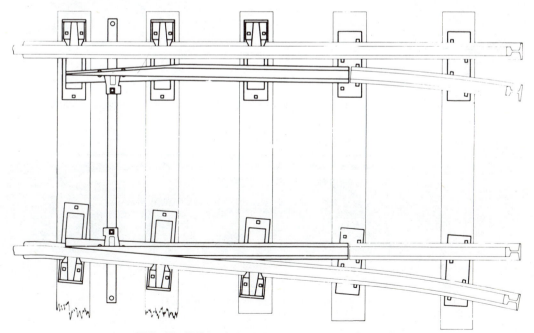

FIG. 37. Split switch, A.M.C. design (Anon., 1980).

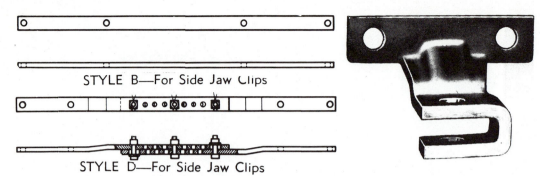

FIG. 38. Switch rods and side jaw type switch clip (Anon., 1980).

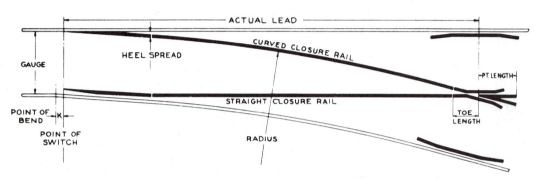

FIG. 39. A.M.C. turnout (Anon., 1980).

FIG. 40(a) Mechanical (Jim Crow screw type) rail bender (b) hydraulic rail bender (Anon., 1980).

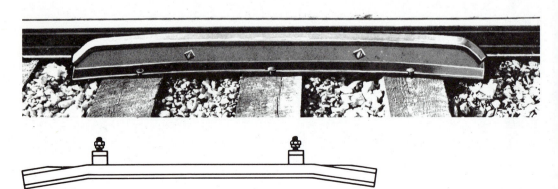

FIG. 41. Guard rail (Anon., 1980).

FIG. 42. Switch stand and exploded view (Anon., 1980).

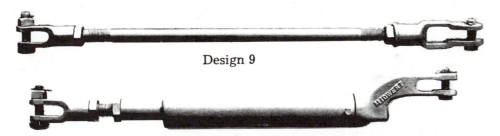

FIG. 43. Adjustable and spring connecting rods (Anon., 1980).

FIG. 44. Riveted clip type steel switch ties (Anon., 1980).

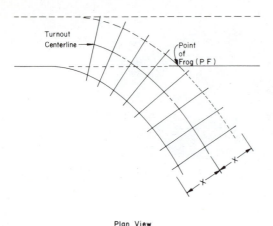

FIG. 45. Roof spads are used to locate the point-of-frog and turnout centerline.

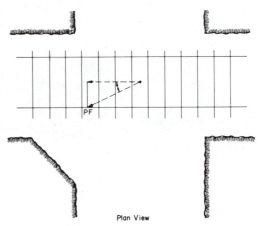

FIG. 46. One method of sighting in the point-of-frog. Notice the PF is located on inner rail, closest to the turnout.

The first table includes rail from 9.1 to 27.2 kg (20 to 60 lb) while the second is from 31.8 to 45.4 kg (70 to 100 lb) rail. Turnouts, once designed, can readily be ordered from manufacturers with specific features needed for individual operations. The average cost for a complete turnout in 1980 was $2200.

Turnout Preparation

Turnouts are usually put into the rail system after the track has been laid past the turnout area. There are several reasons for retrofitting the switch to the track system; one of these is having the position of the frog fixed with respect to the straight trackway. A second reason is associated with turnout preparation.

In most respects, turnout preparation is identical to the preparation for track laying discussed earlier (i.e., orientation, material delivery, etc.), and so will not be repeated. The important difference is in site orientation for the frog and turnout radius. This point is covered briefly in the following section.

Site Orientation: One of the most important preparation steps necessary for installing a turnout is site orientation. This includes a roof spad marking the point-of-frog (PF), as well as the centerline of the turnout (Fig. 45). The radius of the turnout is the frog angle specified by its number.

The point-of-frog is the spot at the apex of the frog angle. Surveyors can establish the PF in the roof by sighting from a centerline spad at the correct angle and distance, as illustrated in Fig. 46. The centerline of the turnout is obtained in a similar manner from the centerline sight spad. From the turnout data presented earlier, a lead distance is known and marked off the centerline. The curve is established by laying a horizontal curve out with a radius equal to the frog angle (Fig. 47). It is important to have centerline spads for the turnout on 1.5 to 2.4 m (5 to 8 ft) centers. An estimate of transit or theodolite turning degrees is approximately 2° each spad. An illustrative discussion of laying out a horizontal curve for mine track is found in Chapter 11 of *Mining Engineer's Handbook*, 3rd ed., by Peele, published by John Wiley and Sons. Also a general discussion of horizontal curves is included in *Surveying Practice*, 3rd ed., by Kissom, published by McGraw-Hill.

The proper positioning of site orientation for turnouts is crucial for standardized and well in-

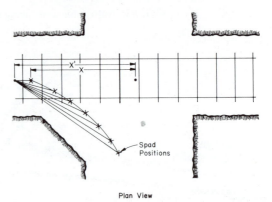

FIG. 47. Laying out a horizontal curve with a radius equal to the frog number.

TABLE 10. AMC Turnout Data: 20-60 Lb Rail, American Mining Congress Design

42 in. GAGE

		FROG			SWITCH			Actual Lead, ft-in.	CLOSURE RAILS		
						Heel			Curved		Straight
No.	Angle	Type	Rail, lb	Toe, in.	Length, ft-in.	Spr., in.	K, in.		Radius, ft-in.	Length, ft-in.	Length, ft-in.
2	28°-04-21	CAST	20-40	6 ¹¹/₁₆	3-6	5	4	12-7	22-0	9-1½	8-6¼
		RPF	20 & 30	17	3-6	5	4	12-7	22-0	8-3¼	7-8
		RPF	40	20	3-6	5	4	12-7	22-0	8-0¼	7-5
2½	22°-37-12	CAST	20-40	8¼	5-0	5	5	16-11	35-0	11-8¼	11-2¾
		RPF	20 & 30	16½	5-0	5	5	16-11	35-0	10-11⅞	10-6½
		RPF	40	20	5-0	5	5	16-11	35-0	10-8½	10-3
3	18°-55-29	CAST	20-60	8⅜	5-0	5	5	19-0	52-0	13-8½	13-3½
		RPF	20 & 30	16½	5-0	5	5	19-0	52-0	13-0¼	12-7½
		RPF	40-60	2-0	5-0	5	5	19-0	52-0	12-5	12-0
4	14°-15-00	CAST	40-60	11 ¹/₁₆	5-0	5	5	23-0	96-0	17-4½	17-1
		RPF	40-60	2-3	5-0	5	5	23-0	96-0	16-0¾	15-9
5	11°-25-16	CAST	40-60	13 ¹³/₁₆	7-6	5	8	31-0	148-0	22-7	22-4¼
		RPF	40-60	2-6	7-6	5	8	31-0	148-0	21-2¾	21-0
6	9°-31-38	CAST	40-60	16 ⁹/₁₆	7-6	5	8	34-9	220-0	26-0¾	25-10½
		RPF	40-60	3-0	7-6	5	8	34-9	220-0	24-5½	24-3
6	9°-31-38	CAST	40-60	16 ⁹/₁₆	10-0	5	10	38-10	210-0	27-7¾	27-5½
		RPF	40-60	3-0	10-0	5	10	38-10	210-0	26-0¼	25-10

44 in. GAGE

		FROG			SWITCH			Actual Lead, ft-in.	CLOSURE RAILS		
						Heel			Curved		Straight
No.	Angle	Type	Rail, lb	Toe, in.	Length, ft-in.	Spr., in.	K, in.		Radius, ft-in.	Length, ft-in.	Length, ft-in.
2	28°-04'-21"	CAST	20-40	6¹¹/₁₆	3-6	5	4	13-2	23-0	9-8⅜	9-1¼
		RPF	20 & 30	17	3-6	5	4	13-2	23-0	8-10½	8-3
		RPF	40	20	3-6	5	4	13-2	23-0	8-7½	8-0
2½	22°-37'-12"	CAST	20-40	8¼	5-0	5	5	17-7	36-0	12-4½	11-10¾
		RPF	20 & 30	16½	5-0	5	5	17-7	36-0	11-8¼	11-2½
		RPF	40	20	5-0	5	5	17-7	36-0	11-5	10-11
3	18°-55'-29"	CAST	20-60	8⅜	5-0	5	5	20-0	55-0	14-8½	14-3½
		RPF	20 & 30	16½	5-0	5	5	20-0	55-0	14-½	13-7½
		RPF	40-60	2-0	5-0	5	5	20-0	55-0	13-5	13-0
4	14°-15'-00"	CAST	40-60	11¹/₁₆	5-0	5	5	24-0	102-0	18-4¾	18-0¾
		RPF	40-60	2-3	5-0	5	5	24-0	102-0	17-0¾	16-9
5	11°-25'-16"	CAST	40-60	13¹³/₁₆	7-6	5	8	32-3	156-0	23-10	23-7¼
		RPF	40-60	2-6	7-6	5	8	32-3	156-0	22-5¾	22-3
6	9°-31'-38"	CAST	40-60	16⁹/₁₆	7-6	5	8	36-3	233-0	27-6¾	27-4¼
		RPF	40-60	3-0	7-6	5	8	36-3	233-0	25-11½	25-9
6	9°-31'-38"	CAST	40-60	16⁹/₁₆	10-0	5	10	40-5	223-0	29-2¾	29-0¼
		RPF	40-60	3-0	10-0	5	10	40-5	223-0	27-7½	27-5

TABLE 10. AMC Turnout Data: 20-60 Lb Rail, American Mining Congress Design (continued)

48 in. GAGE

		FROG			SWITCH			Actual Lead, ft-in.	CLOSURE RAILS		
									Curved		Straight
No.	Angle	Type	Rail, lb	Toe, in.	Length, ft-in.	Heel Spr., in.	K, in.		Radius, ft-in.	Length, ft-in.	Length, ft-in.
		CAST	20-40	6¹¹/₁₆	3-6	5	4	14-3	26-0	10-10¼	10-2¼
2	28°-04'-21"	RPF	20 & 30	17	3-6	5	4	14-3	26-0	10-0¼	9-4
		RPF	40	20	3-6	5	4	14-3	26-0	9-9¼	9-1
		CAST	20-40	8¼	5-0	5	5	19-0	41-0	13-10¼	13-3¾
2½	22°-37'-12"	RPF	20 & 30	16½	5-0	5	5	19-0	41-0	13-2	12-7½
		RPF	40	20	5-0	5	5	19-0	41-0	12-10½	12-4
		CAST	20-60	8⅜	5-0	5	5	21-7	62-0	16-4	15-10½
3	18°-55'-29"	RPF	20 & 30	16½	5-0	5	5	21-7	62-0	15-8	15-2½
		RPF	40-60	2-0	5-0	5	5	21-7	62-0	15-0½	14-7
4	14°-15'-00"	CAST	40-60	11¹/₁₆	5-0	5	5	26-0	114-0	20-5	20-1
		RPF	40-60	2-3	5-0	5	5	26-0	114-0	19-1¼	18-9
5	11°-25-16	CAST	40-60	13¹³/₁₆	7-6	5	8	35-0	174-0	26-7½	26-4¼
		RPF	40-60	2-6	7-6	5	8	35-0	174-0	25-3¼	25-0
6	9°-31'-38"	CAST	40-60	16⁹/₁₆	7-6	5	8	39-4	260-0	30-8¼	30-5½
		RPF	40-60	3-0	7-6	5	8	39-4	260-0	29-0¾	28-10
6	9°-31'-38"	CAST	40-60	16⁹/₁₆	10-0	5	10	43-8	248-0	32-6¼	32-3½
		RPF	40-60	3-0	10-0	5	10	43-8	248-0	30-10¾	30-8

stalled turnouts. Poor or incorrectly marked sights will contribute to derails, poor performance, and lower safety. In addition, a new or inexperienced supervisor and crew can struggle for an inordinate amount of time trying to install a turnout without sights. Mine management should be well aware of this and take measures to correct any problems.

Turnout Installation

After the turnout preparation has been completed, and the correct materials for installation delivered, the actual installation proceeds. Only a supervisor and two workers are necessary for installation, which usually takes two to three manshifts to complete.

The supervisor and crew begins the shift by gathering the tools necessary to complete the job. Table 12 summarizes the turnout tools required.

Then, the work party proceeds to the site. While the crew sorts their tools and begins unclipping the ties, the supervisor drops a string line from the PF spad to the top of the rail and marks the spot. The appropriate distances on both sides of the frog point are marked and the

TABLE 10. AMC Turnout Data: 20-60 Lb Rail, American Mining Congress Design (continued)

56½ in. GAGE											
	FROG				SWITCH				CLOSURE RAILS		
									Curved		Straight
No.	Angle	Type	Rail, lb	Toe, in.	Length, ft-in.	Heel Spr., in.	K, in.	Actual Lead, ft-in.	Radius, ft-in.	Length, ft-in.	Length, ft-in.
		CAST	20-40	6¹¹⁄₁₆	3-6	5	4	16-7	33-0	13-4	12-6¼
2	28°-04'-21"	RPF	20 & 30	17	3-6	5	4	16-7	33-0	12-5¾	11-8
		RPF	40	20	3-6	5	4	16-7	33-0	12-2¾	11-5
		CAST	20-40	8¼	5-0	5	5	22-0	51-0	16-11¼	16-3¾
2½	22°-37'-12"	RPF	20 & 30	16½	5-0	5	5	22-0	51-0	16-3	15-7½
		RPF	40	20	5-0	5	5	22-0	51-0	15-11½	15-4
		CAST	20-60	8⅜	5-0	5	5	25-0	76-0	19-10	19-3½
3	18°-55'-29"	RPF	20 & 30	16½	5-0	5	5	25-0	76-0	19-1¾	18-7½
		RPF	40-60	2-0	5-0	5	5	25-0	76-0	18-6½	18-0
4	14°-15'-00"	CAST	40-60	11¹⁄₁₆	5-0	5	5	30-4	140-0	24-9¾	24-5
		RPF	40-60	2-3	5-0	5	5	30-4	140-0	23-6	23-1
5	11°-25'-16"	CAST	40-60	13¹³⁄₁₆	7-6	5	8	40-7	212-0	32-3	31-11¼
		RPF	40-60	2-6	7-6	5	8	40-7	212-0	30-10¾	30-7
6	9°-31'-38"	CAST	40-60	16⁹⁄₁₆	7-6	5	8	45-10	316-0	37-2¾	36-11½
		RPF	40-60	3-0	7-6	5	8	45-10	316-0	35-¼	35-4
6	9°-31'-38"	CAST	40-60	16⁹⁄₁₆	10-0	5	10	50-7	302-0	39-5½	39-2½
		RPF	40-60	3-0	10-0	5	10	50-7	302-0	37-10¼	37-7

Source: Midwest Steel Catalogue S-180.
Note: RPF denotes "Riveted Plate Frog."
Metric equivalents: in. × 25.4 = mm; ft × 0.3048 = m; lb × 0.453 592 4 = kg.

rail is cut. Bolt holes are torch cut (blown) through the rail where they will correspond to the bolt holes already in the ears of the frog. After the rail cools, the frog is bolted to the rail in position with the PF spad. Holes are marked in the opposite rail end and blown through to attach a filler rail (prebent rail). When the rail cools, the prebent rail used as the outside rail of the turnout is bolted to the straight rail as shown in Fig. 48.

The next step after bolting the frog and prebent rail to the turnout is positioning and clipping the switch ties to the turnout. These six key ties are placed in particular locations to allow a predesigned turnout to be assembled with repetitive tasks in a standardized manner. Tie Nos. 1 and 2 (long ties) are the ties supporting the switch points and also form the switch stand. Tie Nos. 3 and 4 (short ties) support the switch points and form the "heel" of the switch point. Tie Nos. 5 and 6 (frog ties) are placed between the frog and the outside rails of the straight and

TABLE 11. AMC Turnout Data: 70–100 Lb Rail, American Mining Congress Design

36 in. GAGE

No.	Angle	Type	Rail Weight, lb	Point, in.	Actual Toe Length, in.	Length, ft-in.	Angle	Point, in.	Heel Spread, in.	Vertex Distance, in.	Actual Lead, ft-in.	Radius, ft-in.	M.O., in.	Length, ft-in.	Length, ft-in.
													Curved		Straight
3	18°-55′-29″	CAST	70 to 80	1/2	9 1/2	7-6	3° 35 00	3/8	6	6	19-1	39-0	4 3/16	11-1 7/16	10-9 1/2
		CAST	85 to 100	1/2	11	7-6	3° 44 35	3/8	6 1/4	6	18-10 5/8	37-7 3/16	4 5/8	10-9 9/16	10-5 5/8
		RBF	70 to 80	1/2	31	7-6	3° 35 00	3/8	6	6	19-1	39-0	3 3/8	9-3 15/16	9-0
		RBF	85 to 100	1/2	31	7-6	3° 44 35	3/8	6 1/4	6	18-10 5/8	37-7 3/16	3 3/8	9-1 9/16	8-9 5/8
4	14° 15′ 00″	CAST	70 to 80	1/2	12 9/16	7-6	3° 35 00	3/8	6	6	22-5	73-0	4 1/8	14-1 1/2	13-10 1/2
		CAST	85 to 100	1/2	14 5/8	7-6	3° 44 35	3/8	6 1/4	6	22-2 1/2	70-4 5/8	4	13-8 7/8	13-5 7/8
		RBF	70 to 80	1/2	40	7-6	3° 35 00	3/8	6	6	22-5	73-0	2 7/8	11-10	11-7
		RBF	85 to 100	1/2	40	7-6	3° 44 35	3/8	6 1/4	6	22-2 1/2	70-4 5/8	2 7/8	11-7 1/2	11-4 1/2
5	11° 25′ 16″	CAST	70 to 80	1/2	15 11/16	10-0	2° 41 12	3/8	6	8	29-0	114-0	4 1/8	17-10 11/16	17-8 5/16
		CAST	85 to 100	1/2	18 3/16	10-0	2° 48 23	3/8	6 1/4	8	28-9 1/8	110-6	4 1/8	17-5 3/8	17-2 15/16
		RBF	70 to 80	1/2	42 1/2	10-0	2° 41 12	3/8	6	8	29-0	114-0	3 1/4	15-7 7/8	15-5 1/2
		RBF	85 to 100	1/2	42 1/2	10-0	2° 48 23	3/8	6 1/4	8	28-9 1/8	110-6	3 1/4	15-5 1/16	15-2 5/8
6	9° 31′ 38″	CAST	70 to 80	1/2	18 13/16	10-0	2° 41 12	3/8	6	8	32-2	169-0	3 13/16	20-9 3/8	20-7 3/16
		CAST	85 to 100	1/2	21 13/16	10-0	2° 48 23	3/8	6 1/4	8	31-10 1/2	165-4 3/16	3 3/4	20-2 3/4	20-0 11/16
		RBF	70 to 80	1/2	45	10-0	2° 41 12	3/8	6	8	32-2	169-0	3 1/16	18-7 1/4	18-5
		RBF	85 to 100	1/2	45	10-0	2° 48 23	3/8	6 1/4	8	31-10 1/2	165-4 3/16	3 1/16	18-3 9/16	18-1 1/2
7	8° 10′ 16″	CAST	70 to 80	1/2	21 15/16	10-0	2° 41 12	3/8	6	8	35-4 11/16	238-11 11/16	3 9/16	23-8 1/2	23-6 3/4
		CAST	85 to 100	1/2	25 7/16	10-0	2° 48 23	3/8	6 1/4	8	34-9 3/8	234-7 15/16	3 5/16	22-9 11/16	22-7 15/16
		RBF	70 to 80	1/2	56 1/2	10-0	2° 41 12	3/8	6	8	35-4 11/16	238-11 11/16	2 3/4	20-9 15/16	20-8 3/4
		RBF	85 to 100	1/2	56 1/2	10-0	2° 48 23	3/8	6 1/4	8	34-9 3/8	234-7 15/16	2 5/8	20-2 5/8	20-0 7/8
8	7° 9 10	CAST	70 to 80	1/2	25 1/16	15-0	1° 47 27	3/8	6	12	45-8 7/8	298-8 11/16	4 3/16	28-9 1/2	28-7 13/16
		CAST	85 to 100	1/2	29 1/16	15-0	1° 52 14	3/8	6 1/4	12	45-0 3/8	291-9 1/4	3 15/16	27-8 13/16	27-7 5/8
		RBF	70 to 80	1/2	61	15-0	1° 47 27	3/8	6	12	45-8 7/8	298-8 11/16	3 3/8	25-9 11/16	25-7 7/8
		RBF	85 to 100	1/2	61	15-0	1° 52 14	3/8	6 1/4	12	45-0 3/8	291-9 1/4	3 1/4	25-0 7/8	24-11 3/8

42 in. GAGE

No.	Angle	Type	Rail Weight, lb	Point, in.	Actual Toe Length, in.	Length, ft-in.	Angle	Point, in.	Heel Spread, in.	Vertex Distance, in.	Actual Lead, ft-in.	Radius, ft-in.	M.O., in.	Length, ft-in.	Length, ft-in.
													Curved		Straight
3	18° 55′ 29″	CAST	70 to 80	1/2	9 1/2	7-6	3° 35 00	3/8	6	6	21-7	49-0	5 3/4	13-8 1/16	13-3 1/2
		CAST	85 to 100	1/2	11	7-6	3° 44 35	3/8	6 1/4	6	21-4 9/16	47-2 3/4	5 5/8	13-4 3/16	12-11 9/16
		RBF	70 to 80	1/2	31	7-6	3° 35 00	3/8	6	6	21-7	49-0	4 5/16	11-10 9/16	11-6
		RBF	85 to 100	1/2	31	7-6	3° 44 35	3/8	6 1/4	6	21-4 9/16	47-2 3/4	4 5/16	11-8 3/16	11-3 9/16
4	14° 15′ 00″	CAST	70 to 80	1/2	12 9/16	7-6	3° 35 00	3/8	6	6	25-7	90-0	5	17-4	17-0 1/2
		CAST	85 to 100	1/2	14 5/8	7-6	3° 44 35	3/8	6 1/4	6	25-4 3/8	87-10 3/16	4 7/8	16-11 5/16	16-7 3/4
		RBF	70 to 80	1/2	40	7-6	3° 35 00	3/8	6	6	25-7	90-0	3 3/4	15-0 1/2	14-9
		RBF	85 to 100	1/2	40	7-6	3° 44 35	3/8	6 1/4	6	25-4 3/8	87-10 3/16	3 3/4	14-9 15/16	14-6 3/8
5	11° 25′ 16″	CAST	70 to 80	1/2	15 11/16	10-0	2° 41 12	3/8	6	8	33-1	140-0	5 3/16	22-0 1/16	21-9 5/16
		CAST	85 to 100	1/2	18 3/16	10-0	2° 48 23	3/8	6 1/4	8	32-9 1/4	137-4 9/16	5	21-5 7/8	21-3 1/16
		RBF	70 to 80	1/2	42 1/2	10-0	2° 41 12	3/8	6	8	33-1	140-0	4 3/16	19-9 1/4	19-6 1/2
		RBF	85 to 100	1/2	42 1/2	10-0	2° 48 23	3/8	6 1/4	8	32-9 1/4	137-4 9/16	4 1/8	19-5 9/16	19-2 3/4
6	9° 31′ 38″	CAST	70 to 80	1/2	18 13/16	10-0	2° 41 12	3/8	6	8	36-10	209-0	4 5/8	25-5 9/16	25-3 3/8
		CAST	85 to 100	1/2	21 13/16	10-0	2° 48 23	3/8	6 1/4	8	36-6	205-0 3/4	4 1/2	24-10 5/8	24-8 3/8
		RBF	70 to 80	1/2	45	10-0	2° 41 12	3/8	6	8	36-10	209-0	3 7/8	23-3 3/8	23-1
		RBF	85 to 100	1/2	45	10-0	2° 48 23	3/8	6 1/4	8	36-6	205-0 3/4	3 7/8	22-11 7/16	22-9
7	8° 10′ 16″	CAST	70 to 80	1/2	21 15/16	10-0	2° 41 12	3/8	6	8	40-7 13/16	294-1 3/4	4 5/8	28-11 7/8	28-9 7/8
		CAST	85 to 100	1/2	25 7/16	10-0	2° 48 23	3/8	6 1/4	8	39-11 3/4	290-6 1/8	4 1/16	28-0 7/16	27-10 5/8
		RBF	70 to 80	1/2	56 1/2	10-0	2° 41 12	3/8	6	8	40-7 13/16	294-1 3/4	3 1/2	26-1 5/16	25-11 5/16
		RBF	85 to 100	1/2	56 1/2	10-0	2° 48 23	3/8	6 1/4	8	39-11 3/4	290-6 1/8	3 5/8	25-5 3/8	25-3 1/4
8	7° 9 10	CAST	70 to 80	1/2	25 1/16	15-0	1° 47 27	3/8	6	12	52-1 5/8	367-3 3/4	5 1/16	35-2 1/2	35-0 9/16
		CAST	85 to 100	1/2	29 1/16	15-0	1° 52 14	3/8	6 1/4	12	51-4 7/16	360-8 7/8	4 13/16	34-1 1/8	33-11 3/8
		RBF	70 to 80	1/2	61	15-0	1° 47 27	3/8	6	12	52-1 5/8	367-3 3/4	4 1/4	32-2 9/16	32-0 5/8
		RBF	85 to 100	1/2	61	15-0	1° 52 14	3/8	6 1/4	12	51-4 7/16	360-8 7/8	4 1/16	31-5 3/16	31-3 7/16

TABLE 11. AMC Turnout Data: 70–100 Lb Rail, American Mining Congress Design (continued)

44 in. GAGE

No.	Angle	Type	Rail Weight, lb	Frog Point, in.	Frog Actual Toe Length, in.	Switch Length, ft-in.	Switch Angle	Switch Point, in.	Switch Heel Spread, in.	Switch Vertex Distance, in.	Actual Lead, ft-in.	Closure Curved Radius, ft-in.	Closure Curved M.O., in.	Closure Curved Length, ft-in.	Closure Straight Length, ft-in.
3	18° 55′ 29″	CAST	70 to 80	1/2	9 1/2	7-6	3° 35 00	3/8	6	6	22-6	52-0	6 1/8	14-7 1/4	14-2 1/2
		CAST	85 to 100	1/2	11	7-6	3° 44 35	3/8	6 1/4	6	22-2 9/16	50-5 5/16	5 15/16	14-2 3/8	13-9 9/16
		RBF	70 to 80	1/2	31	7-6	3° 35 00	3/8	6	6	22-6	52-0	4 3/4	12-10	12-5
		RBF	85 to 100	1/2	31	7-6	3° 44 35	3/8	6 1/4	6	22-2 9/16	50-5 5/16	4 11/16	12-6 3/8	12-1 9/16
4	14° 15′ 00″	CAST	70 to 80	1/2	12 9/16	7-6	3° 35 00	3/8	6	6	26-10	96-0	5 3/8	18-7	18-3 1/2
		CAST	85 to 100	1/2	14 5/8	7-6	3° 44 35	3/8	6 1/4	6	26-5	93-8 1/1	5 3/16	18-0 1/8	17-8 3/8
		RBF	70 to 80	1/2	40	7-6	3° 35 00	3/8	6	6	26-10	96-0	4 3/16	16-3 3/4	16-0
		RBF	85 to 100	1/2	40	7-6	3° 44 35	3/8	6 1/4	6	26-5	93-8 1/1	4 1/16	15-10 3/4	15-7
5	11° 25′ 16″	CAST	70 to 80	1/2	15 11/16	10-0	2° 41 12	3/8	6	8	34-7	149-0	5 9/16	23-6	23-3 1/4
		CAST	85 to 100	1/2	18 3/16	10-0	2° 48 23	3/8	6 1/4	8	34-1 1/4	146-4 1/16	5 5/16	22-10 1/16	22-7 1/16
		RBF	70 to 80	1/2	42 1/2	10-0	2° 41 12	3/8	6	8	34-7	149-0	4 9/16	21-3 1/2	21-0 1/2
		RBF	85 to 100	1/2	42 1/2	10-0	2° 48 23	3/8	6 1/4	8	34-1 1/4	146-4 1/16	4 7/16	20-9 3/4	20-6 3/4
6	9° 31′ 38″	CAST	70 to 80	1/2	18 13/16	10-0	2° 41 12	3/8	6	8	38-8	222-0	5 1/16	27-3 1/2	27-1
		CAST	85 to 100	1/2	21 13/16	10-0	2° 48 23	3/8	6 1/4	8	38-0 1/2	218-3 5/8	4 13/16	26-5 5/16	26-2 11/16
		RBF	70 to 80	1/2	45	10-0	2° 41 12	3/8	6	8	38-8	222-0	4 1/4	25-1 1/2	24-11
		RBF	85 to 100	1/2	45	10-0	2° 48 23	3/8	6 1/4	8	38-0 1/2	218-3 5/8	4 1/8	24-6 1/8	24-3 1/2
7	8° 10′ 16″	CAST	70 to 80	1/2	21 15/16	10-0	2° 41 12	3/8	6	8	42-4 7/8	312-7 7/16	4 9/16	30-9 1/8	30-6 15/16
		CAST	85 to 100	1/2	25 7/16	10-0	2° 48 23	3/8	6 1/4	8	41-8 5/8	309-1 1/2	4 5/16	29-9 5/16	29-7 3/16
		RBF	70 to 80	1/2	56 1/2	10-0	2° 41 12	3/8	6	8	42-4 7/8	312-7 7/16	3 3/4	27-10 9/16	27-8 3/8
		RBF	85 to 100	1/2	56 1/2	10-0	2° 48 23	3/8	6 1/4	8	41-8 5/8	309-1 1/2	3 5/8	27-2 1/4	27-0 1/8
8	7° 9′ 10″	CAST	70 to 80	1/2	25 1/16	15-0	1° 47 27	3/8	6	12	54-3 3/4	390-2 1/16	5 3/8	37-4 3/16	37-2 1/8
		CAST	85 to 100	1/2	29 1/16	15-0	1° 52 14	3/8	6 1/4	12	53-5 3/4	383-8 11/16	5 1/8	36-2 9/16	36-0 11/16
		RBF	70 to 80	1/2	61	15-0	1° 47 27	3/8	6	12	54-3 3/4	390-2 1/16	4 1/2	34-4 1/4	34-2 3/16
		RBF	85 to 100	1/2	61	15-0	1° 52 14	3/8	6 1/4	12	53-5 3/4	383-8 11/16	4 7/16	33-6 5/8	33-4 3/4

48 in. GAGE

No.	Angle	Type	Rail Weight, lb	Frog Point, in.	Frog Actual Toe Length, in.	Switch Length, ft-in.	Switch Angle	Switch Point, in.	Switch Heel Spread, in.	Switch Vertex Distance, in.	Actual Lead, ft-in.	Closure Curved Radius, ft-in.	Closure Curved M.O., in.	Closure Curved Length, ft-in.	Closure Straight Length, ft-in.
3	18° 55′ 29″	CAST	70 to 80	1/2	9 1/2	7-6	3° 35 00	3/8	6	6	24-1	58-0	6 13/16	16-2 13/16	15-9 1/2
		CAST	85 to 100	1/2	11	7-6	3° 44 35	3/8	6 1/4	6	23-10 1/2	56-10 5/16	6 11/16	15-10 13/16	15-5 1/2
		RBF	70 to 80	1/2	31	7-6	3° 35 00	3/8	6	6	24-1	58-0	5 3/8	14-5 1/4	14-0
		RBF	85 to 100	1/2	31	7-6	3° 44 35	3/8	6 1/4	6	23-10 1/2	56-10 5/16	5 3/8	14-2 13/16	13-9 1/2
4	14° 15′ 00″	CAST	70 to 80	1/2	12 9/16	7-6	3° 35 00	3/8	6	6	28-10	107-0	5 15/16	20-7 9/16	20-3 1/2
		CAST	85 to 100	1/2	14 5/8	7-6	3° 44 35	3/8	6 1/4	6	28-6 1/4	105-3 3/4	5 3/4	20-1 3/4	19-9 5/8
		RBF	70 to 80	1/2	40	7-6	3° 35 00	3/8	6	6	28-10	107-0	4 3/4	18-4 1/16	18-0
		RBF	85 to 100	1/2	40	7-6	3° 44 35	3/8	6 1/4	6	28-6 1/4	105-3 3/4	4 5/8	18-0 3/8	17-8 1/4
5	11° 25′ 16″	CAST	70 to 80	1/2	15 11/16	10-0	2° 41 12	3/8	6	8	37-1	167-0	6 1/8	26-0 1/2	25-9 5/16
		CAST	85 to 100	1/2	18 3/16	10-0	2° 48 23	3/8	6 1/4	8	36-9 5/16	164-3 1/8	5 15/16	25-6 3/8	25-3 1/8
		RBF	70 to 80	1/2	42 1/2	10-0	2° 41 12	3/8	6	8	37-1	167-0	5 1/16	23-9 3/4	23-6 1/2
		RBF	85 to 100	1/2	42 1/2	10-0	2° 48 23	3/8	6 1/4	8	36-9 5/16	164-3 1/8	5 1/16	23-6 1/16	23-2 13/16
6	9° 31′ 38″	CAST	70 to 80	1/2	18 13/16	10-0	2° 41 12	3/8	6	8	41-6	248-0	5 1/2	30-1 15/16	29-11 3/16
		CAST	85 to 100	1/2	21 13/16	10-0	2° 48 23	3/8	6 1/4	8	41-1 1/2	244-9 5/16	5 3/8	29-6 9/16	29-3 11/16
		RBF	70 to 80	1/2	45	10-0	2° 41 12	3/8	6	8	41-6	248-0	4 3/4	27-11 3/4	27-9
		RBF	85 to 100	1/2	45	10-0	2° 48 23	3/8	6 1/4	8	41-1 1/2	244-9 5/16	4 11/16	27-7 3/8	27-4 1/2
7	8° 10′ 16″	CAST	70 to 80	1/2	21 13/16	10-0	2° 41 12	3/8	6	8	45-10 15/16	349-4 7/16	5 1/16	34-3 5/16	34-1
		CAST	85 to 100	1/2	25 7/16	10-0	2° 48 23	3/8	6 1/4	8	45-2 1/4	346-4 1/4	4 13/16	33-3 3/16	33-0 13/16
		RBF	70 to 80	1/2	56 1/2	10-0	2° 41 12	3/8	6	8	45-10 15/16	349-4 7/16	4 1/4	31-4 3/4	31-2 7/16
		RBF	85 to 100	1/2	56 1/2	10-0	2° 48 23	3/8	6 1/4	8	45-2 1/4	346-4 1/4	4 1/8	30-8 1/8	30-5 3/4
8	7° 9′ 10″	CAST	70 to 80	1/2	25 1/16	15-0	1° 47 27	3/8	6	12	58-6 5/16	435-10 3/4	5 5/8	41-7 1/2	41-5 1/4
		CAST	85 to 100	1/2	29 1/16	15-0	1° 52 14	3/8	6 1/4	12	57-8 7/16	429-8 7/16	5 3/4	40-5 3/8	40-3 3/8
		RBF	70 to 80	1/2	61	15-0	1° 47 27	3/8	6	12	58-6 5/16	435-10 3/4	5 1/16	38-7 9/16	38-5 5/16
		RBF	85 to 100	1/2	61	15-0	1° 52 14	3/8	6 1/4	12	57-8 7/16	429-8 7/16	5	37-9 7/16	37-7 7/16

Source: Midwest Steel Catalogue S-180.
Note: RBF denotes "Rigid Bolted Frog."
Metric equivalents: in. × 25.4 = mm; ft × 0.3048 = m; lb × 0.453 592 4 = kg.

TABLE 12. Tools Needed for Turnout Installation

Item	No. Required	Figure No.	Use
Spike maul	2	20	Unclipping ties to rail, clipping ties, and adjusting ties.
Rail tongs	2	22	Move rail and switch parts.
Railroad pick	1	18	Trimming bottom, arranging ties.
Track jack	1 or 2	24	Lift and support rail or switch components.
Track wrench	1 or 2	23	Tighten bolts on frog, switch points and switch stand.
Claw bar	1	21	Clipping or spiking ties to rail and aligning the rail.
Rail bender	1	40	Bending rail to complete turnout.
Hydraulic rail jack	1	—	Power rail bender.
Acetylene torch	1	—	Cut rail at designated spots.

turnout track. Fig. 49 illustrates the approximate location of the key ties.

After the key ties are placed, the curved closure and straight closure rails are bolted to the frog and clipped to the key ties. At this time, regular ties on normal spacing are placed in a criss-crossing pattern to lock the turnout and straight rails together through the switch. These ties are clipped to the rails after being placed. During this step, the filler rails need to be sprung while being clipped.

The switch points are now placed and bolted to the straight and curved closure rails. Specially bent splice bars and bolts are used to allow the motion of the points and the operation of the switch.

The guard rails are the next component to be placed. These are quickly mounted opposite the frog on the outside rails. These are easily placed by two workmen.

The final task to completing the switch is bolting the switch stand to the ties, connecting the rod, and adjusting the "throw" of the switch. Ample lubricants should be used to aid the workmen. Adjustments are made by the throw length of the switch and the spring tension in the connecting rod. Close supervision is necessary to ensure the proper adjustment. The switch should operate smoothly, with little effort.

One final note concerns the inside rail of the turnout. This rail usually must be completely bent by manual methods using the "butterfly." The centerline of the turnout is marked and string lines dropped to the track level. The centerline of the ties are marked after they are clipped to the prebent filler rail (Fig. 50). Once the ties are clipped to the prebent rail and on line with the centerline, then the inside rail can be bent to conform to the tie pattern (Fig. 51). This bending and completion of the turnout is usually accomplished within a shift or two after the switch completion. Along with the final bending task comes the cross-bonding of the switch. This is easily accomplished during the final task by one man. The appropriate pattern of cross-bonding

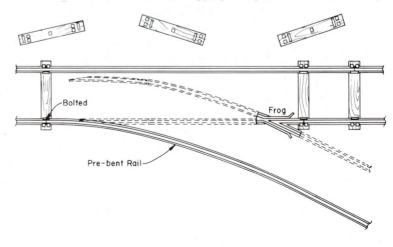

FIG. 48. Prebent rail is bolted to straight rail with standard splice bars.

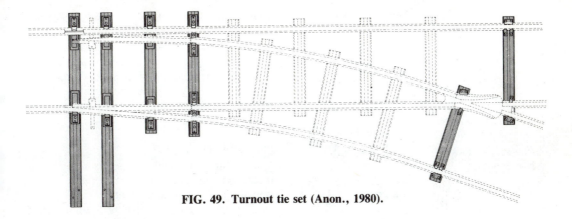

FIG. 49. Turnout tie set (Anon., 1980).

grounds the rails together through the switch, as illustrated in Fig. 52.

In concluding this section on turnout installation, several points need emphasizing:

1) By using a standard approach to installing a turnout, mine management can improve the installation time, training time, and quality control of turnout installation throughout the mine.

2) Switches manufactured to required specifications and then taken underground are necessary for such standardization.

Care should be exercised when installing switches to maintain the integrity and quality of the components and connections. Switch safety is directly dependent upon the installation practices.

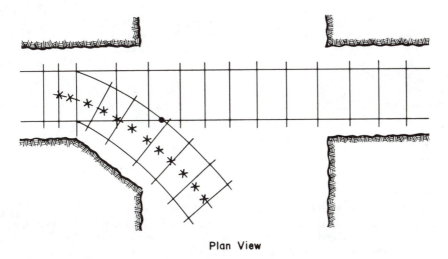

Plan View

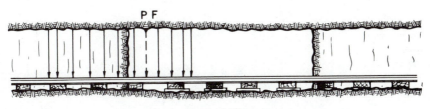

Vertical View

FIG. 50. String lines dropped from the roof mark the centerline of the turnout track.

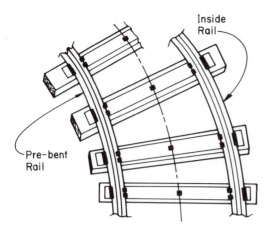

Plan View

FIG. 51. Bending the inside rail to the ties.

INSTALLING TROLLEY WIRE

Once the decision to electrify a railway has been made, the design of such a system begins. Trolley wire is bare copper wire in a number of shapes which carry between 250 and 300 v of direct-current power. There are many varieties of copper trolley wire available to mine operators for powering the system. Figures No. 8 and No. 9 deep-sectioned grooved wire is the most popular wire used underground today.

In discussing trolley wire installation, several topics will be covered. These are:

Preparation—which includes designing and drilling the holes and choosing the materials.

Assembly—which includes hanging the wire and splicing to existing wire.

Cutting a Turnout Switch—which involves all tasks required for installing a trolley switch into the wire already hung for the straight.

Maintenance—which includes lubricating the wire, rehanging or straightening bends, and replacing damaged support hangers.

Each of these will be discussed with regard to the men, materials, and design parameters used in the wire installation. One aspect not under consideration is the initial powering of such a system. Since the chief electrician and a trained crew usually perform this work, it is not considered a typical project beyond the positioning of the power center and rectifier. These will be discussed in later sections. For now, the installation will be discussed.

Preparation

When any construction supervisor begins directing wire hanging activities, there are several steps in preparation to be completed. These include marking the top, drilling the holes, and cutting the support pipe. Preparation activities can take one to two manshifts to complete, and require at least one supervisor and one worker for all preparation activities.

Marking the top for wire installations requires stringing an appropriate sight line along the trackway. Usually this is strung some 152.4 mm (6 in.) to the outside of the wireside rail (Fig. 53). From the string line the foreman measures between the last permanent trolley hanger and the first hole of the new extension (Fig. 54), usually between 3.0 and 4.6 m (10 and 15 ft) away. The foreman marks this spot with chalk or paint. He repeats this procedure for each point where a trolley wire hanger hole is to be drilled. The distance between hanger holes is mainly a function of management desire. As one closes the distance of the holes, say, to 3.0 m (10 ft), more stability and support for the wire is obtained. While this tends to increase the reliability of the system, it also increases cost in terms of materials and installation time.

The next step is drilling the hanger holes by using a rail-mounted air compressor powered by a d-c power source. A stoping drill (or jackleg) and rock bits [25.4 mm (1 in.) diam.] are used to drill holes approximately 203.2 to 304.8 mm (8 to 12 in.) deep. Deeper holes may be necessary if the first few meters (feet) are weak

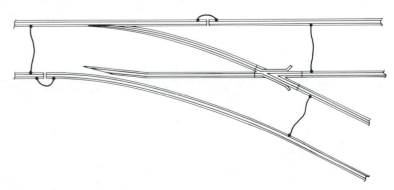

FIG. 52. Typical cross-bonding pattern for a mainline track switch.

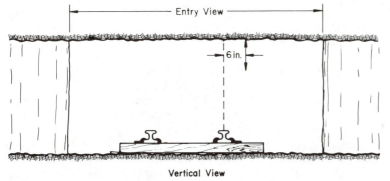

FIG. 53. Marking the wire centerline on a 152.4-mm (6-in.) offset.

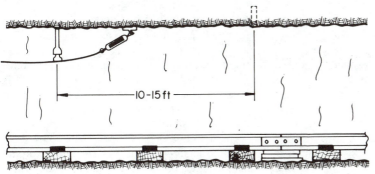

FIG. 54. Measuring the initial distance between the last wire hanger and the first hanger hole. Metric equivalent: ft × 0.3048 = m.

rock. In areas where ten or more holes are being drilled at one time, a second man to handle the air compressor may be necessary to efficiently complete the task.

The final step of preparation is cutting the support pipe. This pipe is usually 1 1/2 in. OD schedule 80 steel pipe. The pipe is needed when the roof of the mine is 2.1 m (7 ft) or greater above the rail because the pipe provides a reinforcing support for the trolley wire hanger to minimize sway and vibration. The foreman will measure from the roof to a point 2.0 m (6 1/2 ft) above the rail level (Fig. 55). This distance, when adjusted for the hanger end caps, is the length needed for the support pipe. The pipe lengths are cut and labeled for each hole. Short distances, less than 127 mm (5 in.) may or may not need a support pipe.

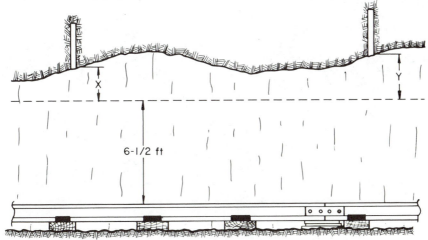

FIG. 55. Establishing a datum line for the wire above the track. In this case, 2.0 m (6 ½ ft) is used.

Assembly

After cutting the support pipe, the assembly task begins. Two workmen and a supervisor are required. The procedure is as follows:

1) At each hanger hole, an expansion bolt (Fig. 56) is inserted and tightened. The bolt has a threaded rod which attaches to an adaptor (Fig. 57) for the support pipe.

2) The support pipe is attached to the adaptor and tightened to solidify the connection (Fig. 58). A second adaptor is bolted to the bottom of the pipe to allow a mine hanger to be installed (Fig. 59).

3) A mine hanger (Fig. 60) is screwed on the end of the adaptor, and a trolley clamp (Fig. 61) is attached to the hanger.

4) Steps 1 through 3 are repeated for each hanger hole.

At the completion of these tasks, the foreman will measure the linear amount of wire needed to hang, allowing an extra 1.5 to 3.0 m (5 to 10 ft) to work with. It is important to mention the typical amount of wire to hang at each setup should be between 91.4 and 152.4 m (300 and 500 ft) for each section since wire splices should not be placed close together. The crew will cut the appropriate amount of wire and stretch it along the trackway. The foreman travels to the outby section insulator (knife) switch, throws it, and tags it out. This de-energizes the wire section where the new piece of wire will be added. The next step is straightening and tying the wire to each hanger with rope. When completed, this gets the wire next to the hanger and at approximately the right height (Fig. 62). The inby wire end is attached to a tip and turnbuckle which is connected to the roof with an expansion bolt and hook, as illustrated by Fig. 63. At this time, the crew again tries to untwist and straighten the wire as much as possible. A chain jack and eccentric clamp are attached to the inby end (end to be attached to existing wire) and secured to the roof. The slack is then jacked out of the wire until the wire becomes taut. The crew now begins at the outby end and mounts the wire into the clamps and tightens them down. They work toward the chain jack end for every hanger. After all hangers are connected and tightened to the wire, the existing wire turnbuckle is loosened and removed so the end can be attached to the new piece of wire. The wire ends are measured and cut with a hacksaw so the ends butt up against one another. A wire splice (Fig. 64) is used to make the connection. A good point to remember is to place the splicing connectors as close to an existing hanger as possible, to minimize sway and vibration of this weak point. After the connection is made, the chain jack is loosened and removed. The inby end is wrapped with insulating material and tagged as a dead-end. A reflector is also hung at this point.

The final task is going back to the outby knife switch and throwing it on. Tools and surplus materials are collected and returned to the appropriate places.

Cutting a Trolley Switch

After a track crew installs a turnout in the track system, the desire for wire servicing the turnout requires a trolley switch to be installed. This task centers around installing a trolley frog (Fig. 65) into the existing wire. Trolley frogs are ordered to be right- or left-handed and with the same frog angle as the track switch. Usually, the trolley frog is mounted on the wire at a point over the "heel" of the turnout switch points. It is recommended the existing wire be pretensioned before installing the frog. This pretensioning helps to support the weight of the trolley

FIG. 62. Tying the wire to each hanger with rope.

Vertical View

TRANSPORTATION—RAILROAD CONSTRUCTION

FIG. 56. Expansion bolt (Anon., 1977).

Number 12810
FIG. 57. Vertical pipe adapter, with stud (Anon., 1977).

Vertical View
FIG. 58. Pipe used to strengthen the wire connector and hanger.

Number 22387
FIG. 59. Vertical pipe adapter, without stud (Anon., 1977).

FIG. 60. Mine hanger (Anon., 1977).

FIG. 61. Trolley clamp (Anon., 1977).

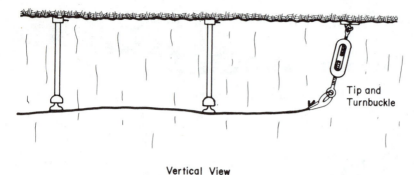

FIG. 63. End of wire is attached to roof by means of a wire tip and turnbuckle.

FIG. 64. Trolley wire splicer (Anon., 1977).

FIG. 65. Trolley frog (Anon., 1977).

frog and also secures the wire to the frog when the tensioning acts upon the frog connections. One additional point in cutting a trolley switch concerns the support. Usually, 9.5 mm (3/8 in.) steel cable is looped through the "eyes" of the frogs and secured to the roof by hanger hooks. This support is provided in a direction generated by the rail traffic as it moves through the switch. In addition, the power to the wire should be off at the outby knife switch while the trolley frog is being installed.

Maintenance

After the installation of a trolley wire system, there are several maintenance tasks to be done. These include lubrication of the wire, rehanging or straightening bends, and replacing damaged support hangers. Except for lubricating the wire, none of these tasks take more than one manshift to complete, but require the power be shut off for repairs. I shall briefly discuss each item.

Lubricating the wire is achieved by the measured use of equipment applicators, as illustrated by Fig. 66. The lubricant is fed into the trolley tank and applied to the wire as the equipment moves along the track. Usually, one or two mine jeeps are fitted with these applicators and filled periodically. Lubricating the wire reduces the friction and hence, wear of the wire.

Rehanging or straightening wire involves adjusting the existing wire. Occasionally, trolley wire vibrates or is knocked from the trolley hanger support. In this case, retightening the trolley clamps around the wire will correct the problem. More often, a bend in the wire which has not been straightened while hanging causes many pole derailings. The wire, therefore, must be hammered to straighten the bend and reten-

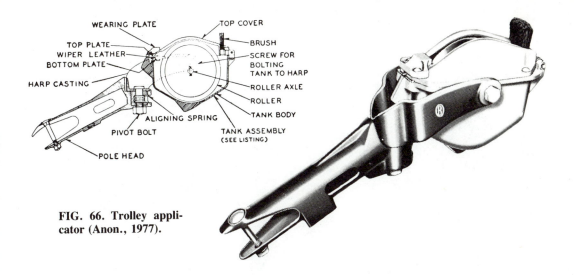

FIG. 66. Trolley applicator (Anon., 1977).

sioned to take up the slack. If this is the case, two men are needed for efficient work.

The final maintenance task is replacing a damaged or pulled support hanger. Many times efficient track systems are plagued with loose support hangers and lose much of their high efficiency. Loose hangers come from weathering top, intense vibrations from moving equipment, and poor installation. To correct the problem involves replacing the expansion bolt with a new one and re-anchoring it 76.2 to 101.6 mm (3 to 4 in.), usually, above the first spot. Sometimes, new holes must be drilled near the existing one, and a complete re-anchoring is required. Again, the power should be shut off before workers are allowed to work around the wire.

This concludes maintenance of the wire system and wire installation. The important aspect of trolley wire to remember is the installation. It should be done correctly the first time. Good wire systems promote safety, efficiency, and are an asset to mining operations.

RAIL EQUIPMENT MANUFACTURERS

The final topic under consideration is one of rail and track vendors. These organizations manufacture the rail, track, and wire components needed for an efficient system. Representatives are well versed in their component capabilities and special care characteristics.

To aid the engineer, superintendent, or purchasing agent in gathering the right materials for their operation, an annotated list of suppliers of various track and wire components is presented in Table 13. These are broken out into component types and alphabetical order. In addition, the address of the companies is listed in Appendix A. This information represents a partial listing of the mining manufacturer as compiled in the 1980 survey by the Mining Information Services group of McGraw-Hill. The complete list is published annually by McGraw-Hill in *Coal Age*.

REFERENCES AND BIBLIOGRAPHY

Anon., *A & K Railroad Catalogue*, A & K Railroad Materials, Inc., Salt Lake City, UT.

Anon., 1976, *Amtrak Specifications for Construction and Maintenance of Track*, National Railroad Passenger Corp., Washington, DC.

Anon., 1977, *The Ohio Brass Company Catalogue No. 70 Mining Equipment*, The Ohio Brass Co., Mansfield, OH.

Anon., 1980, *Midwest Steel Catalogue*, No. S–180, Midwest Corp., Charleston, WV.

Bethlehem Steel Corp., 1980, *Bethlehem Mine and Industrial Trackwork Catalogue*, No. 2341, Bethlehem, PA.

Crickmer, D.F., and Zegeer, D.A., 1981, *Elements of Practical Coal Mining*, 2nd ed., AIME, New York.

Cummins, A.B., and Given I.A., 1973, *SME Mining Engineering Handbook*, AIME, New York.

Peele, R.J., 1943, *Mining Engineer's Handbook*, 3rd ed., John Wiley & Sons, New York.

TABLE 13. Rail and Wire Manufacturers

Jacks, Track
A&K Railroad Materials, Inc.
Anixter Mine & Industrial Specialists
Atlantic Track & Turnout Co.
Big Sandy Electric & Supply Co.
Capital City Industrial Supply Co.
Dresser Industries, Inc.
Duff-Norton Co.
Fairmont Supply Co.
Logan Corp.
Marathon Coal Bit Co., Inc.
Midwest Steel Div., Midwest Corp.
National Mine Service Co.
Ryerson, Joseph & Son, Inc.
Templeton, Kenly & Co.
Tidewater Supply Co.

Rail
A&K Railroad Materials, Inc.
Atlantic Track and Turnout Co.
Atlas Rail Construction Co.
Bethlehem Steel Corp.
CF&I Steel Corp.
Connors Steel Co.
Foster, L.B., Co.
Jennmar Corp.
Midwest Steel Div., Midwest Corp.
Ortner Freight Car Co.
Porter, H.K. Co., Inc.
Ryerson, Joseph T., & Son, Inc.
Schart Co.
Steelplank Corp.
Tidewater Supply Co.
US Steel Corp.

Sheaves
Bolton, R.B.
Card Corp.
Connellsville Corp.
US Steel Corp.

Signals
   1. haulage
   2. rail & highway crossing
Atlantic Track & Turnout Co. (1, 2)
Communications & Control Eng. Co. Ltd. (1)
Davis, John & Son (Derby) Ltd. (1)
Safetran Systems Corp. (1, 2)
Strojex port, pzo (1, 2)
Union Switch & Signal Div., Amer. Standard Inc. (2)

Spike Pullers
Aldon Company
The Atlantic Track & Turnout Co.
Fairmont Supply Co.
Midwest Steel Div., Midwest Corp.
Rexnord Inc., Process Machinery Div.

Spikes Track
(See *Rail Spikes*)

Track Braces
A&K Railroad Materials, Inc.
Anixter Mine & Industrial Specialists
Atlantic Track & Turnout Co.
Card Corp.

Fairmont Supply Co.
Midwest Steel Div., Midwest Corp.

Track Cleaners
American Mine Door Co.
Atlas Railroad Construction Co.
Clayton Mfg. Co.
DuBois Chemicals Div. of Chemical Corp.
Midwest Industrial Supply Inc.
Pettibone Corp.

Track-Switch Heaters
Atlantic Track & Turnout Co.
Bethlehem Steel Corp.
General Electric Co.
Midwest Industrial Supply Inc.
Wiegand, Edwin L., Div., Emerson Elec. Co.

Track Work
Atlas Railroad Contruction Co.
Bethlehem Steel Corp.
Card Corp.
Construction & Mining Serv., Inc.
Fairmont Supply Co.
Jennmar Corp.
Midwest Steel Div., Midwest Corp.
Pettibone Corp.
Porter, H.K. Co., Inc.
Safetran Systems Corp., Mining &
Urban Transit, Signal & Control Div.

Tools
Aldon Company
Atlantic Track & Turnout Co.
Duquesne Mine Supply Co.
Fairmont Supply Co.
Midwest Steel Div., Midwest Corp.
Rexnord Inc.
Tidewater Supply Co.

Trolley Accessories
   1. clamps, connectors, hangers, splicers
   2. contact devices, slide-operated, proximity, etc.
   3. frogs
   4. guards
   5. harps, shoes, sliders, wheels
   6. poles, wood, plastic
   7. signal systems
Anixter Mine & Industrial Specialists (1, 3, 5)
Capital City Industrial Supply Co. (1, 2, 3, 4, 5, 6, 7)
Card Corp. (5)
Communication & Control Eng. Co., Ltd. (7)
Dayco Corp. (4)
Dresser Industries, Inc. (2)
Duquesne Mine Supply Co. (1, 3, 5, 6)
Elreco Corp, The (1, 3, 4, 5, 6)
Fairmont Supply Co. (1, 2, 3, 4, 5, 6)
Gauley Sales Co. (1, 3, 4, 5, 6)
General Equipment & Mfg. Co., Inc. (2, 7)
Goodman Equipment Corp. (5, 6)
Guyan Machinery Co. (1, 4)

Levels, Track
Aldon Co.
Fairmont Supply Co.
Rexnord Inc.

TABLE 13. Rail and Wire Manufacturers (continued)

Lubricators, Rail
Abex Corp.
DuBois Chemicals Div. of Chemical Corp.
Eaton Corp.
E-Power Industries Co.
Lunkenheimer Co., Div. of Conval
Safetran Systems Corp.
Trico Mfg. Corp.

Rail Accessories
1. benders
2. bolts
3. bond terminals
4. bonds
5. braces
6. clamps
7. contact devices
8. dollies
9. drills
10. frogs
11. levelers, spot boards
12. punches
13. signal systems, manual, automatic
14. spike drivers
15. spikes
16. splice bars, plates
17. switch-position indicators
18. switchthrowers
19. switchthrowers, automatic
20. tie plates & rail anchors
21. ties, treated
22. ties, steel
23. trackwork
24. turnouts, switches, stands
25. welding materials
26. guard rails
27. transition rails
28. thermal rail-welding kits

A&K Railroad Materials, Inc. (1, 2, 5, 6, 10, 11, 14, 15, 16, 18, 19, 20, 21, 22, 23, 24, 26, 27)
Abex Corp. (5, 10, 13, 14, 18, 19, 23, 24, 25)
Abex Corp., Railroad Products Group (5, 10, 13, 14, 18, 19, 23, 24, 25)
Aldon Company, The (1, 11)
American Mine Door Co. (3, 19)
Amsco Div., Abex Corp. (10, 23, 24, 25)
Anixter Mine & Industrial Specialists (14)
Armco Inc. (2, 22, 26)
Atlantic Track & Turnout Co. (1, 2, 3, 4, 5, 6, 7, 8, 9, 10, 11, 12, 13, 14, 15, 16, 17, 18, 19, 20, 21, 22, 23, 24, 25, 26, 27, 28)
Atlas Railroad Construction Co. (21, 23, 24)
BTR Trading, Inc. (16, 20, 21, 28)
Bethlehem Steel Corp. (2, 5, 10, 15, 16, 18, 19, 20, 22, 23, 24, 26)
CF&I Steel Corp. (15, 16, 20)
CWI Distributing Co. (2, 15)
Cam-Lok Div., Empire Products, Inc. (25)
Card Corp. (10, 18, 22, 23, 24, 26, 27)
Chemetron Corp., Railway Products (25, 28)
Chicago Pneumatic Equipment Co. (9, 14)
Connors Steel Co. (16, 22)
Co-Ordinated Industries (25)
Deron Corp. (1)
Dresser Industries, Inc. (6, 9, 12)
Duquesne Mine Supply Co. (1, 10, 27)
Dyson, Jos., & Sons Inc. (2)
Electrical Automation (13)
Elreco Corp., The (6, 7)
Ensign Electric Div., Harvey Hubell Inc. (4)
Enrico Products, Inc. (3, 4, 25)
Fairmont Supply Co. (1, 2, 4, 5, 10, 11, 15, 16, 22, 23)
Foster, L.B., Co. (1, 2, 3, 4, 5, 6, 8, 9, 10, 11, 12, 15, 16, 17, 18, 19, 20, 21, 22, 23, 24, 26, 28)
Gauley Sales Co. (4, 10)
General Scientific Equipment Co. (6)
Guyan Machinery Co. (4, 25)
Huwood-Irwin Co. (8)
Janes Manufacturing Inc. (6, 10, 26)
Jennmar Corp.
Jones & Laughlin Steel Corp. (15)
Key Bellevilles, Inc. (7)
Koppers Co., Inc. (20, 21)
Logan Corp. (1, 2, 4, 15, 16, 20, 22)
Marathon Coal Bit Co., Inc. (4)
Microdot, Inc. (6)
Midwest Steel Div., Midwest Corp (1, 2, 5, 6, 10, 11, 15, 16, 18, 20, 21, 22, 23, 24, 26, 27)
Osmose Wood Preserving Co. of America Inc. (21)
Penn Machine Co. (4)
Pettibone Corp. (10, 23, 24, 26)
Porter, H.K. Co., Inc. (3, 4, 6, 16, 22, 23, 24)
Rexnord Inc. (9, 15, 23)
Rexnord Inc., Process Machinery Div. (9, 14)
Safetran Systems Corp., (3, 4, 7, 13, 17, 18, 19)
Service Supply Co., Inc. (2, 15)
Steelplant Corp. (6, 22)
Thor Power Tool Co. (14)
Tidewater Supply Co. (1, 2, 10, 15, 16, 20, 22, 24)
Torque Tension Ltd. (20, 21, 22, 24)
Union Switch & Signal Div., Amer. Standard Inc. (13, 17, 19)
US Steel Corp. (3, 4, 6, 15, 16, 20, 22, 23, 26)
US Thermit Inc. (25, 28)
Wilson, R.M., Co. (18, 19)
Youngstown Sheet & Tube Co., The (15)

Tampers, Mine-Track
Fletcher, J.H., & Co.

Ties
(See *Rail Accessories*)

Wire, Trolley
Anaconda Industries, Wire and Cable Div.
Anixter Bros.
BICC Limited.
Carol Cable Co.
Elreco Corp.
Fairmont Supply Co.
Marathon Coal Bit Co., Inc.
McJunkin Corp.
National Mine Service
Phelps Dodge Industries, Inc.

TABLE 13. Rail and Wire Manufacturers (continued)

Wrenches, Track
Allen Group
Duquesne Mine Supply Co.

Midwest Steel Div., Midwest Corp.
Persingers, Inc.
Tidewater Supply Co.

3

Ventilation Construction

This chapter discusses the necessary construction techniques for installing coal mine ventilation devices. Included in the discussion are the major air controlling devices called overcasts, regulators, stoppings, and doors. Undercasts, although a valid device, are used infrequently in coal mining and are not discussed here.

Typically, coal mines require a great many ventilation devices to control the air movement. Post-1969 mines have had to design their air systems to comply with the Federal Coal Mine Health and Safety Act (CMHSA) as discussed in Chap. 1. Pre-1969 mines have had to redesign major portions of their systems to comply with the basic components of the law such as separation of air courses, single unit split ventilation, and minimum air quantities. It is not uncommon, therefore, to find a large mine (2.0 Mt/a) requiring several hundred overcast installations (permanent, as well as repeat construction), plus doors and regulators placed among a thousand or more stoppings. When each of these devices is subject to leakage from installation and weathering defects, the importance of effective construction becomes evident.

An historic rule of thumb for average mine ventilation says that about 70% of all the air pressure generated by the mine fan is lost between the portal and the working faces. Of course, this rule is subject to many factors, but the message is clear: there is a tremendous inefficiency in controlling leakage through common air devices. This chapter strives to present a standard approach and design alternatives which can go a long way in reducing this leakage, and hence, ventilation costs.

Historically, mine ventilation has progressed from natural, or draft ventilation, to fire-induced ventilation (Fig. 1), furnace-fired centrifugal fans and finally, to today's modern mine fans. Underground controlling devices have been in use in the coal mines since the late 1600s, but the CMHSA of 1969 was the real turning point for mine ventilation. Minimum quantities and velocities of fresh air were required for mining, forcing coal operators and engineers to reevaluate and improve their systems. As a consequence, the cost for ventilating air increased.

Today, companies face a variety of choices among design alternatives, materials, labor, and costs of ventilating devices for their mines. The best approach for any individual company or mine is determined by (1) the needs of the mine, and (2) the local design parameters. In my discussions of ventilation construction I shall be fairly general in terms of device installation steps. Where specifics are warranted for clarification, I will include this information.

BUILDING OVERCASTS

Overcasts, in a strict sense, are devices which allow two air currents to pass one another in direction without mixing. In coal mining, one air current is directed over the other current by constructing a solid "bridge" of incombustible materials (Fig. 2).

This section of the discussion is broken into three broad categories: alternative designs, installing the overcast, and maintaining the overcast. The first discussion, alternative designs, covers design parameters, overcast materials, cost estimates and comparisons, construction time, and required labor. Each will be addressed in terms of the construction foreman's or project engineer's perspective.

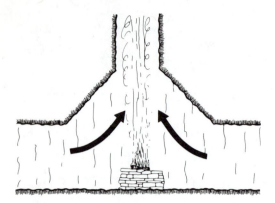

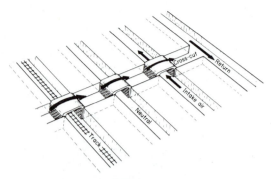

FIG. 1. Typical fire-induced ventilation used in early mines.

FIG. 2. Overcasts direct mine air in modern systems. Notice that most overcasts occur as one of many in a system for air direction.

Alternative Designs

In the majority of modern coal mines, the design of all overcasts put in use is finalized during the premining design stage. Engineers using the best estimation techniques available must choose their design parameters and construction materials and make a cost comparison of their design alternatives. From this comparison comes the type and cost of the overcast to be built during active mining.

Design Parameters: During overcast design, engineers must consider several parameters inherent in the construction of the overcast. These parameters include the required excavation area, necessary air volume, maximum overcast leakage rates, and roof support of the excavated area. In most cases, practical solution of these questions revolve around maximizing the area variable of the equation $Q = VA$, where Q is the required air volume in cubic meters per sec-

FIG. 3. Top cut for overcast. Excavation height equals or exceeds entry height.

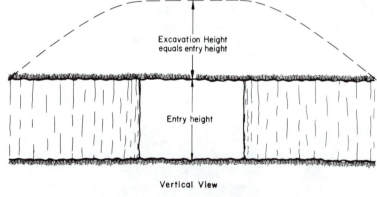

FIG. 4. Banking gob to smooth the airway.

ond, m³/s (cubic feet per minute, cfm), V is the velocity of the air flow in meters per second, m/s (feet per minute, fpm), and A is the effective cross-sectional area meters squared (square feet) of the entry. Installing an overcast (or for that matter, any obstruction) tends to increase the resistance to the air flow unless the design takes this into consideration. For instance, the top of the entry is usually cut or blasted down to provide the same area of airway, as shown in Fig. 3. Coupled with this is the banking of gob material against the support walls and subsequent sloping of this material to allow smoother airflow over the overcast structure (Fig.4). Other aids for maximizing the area (by reducing the frictional resistance) are brushing back the entry top, smoothing the coal ribs, and eliminating trash and materials from the worksite.

Materials: Traditionally, overcasts were built much like houses. Concrete blocks were wet mortared to the design height where steel beams or rail were placed across the gap (Fig. 5). A concrete and wood "floor" was then laid to form the bridge. Concrete block wing walls were then erected to complete the structure. A mortar was applied to the outside of the structure to form the "skin" and complete construction. The blocks, wood, and cement were pretty much standard building materials until recently. Now, in addition to the traditional materials, steel preformed overcasts in several designs are available for construction. For example, one approach is to use a standard aluminum panel design as shown in Fig. 6. This design gives great flexibility in conforming the overcast to the irregularities of the chosen entry. Another popular design consists of preformed sections of corrugated steel or aluminum bolted together (Figs. 7 and 8). In these designs, the number of joints, and hence the amount of potential joint leakage, is reduced over the entire length of the structure.

All three designs (traditional, panel, and sheet) are in popular use today. Personal experience in working with these alternatives tends to favor the prefabricated approach. Prefab overcasts are lightweight, readily portable, quick to erect, and perform excellently over the structure duty life regardless of permanant or temporary life status. Throughout the remainder of the discussions I will emphasize the metal overcast over the traditional approach.

As a general note, a combination approach is available to coal mines with high mining height

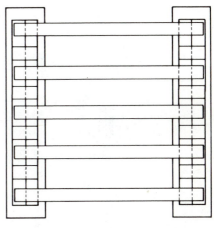

Plan View

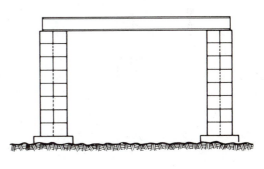

Vertical View

Side View

FIG. 5. Block and steel beams form overcast. As many as 10 beams are placed on the structure walls to form the floor. Beams are often rails cut from 10.1 to 7.3 m (33 to about 24 ft).

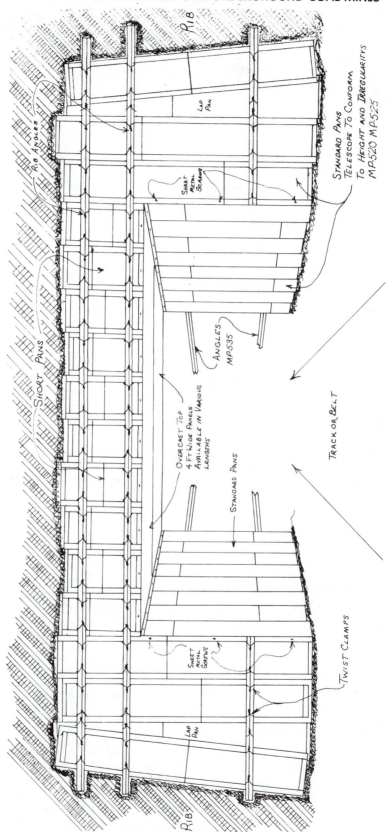

FIG. 6. Aluminum panel overcast (courtesy of Jack Kennedy Metal Products & Buildings, Inc.). Metric equivalent: ft × 0.3048 = m.

VENTILATION CONSTRUCTION

SPAN	RISE		SPAN	RISE		SPAN	RISE	
12 0-1/2	3 2 4 3/4 4 2-3/4 4 6-1/2 4 8-1/2 5 -1/4 5 4 5 9-3/4 5 11-3/4		17 8-3/8	3 2 4 3/4 4 2-3/4 4 6-1/2 4 8-1/2 5 -1/4 5 4 5 9-3/4 5 11-3/4				
12 10-1/4	3 2 4 3/4 4 2-3/4 4 6-1/2 4 8-1/2 5 -1/4 5 4 5 9-3/4 5 11-3/4		19 3-3/4	3 2 4 3/4 4 2-3/4 4 6-1/2 4 8-1/2 5 -1/4 5 4 5 9-3/4 5 11-3/4				
13 7-15/16	3 2 4 3/4 4 2-3/4 4 6-1/2 4 8-1/2 5 -1/4 5 4 5 9-3/4 5 11-3/4		20 0-7/16	3 2 4 3/4 4 2-3/4 4 6-1/2 4 8-1/2 5 -1/4 5 4 5 9-3/4 5 11-3/4				
14 5-5/8	3 2 4 3/4 4 2-3/4 4 6-1/2 4 8-1/2 5 -1/4 5 4 5 9-3/4 5 11-3/4		20 11-1/8	3 2 4 3/4 4 2-3/4 4 6-1/2 4 8-1/2 5 -1/4 5 4 5 9-3/4 5 11-3/4				
15 3-5/16	3 2 4 3/4 4 2-3/4 4 6-1/2 4 8-1/2 5 -1/4 5 4 5 9-3/4 5 11-3/4		21 8-3/16	3 2 4 3/4 4 2-3/4 4 6-1/2 4 8-1/2 5 -1/4 5 4 5 9-3/4 5 11-3/4				
16 1	3 2 4 3/4 4 2-3/4 4 6-1/2 4 8-1/2 5 -1/4 5 4 5 9-3/4 5 11-3/4		24 1-7/8	3 2 4 3/4 4 2-3/4 4 6-1/2 4 8-1/2 5 -1/4 5 4 5 9-3/4 5 11-3/4				

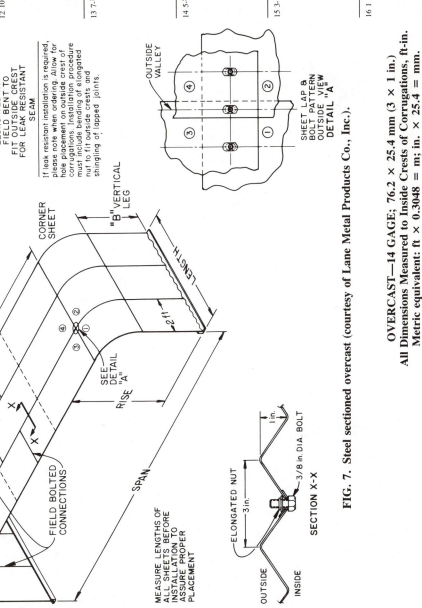

FIG. 7. Steel sectioned overcast (courtesy of Lane Metal Products Co., Inc.).

OVERCAST—14 GAGE; 76.2 × 25.4 mm (3 × 1 in.)
All Dimensions Measured to Inside Crests of Corrugations, ft-in.
Metric equivalent: ft × 0.3048 = m; in. × 25.4 = mm.

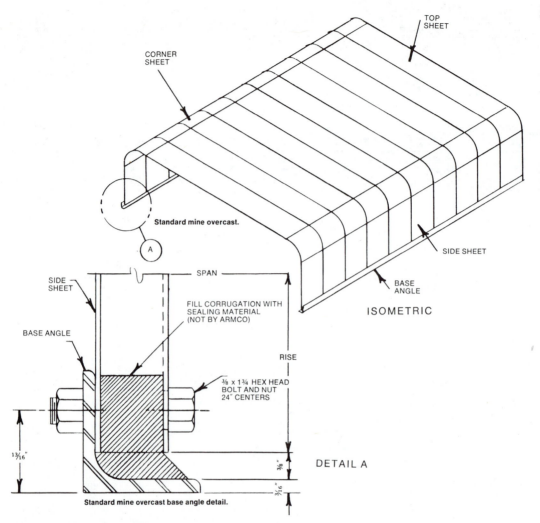

FIG. 8. Aluminum sectioned overcast (courtesy of Armco, Inc.). Metric equivalent: in. × 25.4 = mm.

[greater than 1.8 m (6 ft)]. In this case, structure walls are stacked to a predetermined height. Since most block today is still of nominal dimensions [193.7 × 193.7 × 396.9 mm (7 5/8 × 7 5/8 × 15 5/8 in.)] to allow for 9.5 mm (3/8 in.) mortar joints, planning for stacking a block wall is desirable. Therefore, Table 1 gives a general guide for overcasts, walls, stoppings, and other block structures. The metal overcast is then placed upon the wall and built in the normal manner (Fig. 9). This allows additional passage under the structure, though it requires additional roof material be excavated accordingly.

One of the most innovative changes in materials in recent years has occurred with mine sealants. These sealants are generally a portland cement-based product which combines additional ingredients such as vermiculite, glass fiber, sand, and limestone into a dry cement-type mix. Such mixes have been adopted for mine use from the construction and building trades.

Mine sealants for overcast and other mine work can be troweled on or sprayed. If it is troweled, then a finishing or plastering trowel and plaster hawk is used. For spraying, a machine is used which (1) mixes the sealant with water, (2) pumps the mix through a hose, and (3) induces air at the nozzle of the hose to spray it on. Where speed and bulk handling are required, it is better to spray sealants on. One important note about

FIG. 9. Metal overcast built with structure walls (courtesy of Kaiser Aluminum & Chemical Sales, Inc.).

the mixing machine concerns the type of feed, i.e., either bulk or continuous feed. Generally, more bags of sealant can be mixed and sprayed by a continuous feed method (approximately 60 bags per hr) than with bulk (48 bags per hr).

Today, mine sealants cost about 16¢ to 20¢ per 0.09 m^2 (sq ft) of coverage. Most sealants are shipped in 22.7-kg (50-lb) bags which deliver an average coverage between 4.2 to 5.1 m^2 (45 to 55 sq ft) at a 6.4-mm ($^1/_4$-in.) thickness. Sealants cure in 8 to 24 hr. Sealants can also be either abrasive or nonabrasive, depending upon management's desires. Abrasiveness component is important since some mixing machines can use only nonabrasive mixes.

The performance of any sealant on the market depends on the installation coverage. An even coating of sealant on the structure at approximately 3.2 to 12.7 mm ($^1/_8$ to $^1/_2$ in.) thick is recommended. In some cases, a small percentage (less than 15%) of rock dust or other material can be mixed in the sealant to extend it. However, there may be a possible loss of performance when this is done.

The mixing ratio for these sealants is 18.9 L (5 gal) of water to 2 $^1/_2$ to 3 bags of sealant. The consistency of the mixture must be a pulpy-type slurry for proper spraying. Mixtures too thick will plug the spray hose and dramatically reduce the sealant coverage. Thin mixtures are too runny to stick to the block or metal and usually end up on the floor in a puddle.

Finally, the performance of the sealants depends upon a number of factors working in unison with one another. For example, the air to sealant ratio should support and spray the best sealant mixture on the block for a sticky texture. The mixing machine must mix water and sealant both quickly and evenly to give consistent results.

One final point concerns the mixing machines. Two basic models are available to the mining

TABLE 1. Block Stacking Guide*

Actual Length, ft-in.	Vertical Block Combinations	Horizontal Block Combinations
3-10 1/8	6 Regular	
4-5 3/4	7 Regular	
5-1 3/8	8 Regular	
5-5	8 Regular + 1 half	
5-10 1/8		4 Regular + 1 half
6-5/8	9 Regular + 1 half	
6-6 1/8		5 Regular
6-8 1/4	10 Regular + 1 half	
7-1/4	11 Regular	
7-1 3/4		5 Regular + 1 half
7-3 7/8	11 Regular + 1 half	
7-9 3/4		6 Regular
7-11 1/2	12 Regular + 1 half	
8-5 3/8		6 Regular + 1 half
8-7 1/8	13 Regular + 1 half	
8-11 1/8	14 Regular	
9-1 3/8		7 Regular
9-6 3/4	15 Regular	
9-9		7 Regular + 1 half
9-10 3/8	15 Regular + 1 half	
10-2 3/8	16 Regular	
10-5		8 Regular
10-10	17 Regular	
11-5/8		8 Regular + 1 half
11-5 5/8	18 Regular	
11-8 5/8		9 Regular
12-1 1/4	19 Regular	
12-4 1/4		9 Regular + 1 half
12-8 7/8	20 Regular	
13-1/4		10 Regular
13-1/2	20 Regular + 1 half	
13-4 1/2	21 Regular	
13-7 7/8		10 Regular + 1 half
14-1/8	22 Regular	
14-3 7/8		11 Regular
14-7 3/4	23 Regular	
14-11 3/8	23 Regular + 1 half	
14-11 1/2		11 Regular + 1 half
15-3 3/8	24 Regular	
15-7 1/2		12 Regular
15-11	25 Regular	
16-3 1/8		12 Regular + 1 half
16-6 5/8	26 Regular	
16-11 1/8		13 Regular
17-6 3/4		13 Regular + 1 half
18-2 3/4		14 Regular
18-10 3/8		14 Regular + 1 half
19-6 3/8		15 Regular
20-2		15 Regular + 1 half
20-10		16 Regular
21-5 5/8		16 Regular + 1 half
22-1 5/8		17 Regular

*Based upon: Regular units 193.7 mm high by 396.9 mm long (7 5/8 in. high by 15 5/8 in. long), 9.5 mm (3/8 in.) leveling bed, half block 193.7 mm (7 5/8 in.) long and half high block 85.7 mm (3 5/8 in.) high.
Metric equivalents: in × 25.4 = mm; ft × 0.3048 = m.

TABLE 2. Typical Erection Time for Overcasts* (in Manshifts)

Overcast Type	Foundation	Elements			Skin		Total
		Structure Walls	Floor	Wing Walls	Spray	Travel	
Traditional block and beam	6	3	6-10	2-6	2	4	18-29
Prefab panel	4	1	1	2-4	1	2	9-11
Prefab section†	4	$1/2$	$1/2$	2-4	$1/2$	2	7 $1/2$-11
Prefab with structure wall	4	2	$1/2$	2-4	1	3	9 $1/2$-13 $1/2$

*Assumes structure wall 1.1 m (3 $1/2$ ft) high, wing walls to 2.1 m (7 ft), and overcast is 4.9 m wide by 6.7 m long (16 ft wide by 22 ft long). Skin is sealed inside and out.
†A single prefab section roughly equals four prefab panels bolted together.

industry, each with separate features and prices. The first variety of machine requires the mine to supply a separate power, water, and air supply with a manual water feed for bulk mixing. The cost for this model is about $4500. The second variety of machine is more sophisticated. Here, the machine supplies the air and water (holding tank), while the mine supplies the power for continuous feeding. Cost for this model is approximately $12,000. The price difference between these two machines tends to balance out when considering the annual operating and maintenance costs of each. In looking at both costs, the continuous feed machine is markedly less. The second machine also tends to have a longer useful life, although this is influenced greatly by management.

Labor and Construction Scheduling: Each of the design alternatives has individual labor requirements and erection schedules. These variables are a function of the design and material requirements at each individual installation. Another rule of thumb is that overcast complexity increases both the manpower and construction time variables.

Basically, all overcast construction can be estimated from the required installation elements of the design. For example, a foundation for a prefab section overcast will average four manshifts to construct and pour. Such an estimate includes the necessary siting and assumes the tools and materials have been delivered. Longer times may reflect difficulties with leveling, forming, or curing the foundation. Shorter intervals will indicate fast workers, easy setup, and smooth pouring. The required elements for building an overcast are the following:

The *foundation,* which includes leveling the foundation site, hitching the ribs for wing walls, siting and constructing the foundations forms, and pouring the foundation.

The *structure walls* (optional for prefab) which includes stacking or wet mortaring the concrete block walls to a predetermined height.

The *flooring,* which describes the efforts and materials needed to bridge the two overcast structural walls together and form the floor of the overcast.

The *wing walls* encompasses closing the sides of the overcast from the floor to the roof.

The *skin,* which includes sealing the overcast with appropriate materials to make it airtight and rigid.

The complexity of the overcast design will define the elements in terms of specific labor and construction time.

Table 2 summarizes the estimates of overcast construction. The labor and time is combined into manshifts per construction element. Each element is tied to a particular design approach, i.e., traditional block and beam, prefab panel, prefab section, and prefab with block structure walls. Naturally, these estimates will vary for individual operations, depending upon management desires, mine physical conditions, and system logistics. These estimates also assume there is no problem with supplying tools and the materials to the construction site, and that the excavation is complete.

The optimum number of workers for a construction crew is two to four workers, depending upon the job element. Preferably, one or more workers will have had previous experience.

Cost Comparison: As with any other underground project, there are certain costs associated with overcast construction. These costs reflect the design, materials, and labor needed to install a typical overcast. A trade-off exists between

TABLE 3. Estimated Cost Comparison of Overcast Construction* (1981 Dollars)

Type	Materials	Labor	Total
Block and beam	$1000	$1350–2175	$2350–3175
Prefab panel	$1700	$ 675–825	$2375–2525
Prefab section†	$2600	$ 563–825	$3163–3425
Prefab with structure wall	$2700	$ 713–1013	$3413–3713

*Assumes an overcast 4.9 m (16 ft) wide, 6.7 m (22 ft) long, and 1.8 m (6 ft) high. Individual approaches will alter these dimensions
†A single prefab section roughly equals four prefab panels bolted together.

the higher priced prefabricated design vs. the longer installation time needed for block and beam. The cost estimate presented in Table 3 reflects this trade-off with a standard labor wage averaging $75.00 per manshift. The estimate is broken into two categories, labor and materials. The range given in the table is a result of the manpower requirements for installation, as shown in Table 2.

As the figures indicate, a slight cost advantage can be gained by using the prefabricated panel stopping over the other overcast approaches. Other factors will also temper this advantage. For example, leakage efficiency of the completed stopping is a function of the number of seams and joints needing to be sealed. This point gives an advantage to the sectioned overcast, which has fewer seams. Also, the majority of block and beam stoppings are installed with labor requirements and time closer to the higher end of the manshift range, which again favors metal over the traditional approach. However, individual mines may require a site examination to determine their appropriate needs in terms of the best design approach.

Installing the Overcast

There are five specific tasks required for all overcast construction. These are:

1) *Site preparation*, including the needed excavation, bolting back and cleanup, supplies delivery, and site orientation.
2) *Foundation construction*, encompassing the tasks of forming the foundation, pouring and curing the concrete, as well as hitching the ribs.
3) *Wall construction*, encompassing both structure and wing wall installation.
4) *Floor construction*, including all tasks needed to span the walls and bridge the structure.
5) *Sealing the stopping*, completing by either spraying or troweling the sealant.

To best discuss these tasks, an illustrative example is necessary. The site is assumed to be an intersection with an entry width of 5.5 m (18 ft) and a mining height of 2.1 m (7 ft), as illustrated in Fig. 10. Like track laying, ventilation overcasts are started (i.e., the roof material is cut down or shot) close behind the advancing unit. If cutting the top down, then the work usually begins within 304.8 m (1000 ft) of the working section because of the availability of the miner, bolter, and shuttle car for construction work. This, of course, is not always the case, but it is the predominant approach being used today. For the purposes of the following discussion, a prefabricated sectioned overcast (refer to Fig. 8) mounted on a structure wall will be illustrated. The wall is necessary because track will be laid under the overcast and minimum clearance must be maintained. The engineered design of the overcast for this site is shown in Fig. 11. The first task is site preparation.

Site Preparation: When approaching the site preparation task, there are several subtasks to be identified and completed. These include site orientation, excavation, and supply delivery.

Site Orientation. Site orientation consists of: (1) establishing the minimum excavation requirements, and (2) defining the overcast clearances by establishing a grade line of specific elevation. For our example, to maintain effective cross-sectional area above the overcast, a minimum excavation of 2.1 m (7 ft) is required (Fig. 12). The excavation amount is established for the site by the surveyor and is passed on to the mine construction manager. Concurrently, the surveyors will define a grading line at 2.1 m (7 ft) above the overcast base and mark this line on all four ribs of the intersection, as illustrated in Fig. 13. A construction foreman will then be able to pour his foundations and build the structure wall to exact heights for proper project control.

Site Excavation. After the orientation of the site is finished, the excavation task may begin. Here, the roof is either cut or blasted down. In the United States, the majority of roof is cut by a continuous miner or roadheading machine. For our example, the top will be cut down by a miner and bolted back by a dual boom bolting machine. Cutting and bolting back tasks consist of the following steps:

1) The miner begins the ramping operation approximately 10.7 to 13.7 m (35 to 45 ft) from the intersection in the entry where the air systems will cross over one another (Fig. 14). If, for some reason, the miner ripper head cannot reach the roof to begin the ramp, then the operator must back up an additional 1.5 m (5 ft) and dig some floor material out, pile it up, and use it to reach the roof.

2) The miner will cut approximately 1.2 m (4 linear ft) of top down, then pack it solid by running back and forth over it. He will repeat this advance and pack operation until he has cut down approximately 4.9 m (16 ft) of top. Note: The in-place roof support at the time is usually not recovered (Fig. 15).

3) The miner backs off the ramp and allows the roof bolter to replace the bolts cut down as well as spot bolting in loose areas (such as the brow) and scaling the top and ribs for loose material.

4) The miner and bolter will continue to work and exchange positions until approximately 4.6 m (15 ft) after the center of the entry. Here, the ramping operation begins to slope down. By approximately 10.7 to 13.7 m (35 to 45 ft) past the centerline of the entry, the miner should have finished the cutting operation and the bolter replaced all bolts (Fig. 16).

This operation of cutting and bolting back ideally should not take more than 6 to 12 shifts (one foreman, five miners for one to two shifts). However, problems with equipment movement in setting up and the actual operation sometimes prolongs these activities. One of the possible trouble spots during the cutting operation is getting the machine stuck on a poorly packed ramp. This is often the case when an operator tries to extend his reach at each cutting sequence. The cut material must swell in the entry and be packed by the miner completely before an operator can safely maneuver on the ramp. Premature work on loosely packed ramps will not work.

A second trouble spot is ramping too steeply. If this happens, both machines (miner and bolter) will have a difficult time maneuvering and climbing the ramp, thus losing valuable working time. An efficient ramp should pitch no steeper than a rise of 0.6 m (2 ft) in a run of 3.6 m (12 ft), or approximately 10° (Fig. 17).

After the cutting and bolting back tasks are complete, the load-out (cleanup) task begins. This consists of having the continuous miner load out the cut material into a shuttle car for transport out of the mine (if necessary). Much maneuvering is required during cleanup to ensure a complete job. Cleanup will take approximately one shift to complete, thus making the entire excavation task an excellent idle day task (three shifts or less), since air and equipment diverted to this operation would interrupt regular mine production activities.

Supply Delivery. The supply delivery task consists of transporting the correct amounts of material needed for the overcast construction to

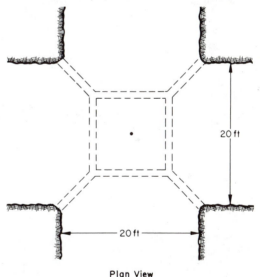

FIG. 10. Overcast site in the intersection.
Metric equivalent: ft × 0.3048 = m.

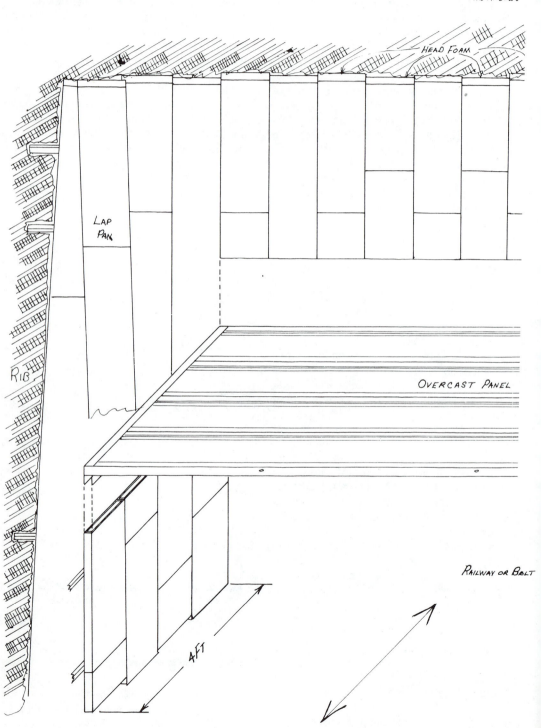

FIG. 11. Site design for the overcast (courtesy of Jack Kennedy Metal Products & Buildings, Inc.). Metric equivalent: ft × 0.3048 = m.

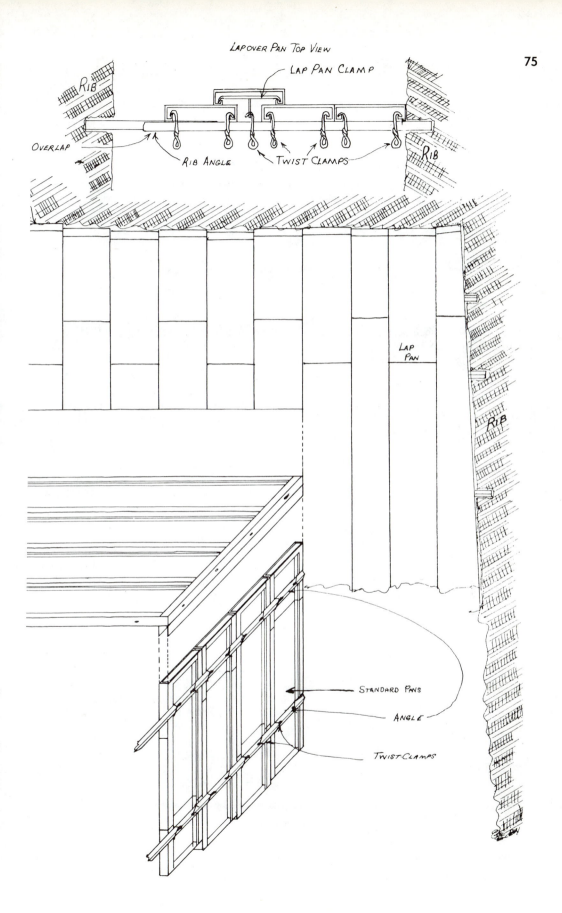

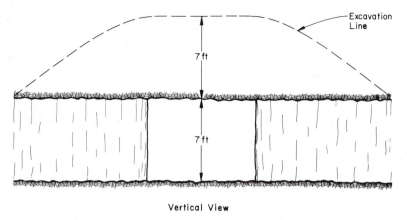

FIG. 12. A minimum of 2.1 m (7 ft) of excavation is required.

the site. This can include any tools such as a concrete mixer (for the foundation), air compressor (for the floor), and hand tools (trowel, axe, pick, shovel, string, and so on). These materials should be placed in a convenient spot for construction use, and in a usable manner (i.e., not having the concrete bags under a pile of cinderblock!). Broken bags of concrete, cracked and broken blocks, ripped bags of bolts and nuts, bent structure plate, and other damaged materials are not usable in many instances. These damaged items are a product of thoughtless handling, so care should be exercised in transporting these materials.

Once the supplies and materials have been delivered to the site, the preparation task ends. The next set of tasks falls under the general step of foundation construction.

Foundation Construction: Foundation construction consists of two tasks: a structural wall footing and wing wall rib hitching. In some mines, the practice of building a foundation is largely ignored. This is unfortunate since an effective overcast should have a strong, level base on which to stand.

A footing foundation does not require extensive time or labor. In many cases, a trench 101.6 to 152.4 mm (4 to 6 in.) deep the length of the structure wall and 203.2 mm (8 in.) wide will suffice for the footing. If the mine floor is uneven, then some combination of trenching and building forms may be required to give a level footing (Fig. 18) at the proper elevation below

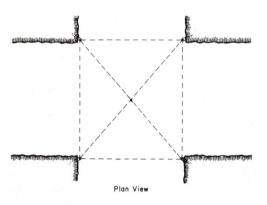

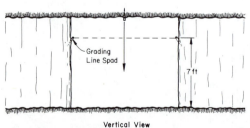

FIG. 13. Orientation sites for overcast construction. Metric equivalent: ft × 0.3048 = m.

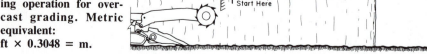

FIG. 14. Start of ramping operation for overcast grading. Metric equivalent: ft × 0.3048 = m.

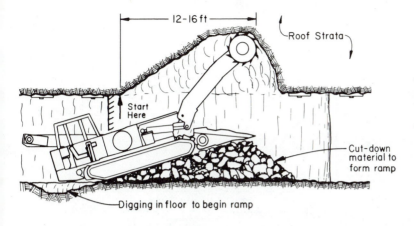

FIG. 15. First cut before bolting back. Metric equivalent: ft × 0.3048 = m.

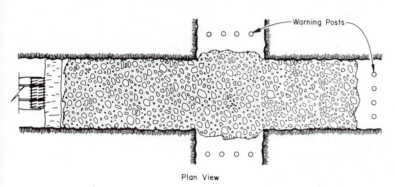

FIG. 16. Miner finishes overcast grading before loading out. Metric equivalent: ft × 0.3048 = m.

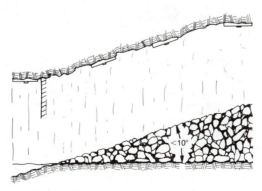

FIG. 17. Ramp operation is always less than 10°.

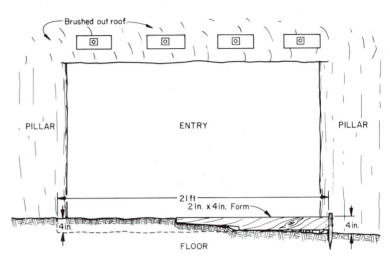

FIG. 18. The foundation is dug out and formed up prior to pouring the concrete. Metric equivalent: ft × 0.3048 = m; in. × 25.4 = mm.

the grade line. The footing is filled and smoothed to the desired level with concrete and allowed to dry for three or more shifts.

The second task of rib hitching is done to anchor the wing walls into the coal ribs. Here the foreman should mark a vertical hitching centerline from the roof to the floor for the worker to use as a guide. Hitchings are picked out 76.2 mm (3 in.) deep by the width of the wall block, usually 203.2 mm (8 in.), as illustrated in Fig. 19. All four corner ribs are hitched on an angle approximately 45° from the end of the structure wall (Fig. 20). In our example, one man can easily hitch all four coal ribs in one shift. As a final task, the hitched areas are cleaned of coal and debris in anticipation of laying wing wall block.

Wall Construction: Wall construction is a two-part task separated by the floor construction task. The first part consists of building the structure and wing walls to a point where the two walls can be bridged. On overcasts without any structure wall (i.e., full prefabricated designs), this first part is eliminated. For our example, the completed block wall would be stacked to the predetermined elevation of 0.8 m (2 3/4 ft) so the overcast will have the required clearance of 2.1 m (7 ft), Fig. 20.

An important point to remember in wall construction is keeping the wall level and plumb. This is best accomplished when the first row of block is leveled and plumb. Difficulties can be minimized by using rock dust or by shaving the

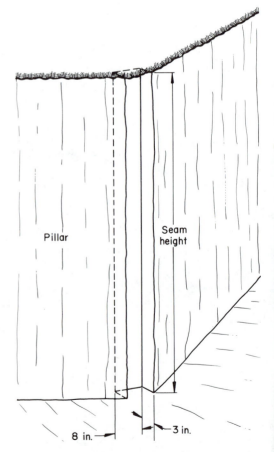

FIG. 19. Rib hitching for overcast wing walls. Note rib is hitched from the floor to roof of the seam. Metric equivalent: in. × 25.4 = mm.

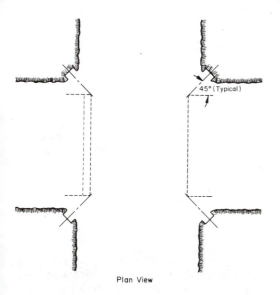

FIG. 20. Rib hitching angle is usually 45° to the structure walls.

concrete block. A leveling tube should be used almost constantly. All too often a wall will begin leaning with the second row, only to fall down under its own weight near completion. This is not only a wasted effort, but a dangerous safety hazard (especially in high coal).

Once the floor is completed (i.e., bridged), the second part of the wall construction begins. This entails stacking or mortaring (if so designed) the block along the floor edges of the overcast and up to the roof of the entry (Fig. 21). Blocks are then shaved and shaped to conform to irregularities with the roof perimeter (Fig. 22). The wing wall is wedged tightly in several places to stiffen the structure. Depending upon the size of the overcast, this task of wall construction can be the most time-consuming of the entire project, since both sides must be done with care. It is in the ease of finishing the task that one really sees the results of carefully keeping the walls level and plumb because such a wall will be completed easily. Out-of-plumb walls, as mentioned before, will compound the problems.

Floor Construction: Floor construction is the general term used to describe the bridging operation of the two walls. For our example, this first involves placing an angle bar (base) on top of each structure wall for the sections to stand in (note: two blocks have been turned up and mortared on each wall to provide a brace for the angle bar), as shown in Fig. 23. In a numbered sequence, the overcast is built from one end to the other by bolting two legs and a floor plate together, thus creating an ''arch'' or bridge (Fig. 24). Approximately 14 to 16 sections are needed for our example; this number would vary for other designs and materials. The bolts are hand tightened until the overcast is finished; thereupon, an air wrench will be used to stiffen the frame. Once the floor bolts are tightened, the wing walls are completed and the overcast is ready for sealing.

Sealing: The sealing of an overcast is accom-

FIG. 21. Two courses of concrete block between overcast structure and ceiling can be seen near left arm of workman in foreground (courtesy of Kaiser Aluminum & Chemical Sales, Inc.).

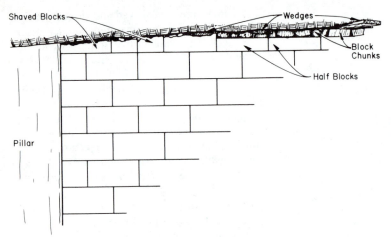

FIG. 22. Close-up view of roof contact area.

plished by either (1) spraying on a sealant from an air-operated machine, or (2) troweling the sealant on from buckets of material, or (3) a combination of the two. Sealing is done from the inside of the overcast first, and since both sides of the overcast should be covered, the inside becomes the most difficult area to reach. With spraying, a key block in the wing wall is carefully removed to allow a hose and nozzle to spray the inside with about 3.2 mm (1/8-in.) of sealant material. Two workers are required to spray an overcast, while any number of workers can trowel the sealant on. Also when spraying, a certain amount of troweling can be done around the edges of the structure to ensure a good seal. Any sealant needs some amount of time to cure and harden.

This completes the section on installing the overcast. Several steps are required for efficient construction and each should be part of an effective design. Various materials are available to the mine operator, permitting a choice among

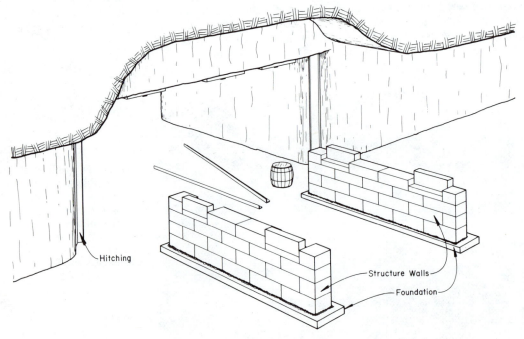

FIG. 23. Overcast structural walls with blocks turned up for angle braces.

VENTILATION CONSTRUCTION

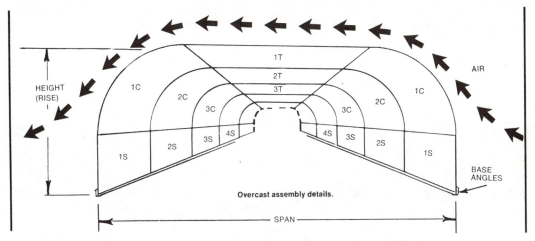

FIG. 24. Overcast assembly details (courtesy of Armco, Inc.).

steel, aluminum, block, and concrete to the design engineer. Our completed example is sketched in Fig. 25.

Maintaining and Inspecting the Overcast

After an overcast has been installed, an operator cannot just walk away from it. The overcast must be inspected and repaired to keep it effectively separating the two air streams from one another. Two factors, weathering and degradation, are constantly attacking the integrity of the overcast. I shall discuss these factors briefly.

FIG. 25. Isometric view of completed overcast (courtesy of Kaiser Aluminum & Chemical Sales, Inc.).

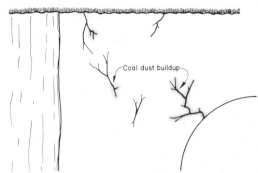

FIG. 26. Leakage through overcast detected by coal dust buildup.

FIG. 27. Hairline cracks show overcast leakage. Some coal dust leakage may be present.

Weathering and Degradation: Weathering and degradation factors are broken into two major categories, i.e., air and water, and age.

Weathering by air and water centers on the changes in ventilating humidity and pressure, and their effect on the overcast and coal roof, floor, and ribs. The roof and ribs tend to loosen and slough off leaving gaps in the sealing while the block walls begin to weather from increased exposure to the humidity. Standing water also takes its toll on the overcast by softening the bottom material. Here, water can allow the overcast to sink (if it has no foundation) or crack its seal.

Age plays a negative role on overcasts. Shrinkage cracks will appear in time on the "skin" of the overcast because of vibrations and hardening of the sealant. Bolts will loosen from age and vibration, especially if trolley or cable hangers are installed in the overcast. Floor heave from ground pressure will invariably cause a shift in the overcast structure. All in all, age can and will reduce the efficiency of the overcast and thereby increase leakage.

Examination and Resealing of the Overcast: One of the best methods for minimizing the effects of weathering and age on an overcast relies on regular inspection. Coal mine operators can institute such a program in conjunction with the inspections of the mine examiner. Daily and weekly inspections will reveal the changing nature of the overcast efficiency. General signs of leakage are the following:

A buildup of coal or other dust around a leakage point showing air movement (Fig. 26).

Hairline cracks on the "skin" of the overcast covering the wing and structure wall areas (Fig. 27).

A slight sagging of the overcast in the middle of the entry floor can be seen upon visual inspection (Fig. 28).

If an examination reveals any of these or other traits, then a general patching program should be initiated. This program could be simply one man with a trowel and bucket of sealant, or a team which completely resprays the skin. In any case, it should be done to prevent further degradation of the overcast to a point where a rebuilding is necessary.

This concludes the discussion on overcasts. These structures are a staple of coal mine ventilation, so their efficient performance is a must in terms of safety, cost, and system operation. Installing these overcasts with careful engineering will help achieve this goal.

CONSTRUCTING A REGULATOR

The second ventilation control device under consideration is the regulator. A regulator is a device which controls the airflow into or out of a split of air by varying the "resistance" to the airflow.

There are two types of regulators under discussion, temporary and permanent. Temporary regulators are placed on short-lived mine workings such as production panels, pillaring sections, or where there are temporary air deflection needs such as battery stations, trolley barns, repair sleds, etc. Permanent regulators are used to direct air from mains and submain entries where the life of the regulator would be greater than five years.

There are three separate topics under this discussion. The first looks at designing a regulator while the second discusses how to build a reg-

FIG. 28. Overcast sagging in the middle due to weight and support problems.

ulator and the third covers the maintenance of the installed regulator. Each topic will be approached from the viewpoint of a mine operator or construction foreman.

Designing the Regulator

A regulator is installed for two types of control demands, namely present and future regulating requirements. Therefore, as the mine grows, management must be able to adjust the regulator to meet its needs.

A regulator is designed from the results of a network analysis of the mine's ventilation system. A simple network is shown in Fig. 29, which was reduced from the mine shown in Fig. 30. Notice the ventilation system has been reduced to a single circuit containing the equivalent resistances found in the mine. From this network analysis, a resistance can be calculated for each split. As a mine grows, some loops (denoting submains, etc.) will increase the natural resistance, thus requiring less from the regulator to keep the mine in balance. A regulator must be built with a measure of flexibility in its design. A good text detailing the analysis and calculation of the regulator size is *Mine Ventilation and Air Conditioning,* 2nd edition, by Howard L. Hartman, published by John Wiley. Interested persons should consult this text for further discussion.

There are several design alternatives available for regulator construction. The first, and easiest, is simply to make a hole in a concrete block or metal panel stopping by removing some of the blocks or panels (Fig. 31). This gives only an estimated amount of regulation at any one point, but usually is good enough for temporary mining purposes. The design's chief advantages lie in its flexibility and cost. Many mines employ this type for temporary installation.

A second design is the use of a mine door installed across the entry; usually a man door is adapted for this purpose. (Fig. 32). These types of regulators give better control over the airflow for a little more expense. This door design, especially the man door, finds a great deal of use in coal mines for both temporary and permanent installations.

The third regulator design seen in coal mining is the automatic vane type of device. This is placed near the mine fan to give greater control over the mine resistance. It is not a common device of mine support construction and so will not be reviewed further.

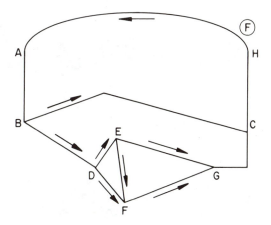

FIG. 29. Ventilation network in simplified form.

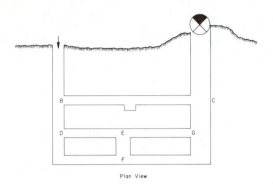

FIG. 30. Ventilation system of coal mine as shown in Fig. 29.

Once mine management has decided which type of regulator to build, there is an associated cost to be reviewed. To facilitate discussion, Table 4 summarizes the estimates for construction of some of the regulators discussed previously. Notice the costs are relatively close between designs, so a choice is usually left to management desires.

Building the Regulator

Building a regulator is much like building a stopping. The difference comes in choosing the appropriate site for the regulator and then setting the regulator for proper performance. For the purposes of the discussion, a sliding man door type of regulator will be covered. The first step is picking and preparing the site.

A regulator is usually placed in the return entry airstream near the intersection of the airstream with the main return of air (Fig. 32). If possible, it should also be located in an accessible and relatively stable area. Blocks and the door are transported to the site and carefully stacked to one side.

The perimeter of the site is then shoveled and scraped clean and level. The roof is made secure by scaling down the loose material. The coal ribs are hitched from roof to floor approximately

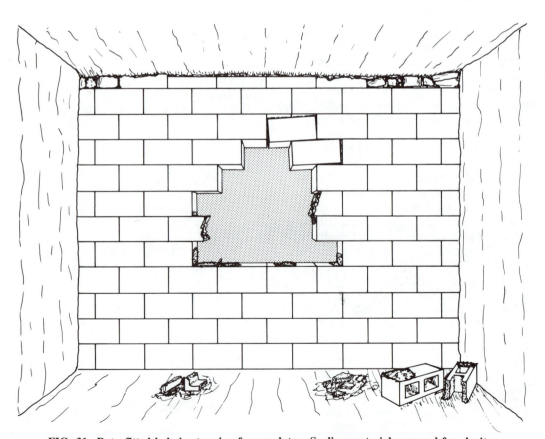

FIG. 31. Retrofitted hole in stopping for regulator. Sealing material removed for clarity.

TABLE 4. Estimates of Regulator Cost and Labor Requirements* (1981 Dollars)

Regulator Type	Material Cost	Manshifts Required	Labor Cost	Regulator Total Cost
Block stopping with hole	$20	2	$150	$170
Metal stopping with panel	$200	1/2	$38	$238
Block stopping with man door	$75	2	$150	$225
Full-entry door*	$200	3	$225	$425

*Assumes the door is constructed from mine materials for the express purpose of air regulation.

76.2 mm (3 in.) deep by the width of the block. The two workmen are now ready to build the regultor.

When constructing the block frame around the door, it is important to keep the frame level and plumb (Fig. 33). The first row of block is critical for keeping the entire structure in line. The block frame is built to the height of the regulator bottom, which is usually one-half the regulator door height subtracted from the entry centerline, as illustrated in Fig. 34. The door is placed on the block frame with the door opening of the regulator centered in the entry. The block frame is then finished and wedged at the roof to stiffen the structure. Sealant is applied to the inby side of the regulator frame to form a tight skin.

Setting the regulator requires a knowledge of the quantity of air needed to pass through. From the equation $Q = VA$, the area of the regulator setting can be varied to produce the required velocity, and thus regulate the quantity of air. The measurement of the velocity is accomplished by using a high speed vane anemometer just behind the regulator opening. The opening is traversed with an appropriate method for 30 sec and multiplied by 2 to give the approximate foot per minute velocity. A mine supervisor should be familiar with this procedure before setting a regulator.

Maintaining the Regulator

Once the regulator is set and operating, it cannot be forgotten. Periodic checks and maintenance are required to keep the regulator functioning correctly. At least once every three months the regulator should be examined for weathering and aging effects. If hairline cracks or shrinkage cracks are discovered, then resealing with new material will be necessary. At least every six months the regulator should be checked for its setting vs. the needs of the mine. Discrepancies must be resolved to maintain an efficient and cost-effective system.

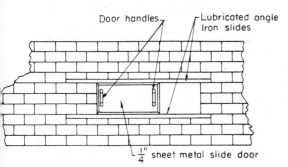

FIG. 32. Man door type of regulator. Metric equivalent: in. × 25.4 = mm.

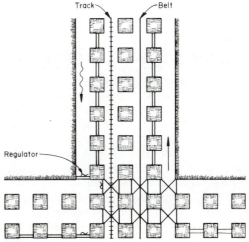

FIG. 33. Regulator placement near the main airways.

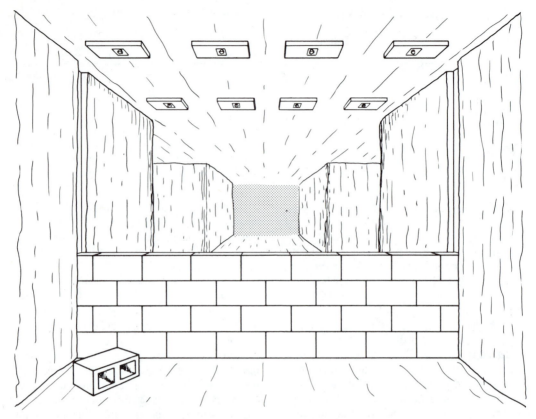

FIG. 34. Half-finished stopping ready for door-type of regulator.

This concludes the discussion on regulator construction. These structures do not require a great deal of technical skill, but are very important for balancing the mine ventilation system.

CONSTRUCTING A STOPPING

A stopping, in the strictest sense, is a device which prevents air from passing through a particular partition. Stoppings prevent the short circuiting of air currents and can seal off any desired areas from the rest of the mine.

There are two types of stoppings, permanent and temporary. Permanent stoppings are made of incombustible materials, usually concrete block (Fig. 35), but other materials such as steel panels are also used. Temporary stoppings are usually built from a combination of wood and brattice cloth, or the more popular metal panel (Fig. 36). For our discussions, the temporary stopping will center on the metal prefabricated type.

Permanent Stoppings

There are two factors which classify a stopping as permanent. The first is location and the second is useful life. Permanent stoppings are found in the main and submain workings of a mine where their life is expected to be greater than three years. Most of these stoppings are constructed with concrete block and sealed on both sides.

The cost of building a stopping varies because of (1) the size required, (2) whether the stopping is wet-mortared or dry-stacked, (3) what type of sealant is applied, and (4) the number of men used on the task. As a rough rule of thumb, a stopping can be constructed (in 1981 dollars) for $1.50 to $2.50 per sq ft. High coal mines, over 2.1 m (7 ft), will experience higher costs, while low coal mines will be on the low end of the scale. This rule of thumb assumes two men are working on the stopping and the supplies are delivered and ready to go.

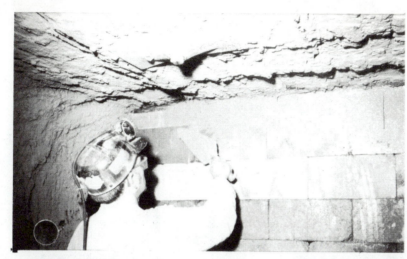

FIG. 35. A stopping constructed with concrete block. The blocks can be either hollow or solid (courtesy of BBond Industries).

In building a permanent stopping, the following procedures are required:

A site is selected which has relatively straight ribs, solid roof, and level floor. The site is scaled and made secure.

Both ribs are hitched from roof to floor approximately 101.6 mm (4 in.) deep and the width of the block, usually 203.2 mm (8 in.) (Fig. 37). It is important to square the corners (using a pick) and level the floor by shoveling the loose material to one side.

The first row of block should be laid (wet or dry) as level and plumb as possible. Rock dust, sand, or more shoveling will help align the block.

Alternating the joint pattern, the block should be stacked or mortared from the first row to the roof of the entry (Fig. 38).

Shaved or sliced blocks are fitted into the roof-stopping seam to fill any gaps. The stopping is wedged to stiffen the structure.

Sealant is sprayed or troweled on both sides of the stopping for maximum performance. The

FIG. 36. A metal type stopping which is gaining in popularity (courtesy of Jack Kennedy Metal Products & Buildings, Inc.).

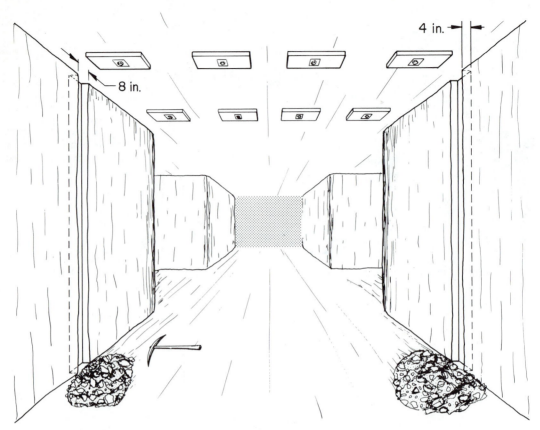

FIG. 37. Rib hitchings for putting in a stopping. Metric equivalent: in. × 25.4 = mm.

FIG. 38. Stacked blocks before sealing.

edges of the stopping are troweled to ensure a good seal.

After the stopping is completed, periodic inspections and maintenance (every six months) should help keep the stopping in good condition with a minimum of leakage. Signs of weathering and age to look for are: a buildup of coal or other dust around a leakage point showing air movement, hairline cracks on the "skin" of the stopping, spaulling of loose rib or roof material, and heaving of floor material. All leakage should be taken care of as soon as possible, since the overall cost of the mine ventilation will increase with time.

Temporary Stoppings

These stoppings are characterized by a short mine life and the ability to be reused many times. Traditionally, wood and brattice cloth stoppings were used, but these are now giving way to the more efficient metal stoppings (Fig. 39). Although the cost of a metal stopping is greater than the homemade variety, the difference in performance, safety, and durability makes these stoppings far superior to their wooden counterparts. The cost of a metal stopping is estimated at $240 installed (1981 dollars). A wooden counterpart averages approximately half of this cost, but the metal stopping has more advantages.

Metal stopping panels come in a variety of sizes based on seam height. The most common size of panel is 0.3 m [12 in. (1 ft)] wide, with a length of 1.5 m (5 ft) collapsed which telescopes to 2.4 m [8 ft (hence the designation 5-8 ft)]. One manufacturer, Jack Kennedy Metal Products, Inc., stocks sizes from 0.6-0.9 to 2.4-4.3 m (2-3 ft to 8-14 ft). The weight of one

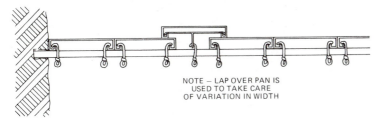

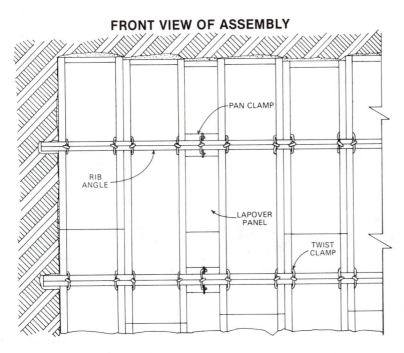

FIG. 39. A high efficiency metal stopping (courtesy of Jack Kennedy Metal Products & Buildings, Inc.).

STANDARD PACKAGE STOPPINGS

May be ordered in any width and height. Galvanized pans are one foot wide and will telescope to any height coal seam.

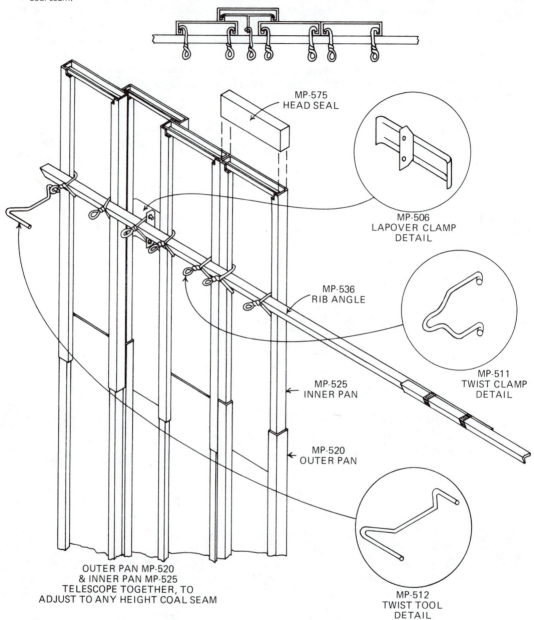

FIG. 40. Closeup of the twist-tie clamp for using metal stoppings (courtesy of Jack Kennedy Metal Products & Buildings, Inc.).

VENTILATION CONSTRUCTION

FIG. 41. Worker readying a place to erect the stopping (courtesy of Jack Kennedy Metal Products & Buildings, Inc.).

entire aluminum stopping, 5.5 m (18 ft) wide, averages 218 kg (480 lb), underscoring the fact these stoppings are lightweight as well as durable.

The installation of these metal stoppings is very quick. The procedure hinges upon using the telescoping panels together with the twist clamp which binds the panels together (Fig. 40) against an angle brace. The steps involved are the following:

A site is chosen for the stopping; scaled and cleaned of debris.

The coal ribs are picked flat by the worker until a good fit between the rib and first panel is obtained (Fig. 41); note: it need not be plumb.

Rib holes are picked on both ribs approximately 50.8 mm (2 in.) deep on the centerline of the stopping. These holes, as a set (one on each rib), will hold an angle brace. Rib holes should be spaced from the middle of the mining height on 0.6-0.9 m (2-3 ft) distances (Fig. 42)

Angle brace ends are placed in the holes and secured in the middle as shown in Fig. 43.

Telescoping panels are placed end to end to form a wall (Fig. 44). Head seals of styrofoam are placed on each panel before being placed on the angle. Twist clamps are used to hold the panel in place.

At the completion of the stopping, sealant is applied to the edges of the panels on the smooth side and to the seams of the panels, also on the smooth side. This makes the stopping airtight.

Once the stopping is installed, it becomes very rigid because of the pressure it exerts between the roof and floor. As mentioned earlier, these stoppings can be used as permanent types, although the majority of use is as a temporary stopping.

In conclusion, the discussion on stoppings stresses the following points:

Stoppings are a common but important device in underground ventilation.

The two most popular choices of stoppings are the concrete and steel panel stoppings, both types of which can be used as permanent or temporary devices.

Metal stopping, although costing more initially, appears superior for construction purposes. That is, it is lightweight, easy to transport and build, durable, recoverable, and so on.

New mines analyzing their future ventilation needs should review their expected power costs vs. the overall leakage they expect to incur. Well installed and maintained stoppings will help keep leakage and subsequent leakage costs in line.

CONSTRUCTION OF A DOOR

Earlier in the discussion it was mentioned that there were two types of doors; man doors and full-entry doors. Man doors (Fig. 45) are used as access points between stopping lines for men, material, and supplies. Full entry doors also allow passage of men, material, and supplies, but on a much larger scale. Full-entry doors are used to block air from entering certain parts of the mine. Today, many full-entry doors are used to control air flow into the neutral entries of a mine by building air locks (two full-entry doors close together).

Man Doors

Man doors are constructed from two possible situations. The first is retrofitting the door into the stopping, and the second is planning the stopping and door together. Both methods are commonly used throughout the mining industry.

The man door used today is almost exclusively a steel or aluminum type (Fig. 45). This door is lightweight, durable, easy to transport, and cost effective. A typical door installation costs

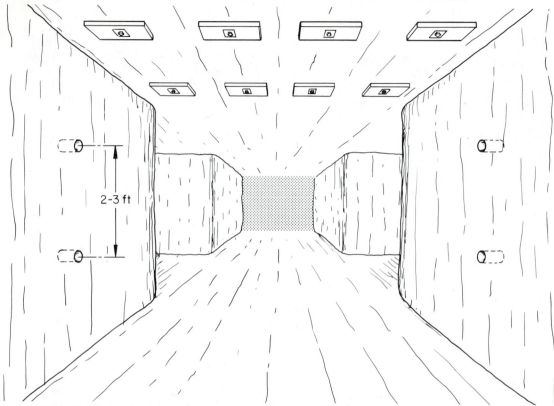

FIG. 42. Rib holes are picked into the ribs for metal stopping angle braces. Metric equivalent: ft × 0.3048 = m.

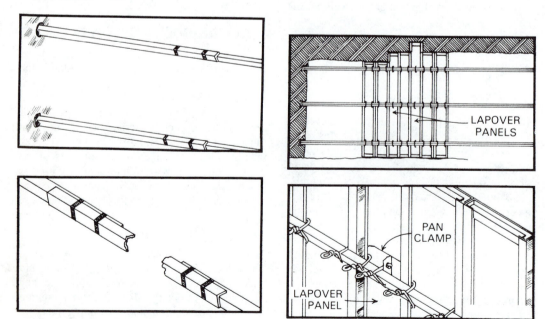

FIG. 43. Angle braces taped in the middle to provide initial support for the stopping panels (courtesy of Jack Kennedy Metal Products & Buildings, Inc.).

FIG. 44. A typical telescoping panel used in metal stoppings (courtesy of Jack Kennedy Metal Products & Building, Inc.).

VENTILATION CONSTRUCTION

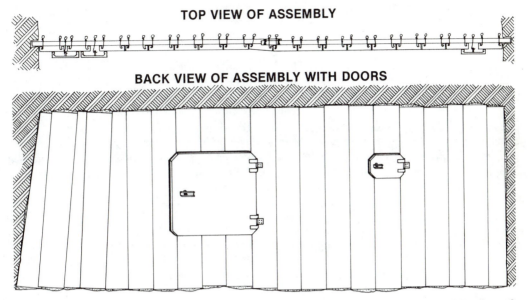

FIG. 45. A typical man door used as an access point (courtesy of Jack Kennedy Metal Products & Buildings, Inc.).

about $100 (1981 dollars), assuming that it takes one worker approximately a half shift to install. Man doors can open from the horizontal (and swing out) or from the vertical (and swing up), depending on management desires. An important installation note to remember is that doors should open up to the pressure side of the stopping so there is always a tendency for the door to close.

Man doors are easily retrofitted into concrete block stoppings by knocking the appropriate hole through the stopping [approximately 762 mm or 0.8 m (30 in.) from the mine floor], and fitting the door in (Fig. 46). The hole is then closed around the door with shaved and split blocks. The stopping and door edges are sealed to prevent leakage.

To retrofit a door in a metal panel stopping, the general procedure requires that the approximate number of panels be removed and replaced with a set of short panels and additional angle braces as shown in Fig. 47. The door is then fitted into the stopping and bolted or riveted in place. Both types of retrofits are easy to complete, but should be done during an idle shift or other scheduled time when the short circuiting of the air will not affect mining activities.

Building man doors into block or metal stoppings during construction is very much like building a door-type regulator into a stopping except that the bottom height of the door is usually 762 mm or 0.8 m (30 in.) off the floor. For the block stopping, the construction is exactly the same procedure as building a regular stopping.

For building a door into a metal stopping, the

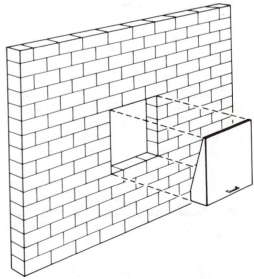

FIG. 46. Exploded view of man door going into the stopping.

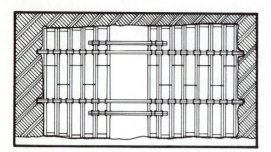

FIG. 47. Metal door being fitted into a stopping (adapted from Jack Kennedy Metal Products & Buildings, Inc. catalogue).

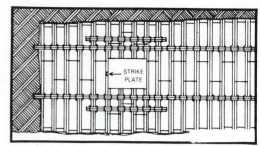

FIG. 48. Stopping ready for door to be inserted (courtesy of Jack Kennedy Metal Products & Buildings, Inc.).

procedure is identical to building a regular stopping, except for replacing two or three regular panels with short panels which are supplied as part of the door kit. The short panels are tied to the regulator structure by angle braces (Fig. 48). The door is bolted or riveted to the panels and is secured by means of a latch, as illustrated in Fig. 49. The structure is sealed in the same manner (on the smooth side) as a regular stopping.

When a man door is placed into a stopping it may experience a great deal of traffic. This tends to create leakage in the stopping and door. Therefore, the mine management should inspect and patch the stopping as necessary to maintain the door's efficiency.

Full-entry Doors

Construction of full-entry doors (also referred to as mine doors) is a difficult task in ventilation control because of the inherent leakage found in the structure. Often, these doors are set in tandem to provide an air lock to further control the airflow.

Materials for full-entry doors have traditionally been wood and brattice cloth. However, these materials have given way to the more cost-effective steel doors and hinges. Steel doors, over their useful life, will better minimize the leakage through the door and can withstand normal mine use longer than their wooden counterparts.

There are several design parameters which must be known before a door is fabricated in the mine shop or ordered from a manufacturer. These include: (1) height of entry, from the mine bottom or top of the tie (if track entry), to the entry roof; (2) width of the entry; (3) the width and height of the largest load required to go through the door; (4) air pressure at the door site; and (5) the track gage and rail weight if the door is to be used for haulage entry. Once these parameters are established, a door size can be chosen which includes the necessary clearances for safe and efficient use.

There are two basic powering devices for opening mine doors: by hand or by air or electric motor (Fig. 50). Both are common installations in current underground mining. Unfortunately, there is no clear-cut advantage to powered doors except they will remain closed in the event the airflow is reversed. Therefore, the cost advantage lies with the hand-powered mine door.

The cost of installing a mine door is relatively straightforward. A fabricated door (including labor) will cost about $375, without the frame. A total fabricated door package will cost around $500. A prefabricated mine door of equal size and weight (gage), including the frame, will cost $850 (1981 dollars). Putting an automatic opening device of air or electric power will boost this price to $1500. However, under heavy traffic conditions, automatic opening and closing devices do have an application, as well as a possible added safety factor.

Assuming the mine door and frame has been delivered to the installation site, construction can begin. Note: if mine track has been laid through the site, then the frame must go under the ties. Two men are required for approximately two shifts to complete this task. The steps for building the mine doors are as follows:

The centerline of the perimeter edge of the door is marked on the roof and ribs.

The ribs are picked smooth and the roof is scaled of loose material.

The placement of the holes in the door frame is measured. These holes are marked on the rib

VENTILATION CONSTRUCTION

FRONT VIEW OF ASSEMBLY WITH DOOR

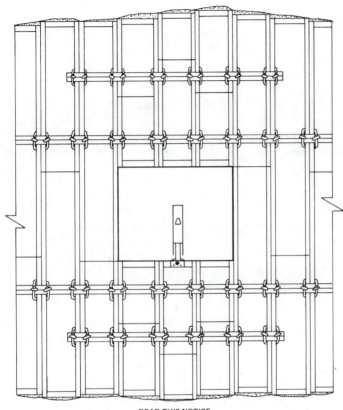

FIG. 49. Door latch (courtesy of Jack Kennedy Metal Products & Buildings, Inc.).

and top (if provided). Using a hand-held auger or stoper drill, bolt holes are drilled approximately 762 mm or 0.8 m (30 in.) into the ribs and top.

The door frame is put in place and secured by tightening the roof bolts 6.0 mm or 0.6 m (24 in.) into the rib (Fig. 51).

The doors are hung on each side and hinged securely.

The frame is sealed between the roof, ribs, and floor (if necessary) with general mine sealant.

The door is sealed by using rubber or other flexible material along the rubbing surfaces, and after installation, may need adjustment for maximum performance.

FIG. 50. Automatic mine door (courtesy of American Mine Door Co.).

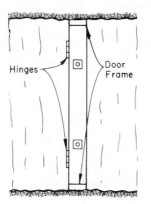
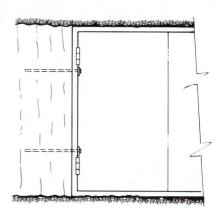

FIG. 51. Door frame bolted into the rib.

Over the life of the door, maintenance work will definitely be necessary. Often, regular mine use will damage the doors in a way that will increase the leakage through it. Therefore, a construction foreman should inspect the door or doors once a month for proper operation. Repairs should be made promptly, since small leakage problems are easier to fix than replacing a whole set of doors.

VENTILATION EQUIPMENT MANUFACTURERS

In an effort to provide as much information as possible, Table 5 lists manufacturers of ventilation equipment. These lists are identified by the type of product, with the manufacturer listed below. This information comes from the 1980 *Buyers' Guide,* published by McGraw-Hill, Inc. as a part of their Mining Information Services. The addresses of these manufacturers are included in Appendix A.

REFERENCES AND BIBLIOGRAPHY

Crickmer, D.F., and Zegeer, D.A., eds., 1981, *Elements of Practical Coal Mining,* 2nd ed., AIME, New York.
Cummins, A.B., and Given, I.A., 1973, *SME Mining Engineering Handbook,* AIME, New York.
Hartman, H.L., 1982, *Mine Ventilation and Air Conditioning,* 2nd ed., John Wiley, New York.
Jack Kennedy Metal Products, Inc., 1981, Technical conversations.
Stefanko, R., 1983, *Coal Mining Technology,* AIME, New York.

TABLE 5. Ventilation Equipment Manufacturers

Coatings (mine roof, ribs, stoppings)
American Minechem Corp.
Austin, J.P., Assoc.
Burrell Construction & Supply
Celite, Inc.
Contractors Warehouse Inc.
DAP Inc.
Eimco Mining Machinery, Envirotech Corp.
Fairmont Supply Co.
Grace, W.R. & Co. (Warehouse Products)
Johnston-Morehouse-Dickey Co.
Michael Walters Industries
Mine Safety Appliances Co.
Cabot, Co.
Preiser/Mineco Div., Preiser Scientific Inc.
Tidewater Supply Co.
West Virginia Belt Sales & Repairs Inc.
Wilson, R.M., Co.

Doors (mine, manual, automatic, air and hydraulic powered)
American Mine Door Co.
Fairmont Supply Co.
Gammeter, W.F., Co.
Kennedy Metal Products & Buildings, Inc.
Leman Machine Co.
Mathews, Abe W., Engineering Co.
Wajax Industries Ltd.

Overcasts (corrugated arch and round)
Armco Inc.
Commercial Shearing, Inc.
Dosco Corp.
Johnston-Morehouse-Dickey Co.
Kennedy Metal Products & Buildings, Inc.
Lane Metal Products, Inc.
Logan Corp.
Norwood, Inc.
Tidewater Supply Co.
US Gypsum Co.

Sealants
Adhesive Engineering Co.
American Cyanamid Co., Industrial Chemicals Div.

Sealants (continued)
Austin, J.P., Assoc.
Banner Bearings
Bearings Inc.
Bowman Distribution, Barnes Group, Inc.
Bruening Bearings, Inc.
Burrell Construction & Supply Samuel, Inc.
DAP Inc.
Darworth Co.
Dixie Bearings, Inc.
Dow Corning Corp.
Dowell Division
du Pont de Nemours, E.I., & Co. Inc.
Fairmont Supply Co.
Firestone Tire & Rubber Co.
Grace, W.R. & Co., Construction Products Div.
Hardman Inc.
Johnston-Morehouse-Dickey Co.
3M Co.
Michael Walters Industries
Midwest Industrial Supply Inc.
Oil Center Research
Pittsburg Corning Corp.
Preiser/Mineco Div., Preiser Scientific Inc.
Rockwell International, Flow Control Div.
Service Supply Co., Inc.
Stonhard, Inc.
Tidewater Supply Co.
Trowelon, Inc.
Uniroyal, Inc.
United McGill Corp.
West Virginia Belt Sales & Repairs Inc.
Wilson, R.M., Co.

Sealant (application machines)
ARO Corp.
Burrell Construction & Supply Co.
Deron Corp.
Dover Conveyor & Equipment Co., Inc.
Johnston-Morehouse-Dickey Co.
Rockwell International, Flow Control Div.
West Virginia Belt Sales & Repairs Inc.
Wilson, R.M., Co.

4

Pump Stations, Sumps, and Drainage Systems

Traditionally, water removal has taken a backseat to many other coal mining functions. However, it is a vital part of the total mine support system. In many US coal mines, more tons of water are removed each year than tons of coal. An excellent discussion of general drainage characteristics can be found in Chapter 8 of *Elements of Practical Coal Mining,* published by SME-AIME.

This chapter is organized into three major sections. These are: designing effective drainage systems; constructing the main mine pumping station (static system); and constructing a portable pumping station (dynamic system). Each topic presents a separate discussion of the elements a construction foreman or engineer should consider for effective installation and operation of the mine drainage system. In addition, a list of pumping equipment manufacturers is available to interested persons and follows these sections.

ANALYZING AND DESIGNING EFFECTIVE UNDERGROUND DRAINAGE SYSTEMS

For each individual mining operation, there must exist an effective water system. What works in one situation may not in another, so it is important the mine engineer or mine foreman be aware of the parameters affecting his drainage requirements.

To analyze any pumping requirements, several design parameters must be available, including (1) the inputs to the mine environment, (2) the required output of the mining operation, and (3) the available system engineering to transport the inputs through the outputs for disposal (Fig. 1). Knowledge of all three of these parameters is necessary to deal effectively with the drainage problems of the underground mine environment.

Inputs to the Mine Environment

There are a number of water inputs to the mine environment. They include: dust sprays from production machinery, dust suppression from mine haulage, coal seam leakage, and surface runoff. Knowing and understanding the sources of water inputs is important when it comes to choosing the best location for the main mine sump or intermediate pumping stations, as well as trying to prevent or control the inputs of water into the mine workings.

In addition to the amount of water input, the construction engineer must look at the makeup and quality of the water input. In many mines, the water makeup contains particles and sediments from the mining operations which can turn clear water into something resembling sludge. The efficiency and useful life of any pump depends on the type of material being handled so cavitation and choking from abrasive particles does not occur. For example, a turbine grinder pump may be better for a water, coal, and rock mixture than a clear water centrifugal pump with a higher head.

The quality of water refers to the pH balance and chemical components of the mine water. Acid mine water (pH less than 5.0) is a real environmental problem which must be controlled before it leaves the mine property. In most cases, the addition of lime into the water output is necessary. Other chemical components (metals, toxins, etc.) are rare, but possible. Design engineers should be aware and check the water makeup and quality to gain the best efficiency and useful life of the pumping system

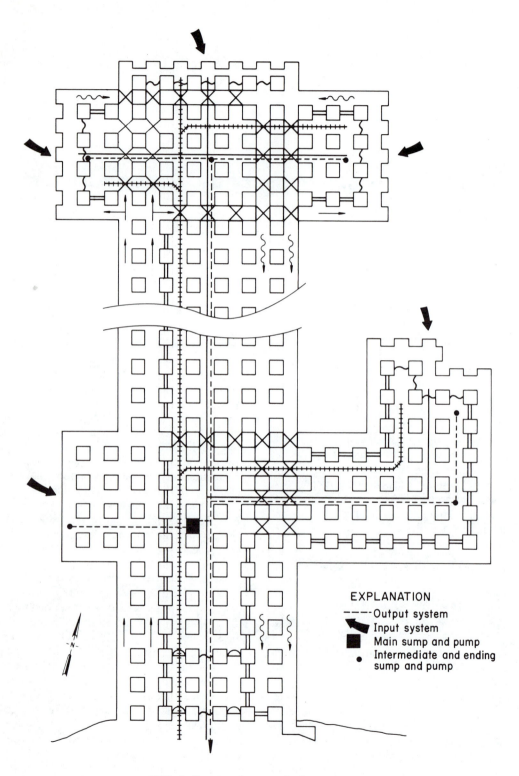

FIG. 1. Input and output drainage patterns.

as well as keeping in compliance with environmental regulations.

Dust Suppression from Production Machinery: Dust suppression from production machinery is a water input to the mine environment at a place where it should not be allowed to accumulate, i.e., the face area. Modern mining machines generally have capabilities of delivering at least 0.9 to 1.6 L/s (15 to 25 gpm) to the working face while mining. Conventional, continuous, and longwall systems all require face sprays for dust suppression, and this water source grows in direct proportion to mine growth over the life of the operation.

In addition to the face sprays, a second and more damaging water input is from the leakage these systems generate over time. Water leaks from valves, splices, adaptors, and pinhole leaks when left unattended can create tremendous water holes, muddy haulways, and soft bottoms on the production section. These leaks on a typical section will release an estimated 0.3 to 0.4 L/s (4 to 7 gpm) to the mine environment. Section foremen can help reduce this input by inspecting their face spray systems and eliminating any system leaks.

Dust Suppression from Mine Haulage: Dust suppression from hauling coal from the face area to the surface starts at the outby haulage system, usually at the section feeder-breaker. Here the coal is crushed and sprayed to allay the dust. Unless the sprays are left on continuously, only a small fraction escapes into the mine environment, since most of the water remains in the crushed coal. Leaks from valves, couplings, and stripped joints are the major concern. Other dust allaying sprays are located at strategic spots along the haulage system such as transfer points and surge bins. Again, the majority of water will pass out of the mine via the crushed run-of-mine material. The leakage from these systems can be the biggest sore spot for drainage, since much of the collected water usually is ignored until it becomes a real problem.

A final note on haulage system dust suppression concerns the fire fighting systems needed along the belt line. These water lines are laid down and then forgotten, making them quite susceptible to unnoticed leaks. These systems should be checked frequently.

Coal Seam Leakage: Leakage from the coal seam itself occurs when the mine expands into virgin territory. Most, if not all, coal seams are aquifers to some degree, and as mining progresses, so does the amount of water entering the operation. Information on how wet the seam is should be gathered during the premining geotechnical studies.

Leakage from old works also falls under this mine input. In this case, water which has collected in pools in previously mined areas enters into the operation from breached ribs, roof, or floor. In many cases, this water violently enters the mine, causing many safety concerns. Water from old works should be avoided whenever possible by leaving an adequate barrier pillar, or by controlled drainage of the old works.

Surface Runoff: The final mine input is surface runoff. Runoff occurs extensively with mines under shallow overburden [less than 18 m (60 ft)]. Here, water from rains and core holes seeps from the roof and entranceways into the mine operation. Where permanent streams and creeks run along the surface, poor development strategies by the mine operator can literally increase the amount of leakage into his operation by as much as 3.2 to 6.3 L/s (50 to 100 gpm) or more (Fig. 2).

In concluding water inputs, one can easily see the prevention of inputs into the mine operation is the prime concern. Mine designs with water inputs taken into account can eliminate the majority of these and reduce the influence of these water inputs which are unavoidable during mining activities.

Outputs from the Mine Environment

If prevention of water inputs into the mining operation cannot and does not stop all this nuisance water, then output systems must be considered. Output systems are the product of the required mine outputs and the necessary system engineering. They contain three separate elements of construction based on the scope of the system. These elements include:

The pumping network which defines the necessary equipment for efficient transport of the water for disposal, including the pumps, piping, motors, valves, and pump controls.

The sump interface encompassing the excavation and reinforcement requirements necessary for storage and surge capacities of the pumping network.

The runoff drainage system which describes the efforts and methods needed to collect the

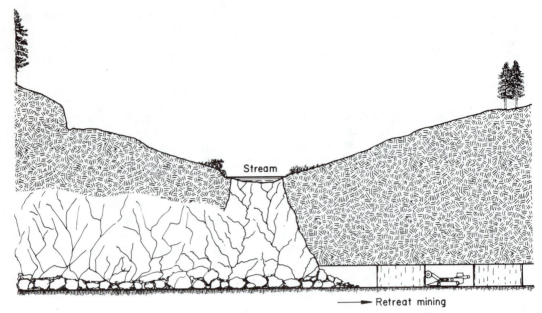

FIG. 2. Fractured strata under stream bed allows water to seep in.

water inputs and channel them into the pumping network.

Of the three elements, the pumping network is the most important consideration in the output system. However, it is the interaction of all the elements which comprises an efficient and successful system. A mine engineer or construction foreman must analyze his mine inputs in relation to his output options in order for the construction of the mine drainage system to properly meet his needs. The first consideration for output systems is the required engineering needed to match the mine inputs.

Engineering the System

In the initial stages of mine drainage design, an engineer can divide drainage construction into two major sections (Fig. 3). The first section encompasses the system from the face to the main sump area. The second section concerns the system from the main sump to the surface. This break in the system denotes a definite change in the scope and construction responsibilities of the mine management in terms of water collection and disposal.

Once the system has been broken into these two parts, the engineer should begin summing up the mine waste water inputs from the working face to the outside to get a total estimate of water to be handled. Parameters from the mine inputs will define minimum pumping rates, power sources and requirements, sump locations, heavy drainage areas, and construction requirements for the final design. As a minimum, detailed drawings of the sump and main pumping station and intermediate pump stations, collection holes, and connecting piping are required. These drawings should specify the placement and quantity of equipment and materials needed to complete the task. Construction prints should be drawn for the life of the mine structures.

The Face to Sump System: The face to sump system is characterized by one word—dynamic. This is the system which must change and evolve each day to meet the varying mine drainage conditions.

Of the three output elements discussed earlier, the pump network is clearly the key component, forming a disposal backbone linking each working face and section together by a series of pumps and intermediate sumps.

The pumping network from the face to sump system is served by a combination of (1) portable or semi-portable pumps; (2) plastic (PVC), steel, fiberglass, cast iron, or aluminum piping; as well as (3) a-c pump controls; and (4) the valves, couplings, hangers, and other hardware needed to hook the system together.

Pump Characteristics. The pumps used in this system are either portable or semi-portable and approved by MSHA for use in gassy environments. These pumps commonly use the turbine

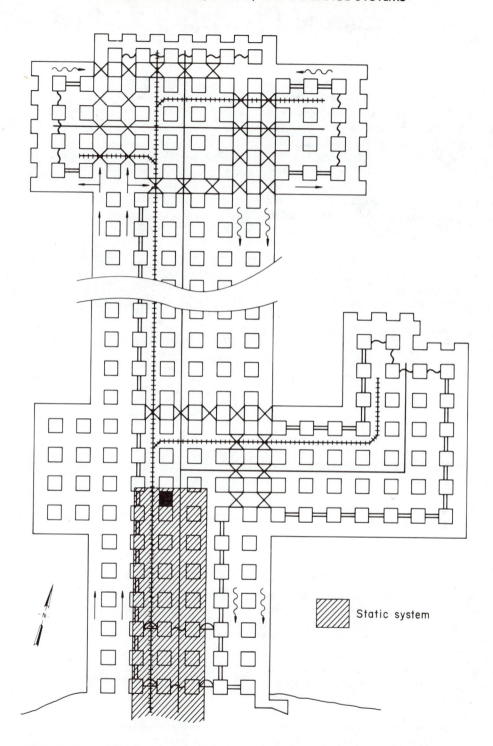

FIG. 3. The static and dynamic components to mine drainage. The dynamic system surrounds the static system. The static system runs from the main sump (solid block) area to the outside.

method of suction, as typified by the Flygt series of permissible mine pumps (Figs. 4 to 8). Here, water is sucked into the discharge line by the action of the impeller, shown in the cutaway view of Fig. 9. A screen aids in keeping abrasive and oversized materials out of the pump and motor areas where damage might occur. The Flygt pumps, like most pumps in the dynamic system area, are light, reliable, rugged, waterproof (submersible), and will not burn out during a dry pumping stage. The range of pumping characteristics is defined by the ability of the pump to deliver water at a certain speed, head, and rate. Tables 1 to 5 illustrate the ranges for the Flygt series. These tables underline the important engineering characteristics necessary to choose a suitable mine pump, namely, the static and friction heads, the capacity rate, and the power characteristics of speed and horsepower.

The static and friction heads define the limits of a pump's capabilities given the rate and power characteristics. The static head is the vertical distance [in meters (feet)] a pump will lift the water. The friction head is the distance lost to the friction of the water traveling in the pipe. The friction head is a real concern during drainage system construction, since it can play a substantial role in the overall performance of the pump. The friction head can be estimated from tables designed for standard pumping characteristics, as shown in Table 6.

The capacity rate [liter per second (gpm)] is the rate at which the water is moved through the system by a pump. It must equal or exceed the water inputs into the mine at all times. For example, a small pump controlling a local water problem may only pump for a few hours each day, but the rate at which it pumps determines how much and how long the pump needs to operate. The mine pump and performance curve shown in Fig. 10 illustrates the relationship between pump rate [liter per second (US gpm)] and total head. At 18.9 L/s (300 gpm), the pump will lift this amount 12 or 24 m (40 or 80 ft) depending upon the nature of the operation (single or two-stage). At 1.6 L/s (25 gpm), it can lift 21 or 43 m (70 or 140 ft) respectively. Construction foremen setting pumps in isolated or local drainage points need to check the pumping rate against the expected head to ensure a proper match.

The speed and horsepower of the pump motor defines the boundaries of the ability of the pump

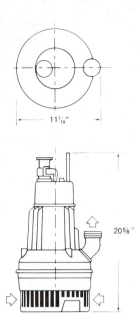

FIG. 4. Flygt Pump B2051P (courtesy of Flygt Corp.).

TABLE 1.

Curve	1	2	3
Discharge	2" St'd		
Phase	3∅		
Rated HP	1.6		
Voltage	460 or 575 V		
Amperage (Max.)	2.3 or 1.85		
Power Input (Max.)	1.6 KW		
RPM	3450		
Weight	44 lbs.		
Rec. Gen. Size	6.5 KW		
Min. Transformer for Service Drop	3 KVA		

[Graph: Total Dynamic Head Feet (0–80) vs. Flow U.S. GPM (0–140), labeled "1 (same curve for all)"]

Metric equivalents: in. × 25.4 = mm; ft × 0.3048 = m; lb × 0.453 592 4 = kg; gpm × 0.063 090 2 = L/s.

PUMP STATIONS, SUMPS, AND DRAINAGE SYSTEMS

FIG. 5. Flygt Pump B2075P (courtesy of Flygt Corp.).

TABLE 2.

Curve	1	2
Discharge	3"/4" H. H.	3"/4" H. Vol.
Phase	3φ	
Rated HP	5	
Voltage	440 or 575 V	
Amperage (Max.)	7.5 or 5.8	
Power Input (Max.)	5.2 KW	(Same as 1)
RPM	3500	
Weight	95 lbs.	
Rec. Gen. Size	10 KW	
Min. Transformer for Service Drop	10 KVA	

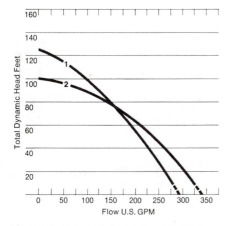

Metric equivalents: in. × 25.4 = mm; ft × 0.3048 = m; lb × 0.453 592 4 = kg; gpm × 0.063 090 2 = L/s.

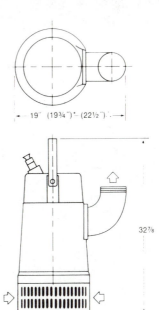

FIG. 6. Flygt Pump B2125P (courtesy of Flygt Corp.).

TABLE 3.

Curve	1**	2
Discharge	3"/4"* H. H.	4"*/6"△ H. Vol.
Phase	3φ	3φ
Rated HP	13	13
Voltage	440 or 550 V	440 or 550 V
Amperage (Max.)	17 or 13	17 or 13
Power Input (Max.)	11.6 KW	11.6 KW
RPM	3400	3400
Weight	203 lbs.	183 lbs.
Rec. Gen. Size	30 or 45 KW	30 or 45 KW
Min. Transformer for Service Drop	20 KVA	20 KVA

**Two stage

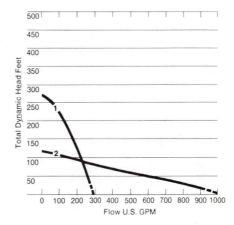

Metric equivalents: in. × 25.4 = mm; ft × 0.3048 = m; lb × 0.453 592 4 = kg; gpm × 0.063 090 2 = L/s.

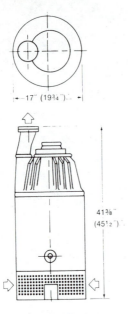

FIG. 7. Flygt Pump B2201P (courtesy of Flygt Corp.).

TABLE 4.

Curve	1	2
Discharge	4" H. H.	6" or 8" △ H.Vol.
Phase	3ø	3ø
Rated HP	58	58
Voltage	460 or 575 V	460 or 575 V
Amperage (Max.)	65 or 52	65 or 52
Power Input (Max.)	47.5 KW	47.5 KW
RPM	3450	3450
Weight	530 lbs.	615 lbs.
Rec. Gen. Size	125/100 KW	125/100 KW
Min. Transformer for Service Drop	75 KVA	75 KVA

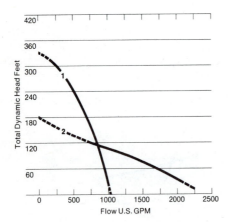

Metric equivalents: in. × 25.4 = mm; ft × 0.3048 = m; lb × 0.453 592 4 = kg; gpm × 0.063 090 2 = L/s.

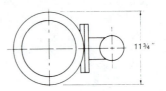
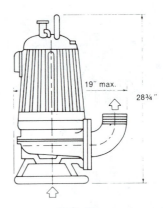

FIG. 8. Flygt Pump DS3080P (courtesy of Flygt Corp.).

TABLE 5.

Curve	1	2	3
Discharge	2" ST	3" H.H.	4" H. Vol.
Phase	3ø	3ø	3ø
Rated HP	9.0	9.0	6.0
Voltage	460 or 575	460 or 575	460 or 575
Amperage (Max.)	12 or 10	12 or 10	9 or 7.5
Power Input (Max.)	9.2	9.2	6.0
RPM	3450	3450	1750
Weight	110 lbs.	110 lbs.	110 lbs.
Rec. Gen. Size			
Min. Transformer for Service Drop	10 KVA	10 KVA	10 KVA

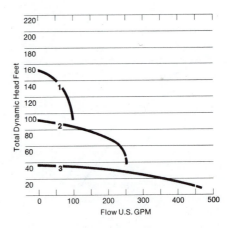

Metric equivalents: in. × 25.4 = mm; ft × 0.3048 = m; lb × 0.453 592 4 = kg; gpm × 0.063 090 2 = L/s.

PUMP STATIONS, SUMPS, AND DRAINAGE SYSTEMS

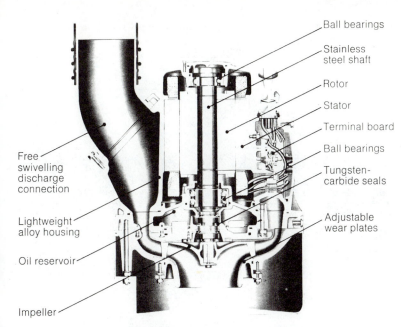

FIG. 9. Cutaway of Flygt turbine pump (courtesy of Flygt Corp.).

to lift the water and the rate at which it does it. The speed is designed in revolutions per minute (rpm), while the horsepower is equal to 745.7 W (33,000 ft-lb per min). The horsepower of a pump installation can be defined simply as:

$$Hp = \frac{\text{Pump rate (gpm)} \times \text{total head (ft)} \times 8.338 \text{ lb per gal}}{33,000 \text{ ft-lb per min} \times \text{efficiency}}$$

where the pump efficiency ranges between 70 and 90%.

Earlier, it was mentioned that pumps could be portable or semi-portable. Portable pumps (such as the Flygt B2051P) usually weigh less than 68 kg (150 lb) and need only one man to transport them. They usually are 220/440 V, with less than 7.5 kW (10 hp) motor, and can always dry pump without burning out. These pumps usually operate with a discharge line of 76 mm (3 in.) or less. Portable pumps are used in the production face areas, areas with access difficulties, and in remote less traveled mine entries with small defined drainage requirements.

Semi-portable pumps, on the other hand, are usually the centrifugal variety. They weigh between 68.5 and 136 kg (151 and 300 lb), need

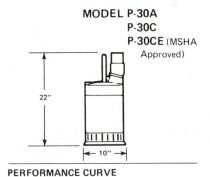

PERFORMANCE CURVE

Discharge	3" NPT	3" NPT
Phase	1 ∅	3 ∅
Rated HP	3	4.6
Voltage	230	230/460/575
Amperage (Max.)	13.5	13/6.5/4.8
Power Input (Max.)	3.1 KW	4.1 KW
RPM	3450	3450
Weight	66 lb.	66 lb.
Rec. Generator Size	7 KW	10 KW

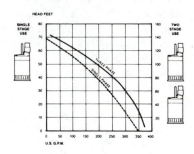

FIG. 10. Stanco mine pump P-30CE with performance curve (courtesy of Linden-Alimak, Inc.). Metric equivalents: in. × 25.4 = mm; ft × 0.3048 = m; lb × 0.453 592 4 = kg; gpm × 0.063 090 2 = L/s.

TABLE 6. Friction Loss of Water in Feet per 30.5 m (100 Ft) Length of Pipe. Based on Williams and Hazen Formula Using Constant 100; Standard Pipe Sizes in In.*

Pipe Size US Gpm	1/2 In. Velocity, Fps	1/2 In. Head Loss in Ft	3/4 In. Velocity, Fps	3/4 In. Head Loss in Ft	1 In. Velocity, Fps	1 In. Head Loss in Ft	1 1/4 In. Velocity, Fps	1 1/4 In. Head Loss in Ft	1 1/2 In. Velocity, Fps	1 1/2 In. Head Loss in Ft	2 Velocity, Fps
2	2.10	7.4	1.20	1.9	....	....	....	....	....	....	....
4	4.21	27.0	2.41	7.0	1.49	2.14	0.86	0.57	0.63	0.26	....
6	6.31	57.0	3.61	14.7	2.23	4.55	1.29	1.20	0.94	0.56	0.61
8	8.42	98.0	4.81	25.0	2.98	7.8	1.72	2.03	1.26	0.95	0.82
10	10.52	147.0	6.02	38.0	3.72	11.7	2.14	3.05	1.57	1.43	1.02
12	....	....	7.22	53.0	4.46	16.4	2.57	4.3	1.89	2.01	1.23
15	....	....	9.02	80.0	5.60	25.0	3.21	6.5	2.36	3.00	1.53
18	....	....	10.84	108.2	6.69	35.0	3.86	9.1	2.83	4.24	1.84
20	....	....	12.03	136.0	7.44	42.0	4.29	11.1	3.15	5.20	2.04
25	....	....	....	....	9.30	64.0	5.36	16.6	3.80	7.30	2.55
30	....	....	....	....	11.15	89.0	6.43	23.0	4.72	11.0	3.06
35	....	....	....	....	13.02	119.0	7.51	31.2	5.51	14.7	3.57
40	....	....	....	....	14.88	152.0	8.58	40.0	6.30	18.8	4.08
45	....	....	....	....	....	....	9.65	50.0	7.08	23.2	4.60
50	....	....	....	....	....	....	10.72	60.0	7.87	28.4	5.11
55	....	....	....	....	....	....	11.78	72.0	8.66	34.0	5.62
60	....	....	....	....	....	....	12.87	85.0	9.44	39.6	6.13
65	....	....	....	....	....	....	13.92	99.7	10.23	45.9	6.64
70							15.01	113.0	11.02	53.0	7.15
75							16.06	129.0	11.80	60.0	7.66
80							17.16	145.0	12.59	68.0	8.17
85							18.21	163.8	13.38	75.0	8.68
90							19.30	180.0	14.71	84.0	9.19
95							....	....	14.95	93.0	9.70
100									15.74	102.0	10.21
110									17.31	122.0	11.23
120									18.89	143.0	12.25
130									20.46	166.0	13.28
140									22.04	190.0	14.30
150									....	....	15.32

two men to handle, and are rarely used for any pumping requirements other than intermediate and main sumps. Centrifugal pumps operate on a different lifting principle than that of the turbine pump. This pump rotates the water with the use of an impeller to throw the water to a height specified by the capacity rate, horsepower, efficiency, speed, and impeller size. Fig. 11 is a photograph and cutaway view of a centrifugal pump manufactured by ITT Marlow. The performance curve of this pump is shown in Table 7. The cost of a pump like this averages $2,500 assembled. These pumps are almost always 440 V a-c power with a discharge line of 76 mm (3 in.) or more, and operate with a horsepower between 10-50 hp (7.5–17 kW). These pumps are not designed to dry pump; therefore, they need more attention and controls (float switches, etc.) for operation. Fig. 12 is an example of a turbine-type semi-portable pump in a typical intermediate sump operation.

Piping Characteristics. The piping used in this dynamic system is chosen from a variety of materials and material characteristics. The basic materials are either steel, cast iron, aluminum, fiberglass, or plastic (PVC), with typical diameters ranging from 51 to 102 mm (2 to 4 in.). Steel piping, usually of schedule 40 to 80 strength, and cast iron piping [57.7 kg (125 lb)] give the advantage of durability, but are heavy and a nuisance to work with. Aluminum pipe, while very light (one man can easily lift a length of

TABLE 6. Friction Loss of Water in Feet per 30.5 m (100 Ft) Length of Pipe. Based on Williams and Hazen Formula Using Constant 100; Standard Pipe Sizes in In.* (Continued)

In.	2 ½ In.		3 In.		4 In.		5 In.		6 In.		Pipe Size
Head Loss in Ft	Velocity, Fps	Head Loss in Ft	Velocity, Fps	Head Loss in Ft	Velocity, Fps	Head Loss in Ft	Velocity, Fps	Head Loss in Ft	Velocity, Fps	Head Loss in Ft	US Gpm
....	....	....	....	....	....	....	....	....	....	....	2
....	....	....	....	....	....	....	....	....	....	....	4
0.20	....	....	....	....	....	....	....	....	....	....	6
0.33	0.52	0.11	....	....	....	....	....	....	....	....	8
0.50	0.65	0.17	0.45	0.07	....	....	....	....	....	....	10
0.79	0.78	0.23	0.54	0.10	....	....	....	....	....	....	12
1.08	0.98	0.36	0.68	0.15	....	....	....	....	....	....	15
1.49	1.18	0.50	0.82	0.21	....	....	....	....	....	....	18
1.82	1.31	0.61	0.91	0.25	0.51	0.06	....	....	....	....	20
2.73	1.63	0.92	1.13	0.38	0.64	0.09	....	....	....	....	25
3.84	1.96	1.29	1.36	0.54	0.77	0.13	0.49	0.04	....	....	30
5.10	2.29	1.72	1.59	0.71	0.89	0.17	0.57	0.06	....	....	35
6.6	2.61	2.20	1.82	0.91	1.02	0.22	0.65	0.08	....	....	40
8.2	2.94	2.80	2.04	1.15	1.15	0.28	0.73	0.09	....	....	45
9.9	3.27	3.32	2.27	1.38	1.28	0.34	0.82	0.11	0.57	0.04	50
11.8	3.59	4.01	2.45	1.58	1.41	0.41	0.90	0.14	0.62	0.05	55
13.9	3.92	4.65	2.72	1.92	1.53	0.47	0.98	0.16	0.68	0.06	60
16.1	4.24	5.4	2.89	2.16	1.66	0.53	1.06	0.19	0.74	0.076	65
18.4	4.58	6.2	3.18	2.57	1.79	0.63	1.14	0.21	0.79	0.08	70
20.9	4.91	7.1	3.33	3.00	1.91	0.73	1.22	0.24	0.85	0.10	75
23.7	5.23	7.9	3.63	3.28	2.04	0.81	1.31	0.27	0.91	0.11	80
26.5	5.56	8.1	3.78	3.54	2.17	0.91	1.39	0.31	0.96	0.12	85
29.4	5.88	9.8	4.09	4.08	2.30	1.00	1.47	0.34	1.02	0.14	90
32.6	6.21	10.8	4.22	4.33	2.42	1.12	1.55	0.38	1.08	0.15	95
35.8	6.54	12.0	4.54	4.95	2.55	1.22	1.63	0.41	1.13	0.17	100
42.9	7.18	14.5	5.00	6.0	2.81	1.46	1.79	0.49	1.25	0.21	110
50.0	7.84	16.8	5.45	7.0	3.06	1.17	1.96	0.58	1.36	0.24	120
48.0	8.48	18.7	5.91	8.1	3.31	1.97	2.12	0.67	1.47	0.27	130
67.0	9.15	22.3	6.35	9.2	3.57	2.28	2.29	0.76	1.59	0.32	140
76.0	9.81	25.5	6.82	10.5	3.82	2.62	2.45	0.88	1.70	0.36	150

pipe), usually cannot last more than five years of typical mine wear and tear. To the contrary, plastic piping gives both durability and ease of handling to the user (Fig. 13), but costs much more than the rest of the piping materials.

The concern with piping for a construction engineer is in minimizing the time required to lay and connect the piping together. This requires some cost and time estimating techniques on the part of the system designer.

In mining, the two basic methods of joining pipe are clamping the ends together with a coupling, or using threaded ends. Of the two, clamping has become the favorite, with threaded piping confined to older mines and systems.

Using welded or flanged piping is generally too time-consuming and costly to find much use underground. To accurately estimate the time and cost of a drainage system, one can use the common pipe and fitting method of calculation. This is a cost estimating procedure which will be discussed in full in Chap. 10. For this discussion, it is appropriate to look at labor estimates.

The pipe and fitting method uses a calculation sheet similar to the one shown in Fig. 14. The foreman or engineer must estimate the total time and labor requirements by breaking down the proposed mine piping system into its component parts (i.e., the joints of pipe, elbows, tees, and

TABLE 6. Friction Loss of Water in Feet per 30.5 m (100 Ft) Length of Pipe. Based on Williams and Hazen Formula Using Constant 100; Standard Pipe Sizes in In.* (continued)

Pipe Size	½ In.		¾ In.		1 In.		1¼ In.		1½ In.		2
US Gpm	Velocity, Fps	Head Loss in Ft	Velocity, Fps	Head Loss in Ft	Velocity, Fps	Head Loss in Ft	Velocity, Fps	Head Loss in Ft	Velocity, Fps	Head Loss in Ft	Velocity, Fps
160											16.34
170											17.36
180											18.38
190											19.40
200											20.42
220											22.47
240											24.52
260											26.55
280											....
300											....
320											....
340											....
360											
380											
400											
450											
500											
550											
600											
650											
700											
750											
800											
850											
900											
950											
1000											
1100											
1200											
1300											
1400											
1500											
1600											
1800											
2000											
2200											

*Metric equivalents: in. × 25.4 = mm; gpm × 0.063 090 2 = L/s; fps × 0.3048 = m/s; ft × 0.3048 = m.
Source: Crickmer and Zegeer, 1981.

valves required). From these components, the estimated man-hours required to install the component can be taken from a table similar to Table 8. Once the time needed to install a system is calculated, the scheduling of the work can be done with confidence.

The controls and miscellaneous couplings (such as starters, float valves, etc.), hangers, and gate valves are usually purchased as separate items for mine use. The pump controls are supplied by the pump manufacturer at the time the pump and motor are purchased. The valves, couplings, and other hardware are supplied by plumbing houses and pump manufacturers based on the specifications of the operator's system. The quantities of these items required by the operator should be developed during the active mining stages so the appropriate design loads for mine drainage can be met and the required installation time estimated.

TABLE 6. Friction Loss of Water in Feet per 30.5 m (100 Ft) Length of Pipe. Based on Williams and Hazen Formula Using Constant 100; Standard Pipe Sizes in In.* (continued)

Pipe Size	2 In.	2½ In.		3 In.		4 In.		5 In.		6 In.		Pipe Size
In.	Head Loss in Ft	Velocity, Fps	Head Loss in Ft	Velocity, Fps	Head Loss in Ft	Velocity, Fps	Head Loss in Ft	Velocity, Fps	Head Loss in Ft	Velocity, Fps	Head Loss in Ft	US Gpm
	86.0	10.46	29.0	7.26	11.8	4.08	2.91	2.61	0.98	1.82	0.40	160
	96.0	11.11	34.1	7.71	13.3	4.33	3.26	2.77	1.08	1.92	0.45	170
	107.0	11.76	35.7	8.17	14.0	4.60	3.61	2.94	1.22	2.04	0.50	180
	118.0	12.42	39.6	8.63	15.5	4.84	4.01	3.10	1.35	2.16	0.55	190
	129.0	13.07	43.1	9.08	17.8	5.11	4.4	3.27	1.48	2.27	0.62	200
	154.0	14.38	52.0	9.99	21.3	5.62	5.2	3.59	1.77	2.50	0.73	220
	182.0	15.69	61.0	10.89	25.1	6.13	6.2	3.92	2.08	2.72	0.87	240
	211.0	16.99	70.0	11.80	29.1	6.64	7.2	4.25	2.41	2.95	1.00	260
	. . . .	18.30	81.0	12.71	33.4	7.15	8.2	4.58	2.77	3.18	1.14	280
	. . . .	19.61	92.0	13.62	38.0	7.66	9.3	4.90	3.14	3.40	1.32	300
	. . . .	20.92	103.0	14.52	42.8	8.17	10.5	5.23	3.54	3.64	1.47	320
	. . . .	22.22	116.0	15.43	47.9	8.68	11.7	5.54	3.97	3.84	1.62	340
		23.53	128.0	16.34	53.0	9.19	13.1	5.87	4.41	4.08	1.83	360
		24.84	142.0	17.25	49.0	9.69	14.0	6.19	4.86	4.31	2.00	380
		26.14	156.0	18.16	65.0	10.21	16.0	6.54	5.4	4.55	2.20	400
		. . . .	. . . .	20.40	78.0	11.49	19.8	7.35	6.7	5.11	2.74	450
		. . . .	. . . .	22.70	98.0	12.77	24.0	8.17	8.1	5.68	2.90	500
		. . . .	. . . .	24.96	117.0	14.04	28.7	8.99	9.6	6.25	3.96	550
				27.23	137.0	15.32	33.7	9.80	11.3	6.81	4.65	600
				. . . .	. . . .	16.59	39.0	10.62	13.2	7.38	5.40	650
				. . . .	. . . .	17.87	44.9	11.44	15.1	7.95	6.21	700
				. . . .	. . . .	19.15	51.0	12.26	17.2	8.50	7.12	750
				. . . .	. . . .	20.42	57.0	13.07	19.4	9.08	7.96	800
				. . . .	. . . .	21.70	64.0	13.89	21.7	9.65	8.95	850
						22.98	71.0	14.71	24.0	10.20	10.11	900
						. . . .	. . . .	15.52	26.7	10.77	11.20	950
						. . . .	. . . .	16.34	29.2	11.34	12.04	1000
						. . . .	. . . .	17.97	34.9	12.48	14.55	1100
						. . . .	. . . .	19.61	40.9	13.61	17.10	1200
						. . . .	. . . .	. . . .	. . . .	14.72	18.4	1300
										15.90	22.60	1400
										17.02	25.60	1500
										18.10	26.9	1600
										. . . .	. . . .	1800
										. . . .	. . . .	2000
										. . . .	. . . .	2200

Sump to Surface System: The sump to surface system, in contrast to the face to sump system, must be designed in the premining plans. Here, the system structure of pumps, controls, surge tanks, and piping remains relatively static (constant) during the life of the mine. Therefore, it is essential all the important design and construction parameters covering the engineering of the system to the mine be considered. As a general rule of thumb, the following parameters should be considered:

System Capacity—The system should well exceed the expected drainage requirements of the mine over its lifetime. As a guide, most wet mines can expect two to three times as much water as the yearly mine production. For example, if a mine is producing 0.9 Mt (1 million st) per year in coal, then it is reasonable to expect to control 1.8 or 2.7 Mt (2 or 3 million st) of water per year by the end of the mine life. However, such estimates are extremely site specific and should be investigated further.

FIG. 11(a). ITT Marlow pump 425EL-3 and motor (courtesy of ITT Marlow).

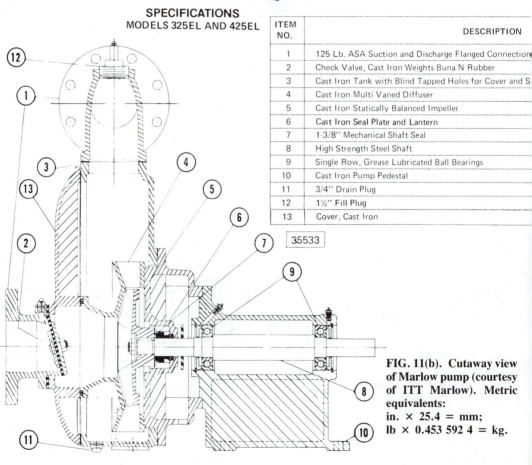

SPECIFICATIONS
MODELS 325EL AND 425EL

ITEM NO.	DESCRIPTION
1	125 Lb. ASA Suction and Discharge Flanged Connection
2	Check Valve, Cast Iron Weights Buna N Rubber
3	Cast Iron Tank with Blind Tapped Holes for Cover and S
4	Cast Iron Multi Vaned Diffuser
5	Cast Iron Statically Balanced Impeller
6	Cast Iron Seal Plate and Lantern
7	1-3/8" Mechanical Shaft Seal
8	High Strength Steel Shaft
9	Single Row, Grease Lubricated Ball Bearings
10	Cast Iron Pump Pedestal
11	3/4" Drain Plug
12	1½" Fill Plug
13	Cover, Cast Iron

35533

FIG. 11(b). Cutaway view of Marlow pump (courtesy of ITT Marlow). Metric equivalents:
in. × 25.4 = mm;
lb × 0.453 592 4 = kg.

TABLE 7. Composite Curves of ITT Marlow Centrifugal Pump.

The performance curves shown here were taken from actual tests of standard production pumps, and reflect an average performance of the pumps indicated.

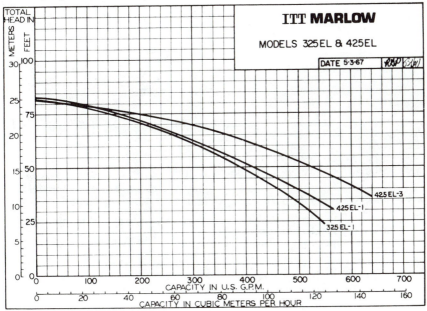

PERFORMANCE:

Pump Model	Speed	TOTAL HEAD IN FEET INCLUDING FRICTION								
		30	35	40	45	50	55	60	70	80
325EL-1	1750	520	490	460	425	390	348	305	205	60
425EL-1	1750	560	535	500	460	420	368	325	225	90
425EL-3	1750	620	620	605	565	525	472	425	290	70

The horsepower for other fluids can be determined by multiplying the horsepower based on water as shown on the curves times the specific gravity of the liquid to be pumped.

Safety Factor—In terms of reliability and backup systems, the safety factor of the system should be considered for all capacity, power, and surge calculations. As a rule of thumb, a factor of 1.25 to 1.50 is used. In terms of construction, the safety factor may mean additional piping, backup pumps, and additional controls.

Surge Capacity—This parameter deals with the ability of the sump to hold a specific amount of waste water from the face to sump system. The average design surge capacity is 36 hr of full discharge into the sump. Additional capacity can be designed in the system by providing surge pumps in the intermediate sump areas amounting to 12 hr or more of capacity.

Accessibility—The main sump areas must be accessible to management for routine maintenance pump and piping changouts, system monitoring, and ease of construction. A logical and convenient site must be chosen to achieve the best construction and maintenance balance with the rest of the mining operations. All too often, the main sump is located in remote or inaccessible areas because of poor planning. Delivery of material, parts, or supplies to badly sited sump areas can be extremely difficult.

Power—The main sump to surface system should have a separate power circuit branching off the mine power system.

Maintenance—A separate and rigorus main-

TABLE 8. Manual "A" Labor Man-hours*, Size, in.†

	2	2.5	3	3.5	4	5	6
Grooved							
Pipe per linear ft	0.06	0.06	0.07	0.08	0.09	0.10	0.14
Elbow	0.32	0.40	0.48	0.56	0.64	0.80	0.96
Tee	0.48	0.60	0.72	0.84	0.96	1.20	1.44
Per joint	0.16	0.20	0.24	0.28	0.32	0.40	0.48
Threaded							
Pipe per linear ft	0.06	0.07	0.10	0.15	0.20	0.30	0.40
Elbow	0.60	0.80	1.00	1.20	1.80	2.40	3.00
Tee	0.90	1.20	1.50	1.80	2.40	3.60	4.50
Per joint	0.30	0.40	0.50	0.60	0.90	1.20	1.50

*Adapted from *Estimator's Workbook* by Victaulic Co. of America.
†Metric equivalents: in. × 25.4 = mm; ft × 0.3048 = m.

tenance schedule should be developed for the system.

Interfacing—The system should adequately interface with all parts (pipe diameters, check valves, etc.) of the drainage system to eliminate any possible bottlenecks which might occur.

Once these parameters have been identified for the system, a better, more cost-effective design can be engineered for a given mining situation. Engineering trade-offs in pumps, excavation, piping and so on will help control drainage costs throughout the mine life.

CONSTRUCTION OF THE MAIN PUMPING STATION

This section discusses the engineering and construction of a main pumping station found at large mines [0.9 Mt/a (1 million stpy) or

FIG. 12. Portable turbine pump in operation (courtesy of Flygt Corp.).

PUMP STATIONS, SUMPS, AND DRAINAGE SYSTEMS

FIG. 13. Plastic PVC piping is durable and easy to handle (courtesy of CertainTeed Corp.).

greater]. Although each mine presents a unique drainage problem, there are certain elements common to the construction of all main pumping stations. This discussion is divided into three main areas: (1) preparing the site, (2) installing the main pumps, and (3) performing operation and maintenance activities. Each area will cover the fundamental points needed to understand the construction of the system.

Preparing the Site

At the start of the construction of the main pumping station, several preparatory tasks are necessary. These include: (1) site orientation, (2) sump excavation, cleaning, and lining, (3) pump foundation construction, and (4) erection and installation requirements at the pump site. When each is completed, these initial tasks will allow the pump installation work to proceed smoothly and efficiently.

Site Orientation: The key individuals in the site orientation task are the mine surveyors, the mine construction engineer, and the construction foreman. Discussions among these individuals will produce answers to the questions of the best site for the sump, the necessary excavation, the labor requirements, the construction schedule, and supply/material deliveries. Each

LABOR COSTS			*Grooved*	*Threaded*
Pipe (linear ft)		@		
Elbows	____	@		
Tees	____	@		
Total Man-hours			_____	_____
$ per Man-hour			X _____	X _____
Total Labor Cost			_____	_____
MATERIAL COSTS				
Pipe (linear ft)		@	($ /linear ft)	($ /linear ft)
Preparation			_____	_____
Elbows	____		($ _____)	($ _____)
Tees	____		($ _____)	($ _____)
Other				
Supplies			_____	_____
Total Materials			_____	_____
+ Labor Cost			+ _____	+ _____
Total Installed Cost			_____	_____

FIG. 14. Pipe and fitting method calculation worksheet. Metric equivalent: ft × 0.3048 = m.

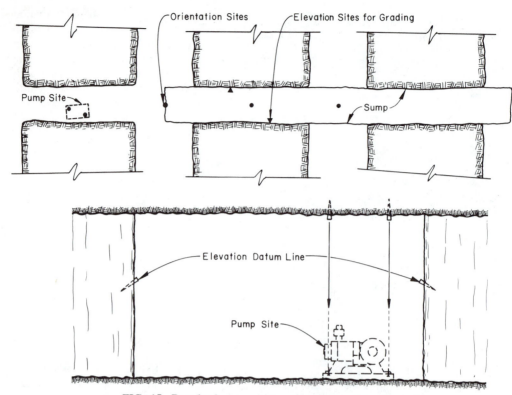

FIG. 15. Required pump station sites for construction.

participant will also gain a complete picture of the project requirements and the necessary construction tasks.

At the end of the discussions, the mine surveyors will provide orientation sites for the important pump station locations. Such location sites are a function of (1) the water volume [3.8 L—0.004 m^3 and 3.78 kg (1 gal—0.134 cu ft and 8.338 lb)]; (2) the sump capacity; (3) pump station design, including backup systems; (4) the required auxiliary systems (i.e, controls, fire protection, explosion-proof housing, hoists, and so on). The important locations, therefore, deal with placing the pump in the correct position and elevation and placing the pump station in a corresponding manner. Fig. 15 illustrates the pertinent sights for a typical main pumping station whose sump is located in the neutral entries behind it. As a minimum, the following sights should be established for this type of project:

An *elevation datum line,* so proper and sufficient excavation work for the sump can be calculated and the position of the pump foundation can be located above the sump (if it is a *dry* type).

The *sump boundaries,* so sufficient excavation can be obtained.

The *pump orientation and motor positions* for

FIG. 16. Centrifugal pump station. Note fireproof walls with metal door for passage to check amount of water in sump behind the wall and for work on the suction line. Note small water line with globe valve in top of pump casing to prime pump from back pressure in discharge line (Crickmer and Zegeer, 1981).

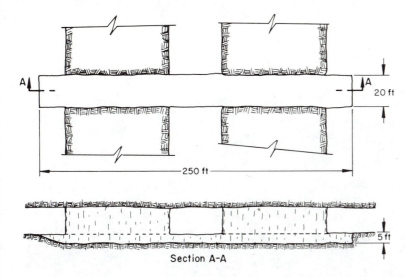

FIG. 17. Excavated sump requirements for a typical operation. Metric equivalent: ft × 0.3048 = m.

calculating the pipe and control requirements and pump foundation dimensions.

Once these sights are established and labeled, the construction foreman can bring in his crews to begin work on the chosen design.

Sump Excavation, Cleaning, and Lining: The initial task of the construction foreman lies with excavating the sump. In coal mines today, this is accomplished by (1) cutting out the bottom, (2) blasting out the bottom, or (3) walling off the necessary area from the mine entry. Several entries and intersections can be dedicated to a sump which has been excavated below the pump site elevation. Fig. 16 is a pump station using a walled entry sump which is placed on a rise above the pump site, letting gravity feed the pump. For mines in which the bottom can not be excavated because there is concern about the wall and floor rock integrity of a shot-out sump, the walled-entry approach provides an alternative. For safety and convenience, however, the best approach for large and long lived sumps handling a great deal of water is almost always a below-grade sump which has been excavated.

In most coal mines, the bottom material is graded out by a miner or roadheading machine. The foreman will string a grade line, locate the starting and stopping points, and mark the required depth on the rib for the workmen to follow. The material, (usually fire clay, shale, or limestone) is loaded into a shuttle car which either dumps into a gob area or loads onto a conveyor belt for transport out of the mine. As a general estimate, an experienced crew of three can excavate 708 m³ (25,000 cu ft) of soft material in one to two shifts. Such a sump can hold over 85 Mg (186,500 gal) of waste water (Fig. 17).

Once the miner is finished rough cutting the sump, a scoop-tractor or other machine is used to clean the bottom of loose gob, while the exposed ribs of the sump are scaled of overhanging and loose material. The cleanup operation usually does not last over one manshift. In some cases, the cleanup can be accomplished during the final retreat maneuvering of the grading machine.

Lining the sump can be accomplished in a number of ways. The two most popular methods are mortared block or wire mesh with sprayed sealant. The reasons for lining a main sump are rather straightforward. First, any exposed bottom strata could allow water to seep out from the sump into the mine environment and cause a recirculation problem. Secondly, and more importantly, waste water will attack the exposed sides of the sump causing sloughing and collapse over time. Neither condition can be tolerated by current mining standards; hence, the main sumps are lined.

In the case of the block walls, wall foundations are poured at a depth of 102 to 152 mm (4 to 6 in.). These foundations should support the lining walls to a height 0.6 to 0.9 m (2 to 3 ft) above the projected waste water level. Shotcrete 6.4 to 13 mm (1/4 to 1/2 in.) thick is then applied to the interior surface of the walls. The floor of the sump is either poured concrete or sprayed shotcrete.

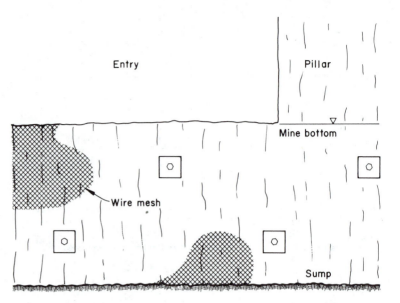

FIG. 18. Staggered bolting pattern for deep sumps.

Mines using wire mesh as a form for the shotcrete begin by bolting the mesh to the sump ribs. The proper bolting pattern for deep sumps is a staggered approach beginning at the top of the mesh with a spacing of 0.9 m (3 ft) or less (Fig. 18). This pattern is necessary to keep the lining from folding in after the shotcrete is applied. Once bolted, shotcrete, approximately 6.4 mm (1/4 in.) thick, is sprayed around the sump perimeter.

Foundation Construction: The next task covers the foundation construction for the main pump station. This task includes the concrete pad for the pump and motor assemblies, plus the foundation for the fireproof control housing.

The pump and motor pad is poured 102 mm (4 in.) thick with 0.3 m (1 ft) clearance around the base of the equipment (Fig. 19). It is important to orient and level the pad to give equal support. If bolts are set in the concrete to secure the pump from moving, the foreman should be familiar with the pump frame and align the bolts accordingly.

The foundation of the control housing is strictly a footing poured to support the block walls. This footing is leveled and smoothed to allow the walls to be level and plumb.

Erection and Installation Requirements at the Site: The final preparatory task concerns the auxiliary equipment needed for the major pump station construction. Here, steel beams and struts are bolted into the roof and ribs to allow the workers to lift and maneuver the pumping equipment at the site. These structures, although optional in many cases, are especially important for operations over the sump area such as laying the suction line and placement of the foot valve (if necesssary).

To aid in maneuvering the equipment with the beams, chain blocks and tackle are mounted on the beam runners. These construction tasks, while taking several manshifts to complete, will save a great deal of time over the mine life. This may be especially important if quick pump changeouts are necessary.

Installing the Main Pump

The tasks needed to install the main pumps in a large coal mine include:

Transporting the pump and other materials to the site.

Anchoring and connecting the pump to the system.

Powering the pump motor and controls.

Priming the pump system (if necessary).

Testing and completing the site.

Each task plays a role in the overall work of making the pump station operational. The first task centers on bringing the pumps and piping to the site.

Transporting the Pump and Other Material to the Site: This task does not cover the *how to* transport, but *what to* transport. In this task, a scoop tractor, shuttle car, rail tram, or other vehicle is used to carry the materials to the site. It is important to deliver the correct

PUMP STATIONS, SUMPS, AND DRAINAGE SYSTEMS

quantity of materials and equipment at the proper time (after the preparation tasks and before construction begins).

The equipment and supplies of the pump station include a variety of items. They are:

Piping—the correct OD piping and fittings (usually standard sizes) for the sump to pump intake lines, the pump to surface connections, and the pump priming and pressure relief systems.

Valves—the necessary gate and butterfly valves to ensure complete control of the water direction and volume.

Pumps—the main and backup pumps and motors as specified in the design.

Materials—includes the block and mortar to construct the fireproof housings and the chains and wire rope needed to maneuver and support the pumps and piping at the site.

Controls—includes the sensors and motor controls to safely operate the pumps and regulate the water flow.

The specific quantities of these items will be identified in the detailed design and engineering of the pump station during the premining stages.

Appropriate lead time for delivery of all items should be planned for and carried through.

Anchoring and Connecting the Pump: Once the pump is delivered to the site, it must be positioned and anchored to the pump foundation. Since the anchor bolts are cemented into the foundation at the right orientation, the task becomes one of setting the pump on the bolts and tightening them down with at least two nuts and a lock washer.

Connecting the pump requires two separate tasks: connecting the intake side and connecting the discharge side. The intake side comes from the sump and usually consists of a 51 to 102 mm (2 to 4-in.) suction pipe laid directly from the center of the sump to the pump inlet. The discharge side is at least one 102 to 152-mm (4 to 6-in.) pipe if the waste water is to travel over 46 m (150 ft) of vertical distance. These connections are usually a flanged coupling to ensure a good seal.

One note on the backup system is appropriate here. It concerns the piping arrangement for switching to a backup pump in times of maintenance or trouble. In this case, gate, or butterfly

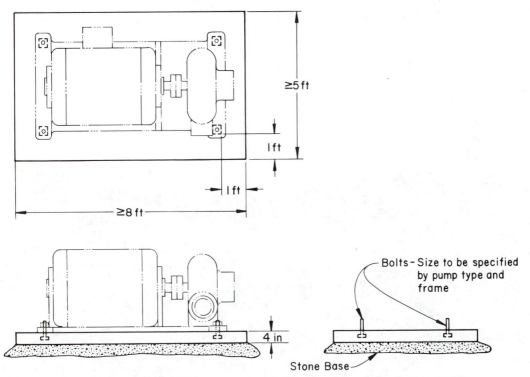

FIG. 19. Working drawing of pump foundation. Metric equivalents: ft × 0.3048 = m; in. × 25.4 = mm.

valves are necesssary to control the water flow into the desired pump (Fig. 20). It is also important to control the discharge flow with a second set of gate valves to prevent back flushing. The proper control of the waste water can be obtained by using a sufficient number of valves.

Powering the Pump Motor: Earlier it was mentioned that mine pumps generally run on 220 or 440 V a-c power. This source comes from an in-line transformer, located near the pump station site, specifically to service the pump. The cables are hung from the roof and the controls are located in the fire- and explosion-proof housing. An electrician should splice the cable leads to the pump leads.

Priming the Pump: This task refers to backfilling the pump and suction line to eliminate the vacuum created during shutdown (if the pump is a centrifugal type). Not all pumps require priming; however, most large general purpose centrifugal pumps used in main waste water pumping must be primed.

The setup for priming is usually fairly straightforward when a vertical rise on the discharge side is present. Here, a small line outby the gate valve taps into the discharge line and leads back to the pump. This line is opened and water drains back through the pump and suction line. Once the vacuum is broken, the pump can be started.

If there is no rise in the discharge line, then an alternative is a small line tapped from a sump feed water pipe to the pump to break the vacuum. If a walled entry sump is used, then the natural flow of the water down to the sump can be utilized, as shown in Fig. 21.

Testing and Completion of the Station: Once the pumps and motors are in place and the controls installed, the filling of the sump can begin. Water is first pumped into the sump for test purposes only. If the mine waste water system is activated before the main station is tested, a minor problem could become a major disaster.

If the pump is the submersible type, such as the Flygt, then the test water should cover the pump completely. Centrifugal pumps located above the sump should have the foot valve submerged for testing.

For practical purposes, the pumping system can gradually be brought on line in successive

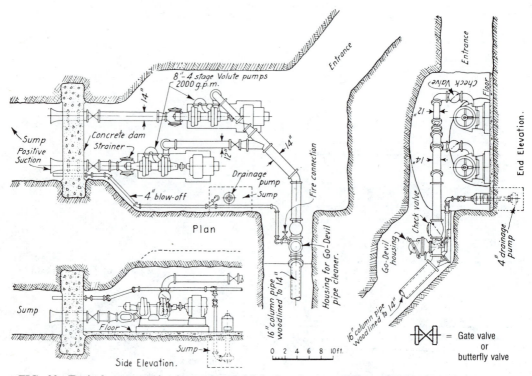

FIG. 20. Typical pump station layout in established coal mines (Tillson, 1938). Metric equivalents: in. × 25.4 = mm; ft × 0.3048 = m; gpm × 0.063 090 2 = L/s.

steps to reduce the water hammer in the pipes. Each test level can be increased until the proper sump level is achieved.

After the test, the final tasks of the pump station are completed. This usually entails constructing a permanent fireproof housing for the pump controls. In addition, the area is identified with warning and alert signs to keep miners from walking in unsafe areas. The sump is checked for leaks and patched accordingly.

Operation and Maintenance

Like any other underground project, the pump station and sump must be maintained over its useful life. These tasks include packing and lubricating the pump as necessary, stopping any pipe leaks, sealing any foundation or lining cracks, and controlling sedimentation. I shall briefly discuss each of these.

Packing and lubricating the pump is necessary to eliminate the frictional heat buildup of the pump shaft bearings and reduce water leakage. In modern pumps, lubrication is self-contained or infrequently required. However, many of the older centrifugal pumps still in use do require lubricating grease at the bearings, as well as repacking the pump glands with the appropriate graphite packing material. The packing eliminates leakage around the pump shaft and the tendency of a vacuum to form in the suction line and stalling out of the pump.

Pipe leaks can happen to the best of systems over time. Joints under stress will loosen from vibration and allow water to leak out of the system. This inefficiency should not be tolerated anywhere in the system and least of all at the main pump station. In this case, the system can be shut down and the old coupling removed. The joint is realigned and a new gasket placed in the coupling. If a leak persists, then the pipes may need to be replaced. As a special note, a new product in the mining industry, called Elbow-Patch from Spate, Inc. is an epoxy-based patching compound for pipe which can seal leaks up to a static pressure of 2.8 MPa (400 psi). This compound may provide a convenient alternative to replacing leaking pipes.

Sealing cracks in the foundation or lining of the sump is an unpleasant task which should be avoided whenever possible. Here, prevention is the key aspect, and using sound installation principles during the initial construction should minimize any potential leakage of the sump.

If leaks are unavoidable and patching is necessary, then a foreman can do one of two things, drain the sump below the leaking level or isolate the leak on the inside of the lining with sealant to cut the flow (Fig. 22). Permanent patching can be accomplished with a grout-type sealant and a trowel. A sufficient amount of time (usually 24 hr) should be available for curing before allowing the water level to return to normal. Again, this is a difficult task and an avoidable one.

The final maintenance task concerns sedimentation. This is caused when the suspended particles in the flowing waste water hit the calm water of the sump and settle out of suspension. Over time, the sump can be reduced from an adequate basin to a mud puddle because of sedimentation.

The task of cleaning a sump is simple. Unfortunately, it is also dirty and wet. One method is for two workers in waders (if it is a shallow sump) and armed with buckets and shovels to hand load the muck from the sump to a gob pile. The work is time-consuming and should be planned for well in advance.

FIG. 21. Permanently mounted centrifugal pump and its motor.
Note bolts in base plate to hold assembly firm to foundation, shielded coupling between pump and motor to protect pumper from getting his clothing caught while servicing the pump, and motor wiring in conduit (Crickmer and Zegeer, 1981).

If the planning of the sump design is done far enough in advance of the construction, then a second method can be considered. This method is an automatic dredging system which, when activated, will clean the sump floor of gob. Several manufactured systems are available to the industry for a price. The reliability of these systems, however, is questionable, and so they can not be recommended as a general rule unless they are specifically built for the system. Also, retrofitting such a system into a working sump is not suggested unless the design shows a decided advantage. All in all, the manual method appears, at this time, to be the best, both economically and feasibly.

This concludes the discussions on construction of a main pumping station. With proper engineering and construction, the system should serve the needs of the mine throughout its life. Maintenance and changeouts will be kept at a minimum with the performance and cost efficiency being maximized.

CONSTRUCTION OF A PORTABLE PUMPING STATION

Often over the life of a coal mine, a need to establish a portable pumping station exists. It can be due to a temporary increase in water seepage in permanent mine entries, or it can be a base to collect water from short-lived works such as production panels or exploratory workings.

A portable pumping station has a life of at least six months to no more than four years. If the need exists over four years, then a permanent station should be considered.

A portable station is located in a low elevation or a small sump cut for the purpose. It is usually established close outby a production unit, or at the head of a production panel. A volume equal to 8–12 hr of collection is designed for the expected full water input. Usually, the sump is excavated by a continuous miner into the floor of the mine by using the stabilizer (stab) jack

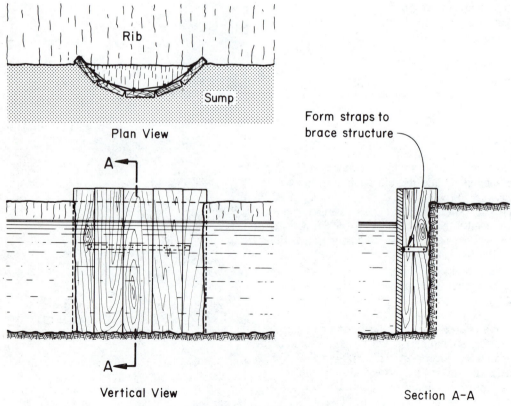

FIG. 22. Isolating pump leak from the inside.

TABLE 9. Expected Seam Runoff as a System Output

Mine Type	Estimated Capacity (Gal per linear ft)	Estimated Flow Rate (Gpm)
Dry	1/2	1/3
Moist	1/4	1
Wet	1/32	2
Saturated	<1/32	>3

to keep the miner head down. Although controlling the total excavated volume is more difficult, the time and cost usually make up for precise measurements. By using a miner, a portable pumping sump can be established in less than a half shift. The labor required consists of a foreman, miner operator and helper, and shuttle car operator.

As the production unit moves further into the reserves, the portable station will serve as an intermediate collection point until the need for the sump (i.e., panel retreat) is eliminated.

Designing for Inputs

Once the decision to establish a portable pumping station has been made, the inputs and outputs to the sump should be evaluated. Since a volume equal to 8 hr of water collection has to be excavated for the sump, an analysis is necessary to determine the minimum expected quantity.

The outputs of the system must be systematically evaluated for capacities and flow rates. Such outputs as drainage ditches, runoff, and face pumping with portable pumps are summed and routed to the sump and pump site. Drainage ditches generally hold 1.9 L (1/2 gal) of waste water per 0.3 m (1 ft) of ditch length at a seam pitch of 3°. Their expected flow rate is under 0.2 L/s (3 gpm). Seams with greater pitch will generally have greater flow rates and a larger capacity.

Seam runoff is a nebulous output because there is little control over its action. However, a simple classification can be estimated for mines with varying water conditions. Table 9 summarizes this rule of thumb for expected seam runoff as a designed output of the drainage system. Specific sites should be reviewed before making any judgement on seam runoff capacity.

Effective face pumping with portable pumps such as the Flygt B2051 P or Stanco 20CF are a function of their performance curve. Necessary pump sizes and pump placement can be estimated from the appropriate total head needed to pump the water from the face to the next sump.

Getting back to the construction aspects, the sump is usually excavated in half a shift by the miner. The construction crew will then clean as much debris as possible from the sump. In the lowest spot of the sump, the workers will build a short base on which the pump shall sit. The base can be made of concrete block, wood, or metal, and should be about 0.3 m (1 ft) high and 0.6 m (2 ft) square. This base is necessary to keep the pump from clogging with sediments over a short period of time.

The chosen pump is then set on the base and water is allowed to run into the sump. The control and power cable to the pump is hung from the roof to the pump to the control box (also off the floor).

The final task is to post signs and warnings about the sump location and where the controls are located. A fire extinguisher should be located with the controls. A barrier is erected to further warn workers of the sump location.

Operation and Maintenance

Once the portable pumping station is started and operating properly, some maintenance is required. These tasks are similar to the main station in many ways. For example, the sump will need to be cleaned of sediment for every four months of use, or more often if it is regularly active. The pump should be checked to ensure the integrity of the power cables, pump fittings, pump housing, and parts. As a routine check, the mine examiner or mine pumper should check its operation at least once a week.

Operation checks also include the outputs of the system to the sump. This check is basically for the face pump feeding the sump. The face pump receives much abuse and relocation in the face area, especially in conventional and continuous mining units. Here, the pump quite likely is being dragged from face entry to face entry in front of the mining machine and draining the water to allow mining. The abuse can put tremendous strain on the pumps, the cable, controls, and discharge lines. Although the section foreman is specifically overseeing the pump's use, a construction foreman should check its operation to maintain the desired efficiency.

This concludes the discussion of main and portable pumping stations. These structures are necessary components of an efficient drainage system. By designing and constructing these stations for efficiency and smooth operation, a coal mine can expect to continue to mine coal without having to shut down because of flooded working faces.

MANUFACTURERS OF PUMPING EQUIPMENT

Table 10 is a listing of pumping equipment manufacturers who offer specific items for coal mining situations. This list comes from the 1980 *Buyer's Guide* as published in *Coal Age* magazine by the McGraw-Hill Mining Information Services. The addresses of these manufacturers are included in Appendix A.

REFERENCES AND BIBLIOGRAPHY

Anon., 1974, *Cost Estimator's Handbook*, Victaulic Co. of America, Easton, PA.
Crickmer, D.F., and Zegeer, D.A., 1981, *Elements of Practical Coal Mining*, 2nd ed., AIME, New York.
Cummins, A.B., and Given, I.A., 1973, *SME Mining Engineering Handbook*, AIME, New York.
Tillson, B.F., 1938, reprinted 1976, *Mine Plant*, AIME, New York.

TABLE 10. Pumping Equipment Manufacturers

Controls	Flow Meters
Allen-Bradley Co.	Air Products & Chemicals, Inc.
American Meter Div., Singer Co.	American Meter Div., Singer Co., The
Automation Products, Inc.	BIF, a unit of General Signal
Babcock & Wilcox	Bristol-Babcock Div., Acco Industries Inc.
Bindicator	Calgon Corp.
Bristol-Babcock Div., Acco Industries Inc.	Capital Controls Co.
CWI Distributing Co.	Crane Co.
Communication & Control Eng. Co. Ltd.	Dresser Industries, Inc.
Compton Electrical Equipment Corp.	Federal Supply & Equipment Co., Inc.
Delavan Electronics, Inc.	Foxboro Co., The
Delavan Corp.	General Electric Co., Instrument Products Operation
Electric Machinery Mfg.	Gustin-Bacon Div., Aeroquip Corp.
Electrical Automation	Halliburton Services-Research Center
Endress & Hauser, Inc.	Hayden-Nilos Conflow Ltd.
Fisher Controls Co.	Honeywell Inc., Process Control Div.
Foxboro Co.	ISCO
General Electric Co.	J-Tec Associates, Inc.
Gould Inc. Distribution & Controls Div., R.B. Denison	Jones & Laughlin Steel Corp.
Guyan Machinery Co.	Kay-Ray Inc.
Honeywell Inc. Process Control Div.	Keene Corp.
Joy Mfg. Co., Denver Equipment Div.	Leeds & Northrup, a unit of General Signal
Leeds & Northrup (div. of General Signal)	Milltronics Inc.
Mag-Con, Inc.	Mining & Transport Engineering B.V.
Milltronics Inc.	Modern Engineering Co.
Ohmart Corp.	National Environmental Instruments, Inc.
Pace Transducer Co., Div of C.J. Enterprises	Pace Transducer Co., Div. of C.J. Enterprises
Preiser/Mineco Div., Preiser Scientific Inc.	Preiser/Mineco Div., Preiser Scientific Inc.
Revere Corp. of America	Senior Conflow
Square D Co.	Stevens International Inc.
Stevens International Inc.	Taylor Instrument Co.
Taylor Instrument Co.	Texas Nuclear
Texas Nuclear	Union Carbide Corp.
WESMAR Industrial Systems	Viking Machinery Co.
Westinghouse Electric Corp.	WESMAR Industrial Systems Div.
	Westinghouse Electric Corp.

TABLE 10. Pumping Equipment Manufacturers (continued)

Pipe
1. aluminum
2. aluminum-plastic
3. aluminum, steam-traced
4. asbestos-cement
5. bronze, copper, red brass
6. cast-iron, wrought iron
7. lined
8. corrosion resistant
9. corrugated
10. drive & driving winches
11. plastic
12. rubber
13. rubber-lined
14. seamless
15. spiral-welded
16. stainless steel
17. steel, steel-welded
18. steel, plastic-coated
19. wood, wood-stave
20. glass fiber reinforced
21. abrasion resistant

Alcoa (1, 3, 14)
Allegheny Ludlum Steel Corp. (8, 16, 17)
Ampco Metal Div., Ampco-Pittsburgh Corp. (5, 8)
Anixter Mine & Industrial Specialists (1, 2, 8, 11, 12, 14, 15, 20)
Armco Inc. (7, 8, 9, 11, 14, 15, 16)
Babcock & Wilcox (8, 14, 16, 17)
Babcock Corrosion Control Ltd. (7, 8, 11, 13, 20, 21)
Berger Industries, Inc. (17)
Bethlehem Steel Corp. (9, 14, 17, 18)
C F & I Steel Corp. (14)
Can-Tex Industries (11)
Capital City Industrial Supply Co.
Carborundum Co., Refractories Div. (7)
CertainTeed Corp., Pipe & Plastics Group (4, 11)
Ciba-Geigy Corp., Pipe Systems Dept. (8, 20)
Clouth Gummiwerke Aktiengesellschaft (7, 8, 12, 13)
Continental Rubber Works, Sub. of Continental Copper & Steel Industries, Inc. (12)
Contractors Warehouse Inc. (15, 17)
Coors Porcelain Co. (7, 8, 21)
Dayco Corp. (11)
Detrick, M. H., Co. (7, 8)
du Pont de Nemours, E. I. & Co. Inc. (11)
Durex Products, Inc. (8, 21)
Duriron Co., Inc., The (8)
ESCO Corp. (6, 8, 16)
Fagersta, Inc. (16)
Fairmont Supply Co. (11, 12, 13, 14, 15, 20)
Fiberglass Resources Corp. (8, 11)
Flexible Valve Corp. (12)
Foamcraft, Inc. (1, 8)
Foster, L. B., Co. (7, 14, 15, 17, 18, 21)
G & W Energy Products Group (7, 8, 13, 14, 16)
Gates Rubber Co., The (13)
General Resource Corp. (1, 6, 8)
General Scientific Equipment Co. (11, 12)
Goodall Rubber Co. (11, 12)
Goodrich, B. F., Co., Engineered Products Group (13)
Goodyear Tire & Rubber Co. (12, 13)

Greenbank Cast Basalt Eng. Co. Ltd. (7, 8)
Greengate Industrial Polymers Ltd. (12)
Gundlach, T. J., Machine Div., Rexnord Inc. (21)
Heil Process Equipment, Fiberglass Equip. Div., Dart Environment & Services Co. (5, 8, 11, 13, 16, 20)
Hercules Inc. (8)
Holz Rubber Co., Inc., a Randtron Sub. (7, 8, 12, 13, 21)
Hydrophilic Industries, Inc. (11)
ITT Grinnell Corp. (11, 14, 15)
ITT Harper (16)
Irathane Systems, Inc. (13)
Jennmar Corp.
Johnston-Morehouse-Dickey Co. (11, 20)
Jones & Laughlin Steel Corp. (7, 8, 14, 17)
Kalenborn (7, 8, 12)
Keene Corp. (4)
Lane Metal Products, Inc. (9)
Linatex Corp. of America (13)
Logan Corp. (11, 17)
McJunkin Corp. (1, 9, 11, 14, 16, 17)
Microdot, Inc. (11)
Midland Pipe & Supply Co. (1, 8, 13, 16)
National Mine Service Co. (1, 2, 11)
National Supply Co., Div. of Armco Inc. (14)
Naylor Pipe Co. (13, 15, 16)
Norwood, Inc. (4, 8, 9, 11, 17, 18)
Peabody International (11)
Peabody Intl. Corp., Peabody ABC Group (11)
Phelps Dodge Industries, Inc. (5, 8)
Phillips Driscopipe, Inc. (8, 11)
Pipe Benders, Inc. (7, 8, 21)
Pipe Systems, Inc. (8, 11, 21)
Poly-Hi/Menasha Corp. (11, 8, 21)
Red Valve Co., Inc. (12)
Reggie Industries, Inc. (14, 15, 16, 17, 18)
Republic Steel Corp. (8, 9, 14, 16, 17, 18)
Rexnord Inc. Resin Systems Div. (7, 8, 11, 18, 21)
Reynolds Metals Co. (1)
Ryerson, Joseph T., & Son, Inc. (1, 8, 11, 14, 16, 17)
Scholten, T.H., & Co. (7)
Smith, A.O.-Inland Inc. Reinforced Plastics Div. (8, 11, 20, 21)
Tidewater Supply Co. (9, 11, 14, 17, 20)
Townley Engineering & Mfg. Co., Inc. (7, 8, 13, 21)
Trelleborg AB (7, 11, 12, 13)
Trelleborg, Inc. (7, 12, 13)
Tricon Metals & Services, Inc. (8, 11, 14, 16, 17, 21)
Tube Turns Div. of Allegheny Ludlum Industries (17)
Union Carbide Corp. (8)
Uniroyal, Inc. (12, 13)
United McGill Corp. (1, 8, 11, 15, 16, 17, 18)
U.S. Steel Corp. (1, 7, 8, 9, 11, 14, 17, 18)
Valley Steel Products Co. (11, 13, 14, 15, 17, 18)
Wheeling Corrugating Co., Div. of Wheeling-Pittsburgh Steel Corp. (8, 9, 18)
Wheeling-Pittsburgh Steel (14, 17)
Wilson, R. M., Co. (2, 7, 8)
Witco Chemical Corp. (11)
Workman Developments, Inc. (7, 8, 11, 12, 13, 14, 21)
XTEK, Inc. (7, 8, 18, 21)
Xtek: Pipe (2, 7, 8, 13, 16, 18, 21)
Youngstown Sheet & Tube Co., The (8, 11, 14, 17)

TABLE 10. Pumping Equipment Manufacturers (continued)

Pipe Accessories
1. couplings
2. couplings, flexible
3. couplings, grooved
4. coverings
5. fittings, brass & bronze
6. fittings, cast-iron
7. fittings, malleable-iron
8. fittings, flanges-fabrication, welding
9. fittings, forged steel
10. fittings, plastic
11. fittings, rubber
12. fittings, stainless steel
13. flanges, forged, stainless, alloy
14. groovers
15. hangers
16. repair clamps, sleeves
17. fittings, cast steel

Aeroquip Corp. (1, 2, 9)
Ampco Metal Div., Ampco-Pittsburgh Corp. (5)
Anchor Coupling Co., Inc. (1, 5, 7, 8, 9)
Anixter Mine & Industrial Specialists (1, 2, 3, 7, 8, 9, 10, 14, 15, 16)
A-T-O Inc. (1, 5, 15)
Babcock & Wilcox (8, 9, 12, 13)
Bethlehem Steel Corp. (9, 13)
Big Sandy Electric & Supply Co., Inc. (3, 6, 7)
Bowman Distribution, Barnes Group, Inc. (1, 6, 7)
C F & I Steel Corp. (1)
Campbell Chain Co. (15)
Can-Tex Industries (10)
Capital City Industrial Supply Co. (1, 2, 3, 4, 5, 6, 7, 8, 9, 10, 11, 12, 13, 14, 15, 16, 17)
CertainTeed Corp., Pipe & Plastics Group (1, 10)
Clouth Gummiwerke Aktiengesellschaft (11)
Conac Corp. (1, 2, 3, 6, 7, 9, 12, 14, 16)
Continental Rubber Works, Sub. of Continental Copper & Steel Industries, Inc. (11)
Contractors Warehouse Inc. (1, 3)
Dresser Industries, Inc. (1, 5, 6, 7, 8, 9, 10, 16)
Dresser Manufacturing, Div. Dresser Industries, Inc. (1, 2, 5, 7, 10, 16)
du Pont de Nemours, E. I. & Co. Inc. (10)
Durex Products, Inc. (1, 10)
Duriron Co., Inc., The (1, 6)
Elreco Corp., The (7, 15)
ESCO Corp. (8, 10, 12, 13)
Fairbanks Co., The (7)
Fairmont Supply Co. (1, 3, 6, 7, 8, 9, 10, 15, 16)
Fiberglass Resources Corp. (1, 10, 16)
Flexible Valve Corp. (11)
Foster, L. B., Co. (1, 8)
G & W Energy Products Group (1, 2, 3, 7, 8, 9, 12, 13)
General Resource Corp. (1)
Goodall Rubber Co. (10, 11)
Gould Inc., Hose and Couplings Div. (1, 5, 10)
Greenbank Cast Basalt Eng. Co. Ltd. (1, 8, 15)
Gustin-Bacon Div., Aeroquip Corp. (1, 2, 3, 6, 7, 9, 12, 14, 16)
Holz Rubber Co., Inc., a Randtron Sub. (2, 11)
Hydrophilic Industries, Inc. (10)
ITT Grinnell Corp. (1, 2, 5, 6, 7, 8, 9, 10, 12, 13, 15, 17)
ITT Holub Industries (15)
Johnston-Morehouse-Dickey Co. (1, 3, 10)
Jones & Laughlin Steel Corp. (1, 3, 4, 6, 7, 8, 9, 10, 12, 14, 15, 16, 17)
Key Bellevilles, Inc. (15)
Ladish Co. (1, 8, 9, 12, 13)
MCC MARPAC, a unit of Mark Controls Corp. (9, 12)
Mark Controls Corp. (1)
McJunkin Corp. (1, 5, 7, 8, 9, 10, 12, 13, 15, 16)
Microdot, Inc. (1, 2, 3, 10)
Midland Pipe & Supply Co. (8, 12, 13)
National Mine Service Co. (1, 3)
Naylor Pipe Co. (8, 12)
Norwood, Inc. (10)
Ohio Brass Co., a Sub. of Harvey Hubbell Inc. (7, 15)
Parker-Hannifin Corp., Hose Products Div. (1, 3, 5, 6, 7, 9, 10, 12)
Parker-Hannifin Corp., Tube Fittings Div. (5, 9, 12, 15, 17)
Phelps Dodge Industries, Inc. (1, 4, 5, 8)
Phillips Driscopipe, Inc. (10)
Plymouth Rubber Co., Inc. (4)
Red Valve Co., Inc. (2, 11)
Rexnord Inc., Resin Systems Div. (10)
Seton Name Plate Corp. (4)
Smith, A. O.-Inland Inc. Reinforced Plastics Div. (10)
Snap-tite, Inc. (1)
Sperry Vickers Div., Sperry Corp. (13)
Spraying Systems Co. (5)
Stockham Valve & Fittings (1, 6, 7)
Stratoflex, Inc. (1, 5, 8, 12, 13)
Thor Power Tool Co. (1)
Tidewater Supply Co.
Townley Engineering & Mfg. Co., Inc. (1, 2, 10, 11)
Transamerica Delaval Inc., Wiggins Connectors Div. (2)
Trelleborg AB (1, 10, 11)
Trelleborg, Inc. (1)
Tube Turns Div. of Allegheny Ludlum Industries (8, 9, 13)
Uniroyal, Inc. (11)
United McGill Corp. (1, 8, 10, 12, 15)
U.S. Steel Corp. (1, 2, 3, 6, 7, 8, 9, 12, 13, 14, 16)
Valley Steel Products Co. (1, 3)
Victaulic Co. of America (1, 2, 3, 6, 7, 12, 14, 16)
Wachs, E. H., Co.
Weatherhead Div., Dana Corp. (1, 5, 9, 12)
Wilson, R. M., Co. (1, 10)
Workman Developments, Inc. (1, 8, 10, 11, 16)
XTEK, Inc. (1, 3, 6)
Xtek: Pipe (1, 3, 6, 10)

Pipe Fabrication, Welding
American Alloy Steel, Inc.
Ampco Metal Div., Ampco-Pittsburgh Corp.
Barber Industries
Dravo Corp.
Foster, L. B., Co.
Greenbank Cast Basalt Eng. Co. Ltd.
Lively Mfg. & Equipment Co.
Manufacturers Equipment Co., The
Marathon Steel Co.
McJunkin Corp.
McLaughlin Mfg. Co.

TABLE 10. Pumping Equipment Manufacturers (continued)

Midland Pipe & Supply Co.
Newport News Industrial, Sub. of Newport News Shipbuilding
Norwood, Inc.
Pipe Benders, Inc.
Reggie Industries, Inc.
Stearns-Roger
United McGill Corp.
Valley Steel Products Co.
Wachs, E. H., Co.
Workman Developments, Inc.

Gages, Liquid-Level
Babcock & Wilcox
Baylor Company
Bindicator
Dresser Industries, Inc.
Foxboro Co., The
Honeywell Inc., Process Control Div.
Kay-Ray Inc.
Lunkenheimer Co., Div. of Conval Corp., Sub. of Condec Corp.
Ohmart Corp.
Preiser/Mineco Div., Preiser Scientific Inc.
Rexnord Inc.
Stevens International Inc.
Texas Nuclear
WESMAR Industrial Systems Div.
Westinghouse Electric Corp.

Gages, Pressure, Vacuum, Flow
Alemite & Instrument Div., Stewart-Warner Corp.
American Meter Div., Singer Co., The
Anixter Mine & Industrial Specialists
Beckman Instruments, Inc.
Bowman Distribution, Barnes Group, Inc.
Deron Corp.
Duriron Co., Inc., The
ENERPAC, Div. of Applied Power Inc.
Foxboro Co., The
Hayden-Nilos Conflow Ltd.
Honeywell, Inc., Process Control Div.
Irad Gage, Inc.
J-Tec Associates, Inc.
Keene Corp.
Martin-Decker
Minnesota Automotive Inc.
Modern Engineering Co.
Pace Transducer Co., Div. of C.J. Enterprises
Preiser/Mineco Div., Preiser Scientific Inc.
Schroeder Bros. Corp.
Senior Conflow
Snap-On Tools Corp.
Sperry Vickers Div., Sperry Corp.
Templeton, Kenly & Co.
TOTCO
Viking Machinery Co.
Westinghouse Electric Corp.

Pumps
1. centrifugal
2. corrosion-resistant
3. diaphragm
4. drum
5. froth-handling
6. metering
7. piston & plunger
8. pressure-testing
9. priming
10. sand & abrasive handling
11. slurry, solids-handling
12. submersible
13. sump
14. transfer
15. trash & sludge
16. vertical centrifugal & turbine
17. power hydraulic
18. explosionproof
19. vacuum
20. gear

AMF Inc. (3, 6, 7)
A-S-H Pump, Div. of Envirotech Corp. (1, 2, 10, 11, 13, 16)
Abex Corp. (2, 6, 10, 11, 15, 17, 18)
Abex Corp., Denison Div. (17)
Acker Drill Co., Inc. (8)
Adams Equipment Co., Inc. (1, 2, 3, 4, 5, 6, 7, 8, 9, 10, 11, 12, 13, 14, 15, 16, 17, 18)
Alemite & Instrument Div., Stewart-Warner Corp. (2, 4, 14)
Allis-Chalmers (1, 2, 10, 11, 12, 15, 16)
Ampco Metal Div., Ampco-Pittsburgh Corp. (1, 2)
Amsco Div., Abex Corp. (2, 10, 11, 15)
Anderson Electrical Connectors, Square D Co. (17)
Armco Inc. (7)
ARO Corp. (2, 4, 6, 7, 14)
Atlas Copco, Inc. (1, 3)
Atlas Copco MCT AB (1, 3, 12, 13)
Aurora Pump, Unit of General Signal (1, 2, 11, 13, 16)
Babcock Hydro-Pneumatics Ltd. (11)
Barrett, Haentjens Co. (1, 2, 5, 10, 11, 12, 13, 16)
Beckman Instruments, Inc. (6)
BIF, a unit of General Signal (3, 6, 7)
Bowman Distribution, Barnes Group, Inc. (4)
Byron Jackson Pump Div., Borg Warner Corp. (1, 2, 12, 13, 14, 16)
C-E Power Systems, Combustion Eng., Inc. (1, 16, 18)
CWI Distributing Co. (10, 12, 13, 15, 18)
Calgon Corp. (2, 3, 6, 7, 14)
Canton Stoker Corp. (2, 7, 10, 11, 14, 15, 18)
Capital Controls Co. (6)
Carborundum Co., Refractories Div. (10, 11)
Carver Pump Co. (1, 2, 8, 10, 11, 13, 14, 15, 16)
Century Hulburt Inc. (4)
Chicago Pneumatic Equipment Co. (12, 13, 15)
Christensen Diamin Tools, Inc. (7)
CompAir Construction & Mining Ltd. (1, 12, 13)
Contractors Warehouse Inc. (10, 11, 12, 13, 15, 18)
Crane Co. (1, 2, 3, 4, 5, 6, 7, 8, 9, 10, 11, 12, 13, 14, 15, 16)
Crisafulli Pump Co., Inc. (1, 11, 12, 13, 15, 16, 17, 18)
Dean Brothers Pumps, Inc. (1, 2, 14, 16, 18)
Deere & Co. (17)
Delavan Corp. (11, 17)
Demco (1)
Dorr-Oliver Inc. (1, 2, 3, 11)

TABLE 10. Pumping Equipment Manufacturers (continued)

Dorr Oliver Long, Ltd. (1, 2, 3)
Dresser Industries, Inc. (1, 2, 3, 7, 11, 12, 13, 14, 16, 17)
Dresser Industries, Inc., Drilling Equipment (3, 7)
Dresser Industries, Inc., Industrial Products Div. (19)
DuBois Chemicals Div. of Chemed Corp. (1)
Duff-Norton Co. (17)
Duriron Co., Inc., The (1, 2, 3, 9, 10, 11, 13, 15, 16)
Dynex/Rivett Inc. (17)
ERL Inc. (12, 13)
Eaton Corp., World Headquarters (17)
ENERPAC, Div. of Applied Power Inc. (7, 8, 17, 18)
English Drilling Equipment Co. Ltd. (7, 18)
E-Power Industries Co. (14)
FMC Corp., Agricultural Machinery Div. (2, 7, 8, 14)
Fairmont Supply Co. (1, 3, 12, 13, 15)
Federal Supply & Equipment Co., Inc. (8, 17)
Fire Protection Supplies Inc. (1, 3, 9, 15)
Flygt AB (2, 11, 12, 13, 15, 18)
Flygt Corp. (2, 10, 11, 12, 13, 15, 18)
Fuller Co., A Gatx Co. (10, 11, 18, 19)
Funk Mfg. Div. of Cooper Industries, Inc. (17)
GEC Mechanical Handling Ltd. (1, 2, 10, 11, 13, 16)
Galigher Co., The (1, 2, 5, 10, 11, 13, 14, 16)
Gardner-Denver, Industrial Machinery Div. (2, 7, 8, 14)
Gauley Sales Co. (1, 2, 3, 4, 5, 6, 7, 8, 9, 10, 11, 12, 13, 14, 15, 16, 17, 18, 19)
General Resource Corp. (10)
General Scientific Equipment Co. (4)
Giant (2, 3, 4, 6, 7, 8, 9, 14, 17, 18)
Gorman-Rupp Co., The (1, 2, 3, 9, 11, 12, 13, 14, 15)
Goulds Pumps, Inc. (1, 2, 9, 10, 11, 12, 13, 14, 15, 16, 18)
Goyne Pumps, Sub. of Goulds Pumps Inc. (1, 2, 10, 11, 13, 16)
Gulf Oil Corp., Dept. DM
Gullick Dobson Intl. Ltd. (1)
Guyan Machinery Co. (1, 2, 3, 7, 10, 11, 12, 15, 19)
Halliburton Services-Research Center (2, 7)
Hardman Inc. (6)
Hartman-Fabco Inc. (1, 10, 11, 13)
Hauhinco Maschinenfabrik (17)
Hayward Tyler, Inc. (2, 12, 18)
Homelite Div., Textron Inc. (1, 2, 3, 10, 11, 12, 15)
Huwood-Irwin Co. (17, 18)
Hydreco, A Unit of General Signal (17)
ITT Marlow Pumps, Pumps and Compressors Div. (1, 2, 3, 5, 7, 9, 10, 11, 12, 14, 15, 17, 18)
Ingersoll-Rand Co. (1, 2, 5, 7, 8, 9, 11, 12, 13, 14, 15, 16, 18)
Jaeger Machine Co. (1, 3, 15)
Jeffrey Mining Machinery Div., Dresser Industries Inc.
Jeffrey Mining Machinery Div., Drilling Equipment (3, 7)
Jennmar Corp.
Johnston Pump Co. (2, 12, 13, 14, 16, 18)
Jones & Laughlin Steel Corp. (13, 14)
Joy Mfg. Co., Denver Equipment Div. (1, 2, 3, 5, 6, 10, 11, 13, 16)
KHD Industrieanlagen AG, Humboldt Wedag (1, 2, 8, 9, 10, 11, 13, 15, 18)
Koppers Co., Inc.
LaBour Pump Co. (1, 2, 5, 9, 11, 13, 14, 16, 18)
Lawrence Pumps, Inc. (1, 2, 9, 10, 11, 13, 14, 16)
LeRoi Div., Dresser Industries, Inc. (3, 10, 11, 12)
Lightning Industries, Inc. (2, 10, 11)
Linatex Corp. of America (1, 10, 11, 13)
Lincoln St. Louis Div. of McNeil Corp. (2, 4, 7, 14)
Line Power Manufacturing Corp. (1, 2, 12, 13, 14, 15, 16)
Logan Corp. (11, 12, 15, 16)
Lucas Fluid Power (17)
McNally Pittsburg Mfg. Corp. (1)
Megator Corp. (2, 3, 6, 9, 13, 14)
Midland Pump, LFE Fluids Control Div. (1, 2, 3, 9, 10, 11, 13, 14, 15)
Midway Equipment, Inc. (1, 3, 12, 15)
Mine Equipment Co. (1, 2, 5, 11, 12, 13, 15)
Mining & Transport Engineering B.V. (1, 10)
Mining Developments Ltd. (13)
Mining Progress, Inc. (7)
Minnesota Automotive Inc. (1, 4, 14)
Mono Group Inc. (6, 10, 11, 15)
Monogram Industries, Inc., Jet-O-Matic Div.
Morris Pumps, Inc. (1, 2, 5, 10, 11, 12, 13, 14, 15, 16)
Muncie Power Products (17)
Myers, F.E., Company (1, 7, 9, 11, 12, 13, 14, 15)
Nagle Pumps, Inc. (1, 2, 9, 10, 11, 12, 13, 16)
Nash Engineering Co. (9, 19)
National Electric Coil Div. of McGraw-Edison Co. (12)
National Environmental Instruments, Inc. (2)
National Supply Co., Div. of Armco Inc. (7, 11, 14)
Peabody Barnes (1, 2, 3, 7, 9, 10, 11, 12, 13, 14, 15)
Peerless Pump (1, 2, 12, 13, 14, 16)
Pekor Iron Works, Inc. (1, 10, 11)
Permco Inc. (17, 19)
Pettibone Corp. (10, 11)
Porto Pump, Inc. (8)
Power Transmission Equipment Co. (17)
Preiser/Mineco Div., Preiser Scientific Inc. (1, 2, 3, 4, 6, 7)
Prosser Industries, Div. of Purex Corp. (1, 2, 12, 13, 14, 17, 18)
Rexnord Inc. (17)
Rish Equipment Co. (1, 3, 15)
Robbins & Myers, Inc. (1, 2, 5, 6, 9, 10, 11, 12, 14, 15)
Sala International (1, 2, 5, 10, 11, 13, 16)
Sala Machine Works Ltd. (1, 2, 5, 11, 13, 16)
Sepor, Inc. (1, 2, 3, 10, 11, 12, 13, 19)
Siemens Corp. (19)
Sperry Vickers Div., Sperry Corp. (17)
Sprague & Henwood, Inc. (7)
Stanadyne/Hartford Div. (7, 14)
Stanco Mfg. & Sales Inc. (1, 2, 10, 11, 12, 13, 15, 18)
Strojexport, pzo (1, 3, 10, 16)
Sundstrand Fluid Handling (1, 2, 8, 14, 17, 18)
T & T Machine Co., Inc. (1, 2, 3, 7, 9, 10, 11, 12, 13, 15, 16, 18)
Taber Pump Co., Inc. (1, 2, 11, 13, 14, 16, 18)
Templeton, Kenly & Co. (17)
Thomas Foundries Inc. (1, 10, 11, 15)
Thor Power Tool Co. (11, 12, 13)

TABLE 10. Pumping Equipment Manufacturers (continued)

Tidewater Supply Co.
Townley Engineering & Mfg. Co., Inc. (2, 10, 11)
TRW Mission Mfg. Co., Div. of TRW Inc. (1, 2, 7, 10, 11)
Turmag (G.B.) Ltd. (1, 2, 7, 9, 10, 11, 12, 13, 14, 15, 16, 19)
Union Carbide Corp. (1, 2)
U.S. Steel Corp.
Valley Steel Products Co. (1, 12, 13, 16, 18)
Viking Machinery Co. (1, 2, 5, 6, 8, 9, 12, 14, 17, 20)
Wachs, E. H., Co. (12, 17)
Wajax Industries Ltd. (1, 3, 7, 10, 11, 12, 13, 16)
Warman International Inc. (1, 2, 5, 9, 10, 11, 13)
Warren Pumps Div., Houdaille Industries, Inc. (1, 2, 7, 10, 11, 14, 15)
Warren Rupp Co., The (1, 2, 3, 7, 9, 10, 11, 12, 13, 14, 15, 18)
WEMCO Div., Envirotech Corp. (1, 2, 5, 10, 11, 12, 13, 14, 15)
Westfalia Lunen (7, 17)
Wheelabrator-Frye, Inc., Materials Cleaning Systems (2, 4, 14)
Wilfley, A. R., & Sons (1, 2, 5, 10, 11, 14, 15, 16)
Wilson, R. M., Co. (1, 2, 7, 11, 12, 13, 15, 16, 18)
Worthington Pump Inc. (1, 2, 7, 11, 12, 13, 15, 16)

5
Belt Drives, Takeups, and Transfer Points

Today's coal mines are turning more and more to belt haulage as the key system to transport coal out of the mine. The advantages of using belt for both intermediate and main haulage are numerous. Belts are reliable high capacity haulage mediums that can be utilized at low operating costs. They are adaptable to rolls in the coal seam and can handle relatively steep grades. They are not noisy, operate with a minimum amount of labor, and are easily extended and retracted. Belts require the least power of any haulage medium, are easily controlled by pushbuttons and interlocks, and present low fire and safety hazards. Belt haulage can also be the largest source of operational problems for mine management when not installed correctly.

This chapter and Chap. 6 concentrate on proper installation of both the static and dynamic components of efficient belt haulage. This chapter looks at the static system component, composed of the belt drive, its installation and operation, as well as belt takeup and transfer point construction. These components do not change readily over the life of the installation, hence the term *static system*. Chap. 6 discusses the dynamic system of belt line construction, which includes belt splicing, breaking down and reanchoring the tailpiece, and extending the belt line, plus operational and maintenance items for belt operation.

This chapter is divided into three major topics. The first is a general discussion covering the points necessary for designing an efficient haulage system; the second looks at installing belt drives, head rollers, and takeups, and the third discussion centers on installing efficient and effective transfer points along the haulage system. Each will address the design and construction issues facing the engineer or construction supervisor in charge of installing the system.

Before looking at the static system, it may be useful to review some of the basic design factors and considerations of all belt haulage systems.

The material usually carried on an underground belt is a mixture of raw coal and rock, commonly known as run-of-mine (ROM) coal. This mixture has certain material characteristics that guide the mining engineer to the final and proper belt selection. The characteristics looked at most frequently are:

1) *Angle of Repose* is the angle at which the material will freely make a pile. Most ROM coal averages a 35°–39° angle of repose.

2) *Angle of Surcharge* is the angle a pile will assume on a moving conveyor. This ranges from 5° to 15° less than the angle of repose in most materials. For ROM coals, 25° is the average angle of surcharge.

3) *Flowability,* which determines the maximum cross-sectional area needed to carry any given material on a belt. It is measured by the angles of repose and surcharge of a material and also serves as an index of the safe incline angle for the belt line.

These characteristics, plus practical engineering research and development work are defined by an organization called the Conveyor Equipment Manufacturers Association (CEMA) for most bulk materials. CEMA also provides a code for each classified material depending on the characteristics above and the subjective considerations of the average size to be encountered, the abrasiveness of the material, and any miscellaneous characteristics (oil resistive, corrosive, etc.) which may be important. For coal, CEMA (composed of member companies in-

TABLE 1. CEMA Classification for Coal

	Anthracite (sized)	Bituminous (ROM)	Subbituminous Lignite
Angle of repose	27°	38°	38°
Angle of surcharge	10°	20°	20°
Flowability	Free Flowing (2)	Ave. Flowing (3)	Ave. Flowing (3)
Size	Granular under 1/2 in.* (C)	Lumpy over 1/2 in. (D)	Lumpy over 1/2 in. (D)
Abrasiveness	Abrasive (6)	Nonabrasive (5)	Abrasive (6)
Average weight (lb/ft³)	55–60	45–55	40–45
Maximum inclination	16°	18°	22°
Miscellaneous characteristics	-----	Mildly Corrosive (T)	Mildly Corrosive (T)
CEMA code	C26	D35T	D36T

*Metric equivalents: in. × 25.4 = mm; lb/ft³ × 16.018 46 = kg/m³.

volved with belt and component manufacturing) provides the information found in Table 1 for conveyor design work. Detailed information on these items can be found in *Belt Conveyors for Bulk Materials* (1979), 2nd ed., published by CEMA and CBI Publishing Company of Boston, or from any CEMA member.

The material characteristics mentioned above largely affect the choice of the conveyor belt and equipment for any given installation. To help evaluate the potential performance of a belt, many mine engineers calculate and use design characteristics. These factors combine the site specific information with the material characteristics to achieve an optimum choice. For example, Table 2 shows the typical design characteristics of *Coalseal* (manufactured by B.F. Goodrich), a popular conveyor belt in the coal industry. A construction engineer will compare the material characteristics in Table 1 against those belt characteristics in Table 2 to achieve the best performance for the application. However, any detailed design efforts of individual or special belt installations should be done in conjunction with a CEMA member's technical staff.

The conveyor belt generally consists of one or more tension-carrying plies sealed within covers which protect the plies from impact and abrasion (Fig. 1). The load carrying ply(s), which before the late 1960s had been made of cotton duck, now consist of stronger materials such as polyester, nylon, and other synthetics. Because these materials are thinner and stronger than cotton duck, more flexible belts now can be manufactured. Belts used in high-tension applications (such as slope belts) utilize steel cable in the ply(s).

Belt covers, once composed of rubber, now consist of stronger and lighter fire-resistant materials. Polyvinyl chloride (PVC) was the first such material, and it was used mainly in the single-ply, woven-carcass type of belt. Neoprene is often used as a cover for multi-ply belts, and special fire-resistant rubber compounds have been introduced and used in belt covers. Common cover thicknesses are 3.2 mm (1/8 in.) for the top cover and 1.6 mm (1/16 in.) for the bottom.

Panel belts are often two-ply or woven carcass (PVC) while many mainline belts are three plies or more. A typical length for a two-ply 914.4 mm or 0.9 m (36 in.) panel belt is 914.4 m (3000 ft), while three-ply 1066.8 mm or 1.1 m (42 in.) mainline belts are normally 1219.2 to 1524 m (4000 to 5000 ft) long.

The type of belt used in underground situations must be pliable enough to trough properly even when empty. Troughing normally is not a problem with most special weave belt constructions, but it can be a problem with multi-ply belts. The belt must also be strong enough to support the material load over the gaps between the idler rollers, as shown in Fig. 2. If it fails to support the load at these gaps, then adhesion failure between plies or breaks in the belt may result.

Belt capacity is a function of belt width, speed, and the troughing angle. Recently, there has been a trend toward increasing belt speeds. By

BELT DRIVES, TAKEUPS, AND TRANSFER POINTS

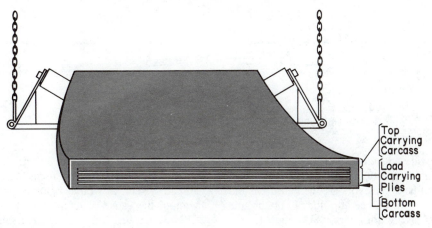

FIG. 1. Cross-section of a typical conveyor.

increasing the belt speed (while belt width and material loading rate remain constant), belt operating tension is lowered because the amount of material per unit length of belt is reduced. This allows the utilization of a lower-rated and less costly belt. It should be noted that higher belt speeds (but below the maximum recommendation) do not appear to cause serious operational problems. Idler roller bearing ratings should be checked before a decision to increase belt speed is made. Also, to avoid excessive belt wear, belt speeds should not be increased to the point where a very small volumetric load exists during belt operation. The maximum belt speeds in 0.3 m/s (feet per second) for coal mining can be summarized by the appropriate belt width:

Belt Speed, m/s (fps)	Belt Width, mm (in.)
182.88 (600)	609.6–914.4 (24–36)
243.84 (800)	1066.8–1524 (42–60)
304.8 (1000)	1828.8–2438.4 (72–96)

Normal production speeds for underground belts are in the neighborhood of the 167.6 m/s (550 fps) range.

DESIGNING FOR EFFICIENT SYSTEMS

This major discussion focuses on the haulage designs amenable to efficient systems by analyzing some of the causes of common haulage problems and then reviewing current design solutions at static haulage points (i.e., at drives, chutes, transfers, etc.).

Typically, panel belts are 914.4 mm or 0.9 m

TABLE 2. *Coalseal* Data Chart

Grades	220	300	330	440	450	550	600
No. of plies	2	2	3	4	3	5	4
Tension Rating PIW							
mechanical	220	300	330	440	450	550	600
vulcanized	220	300	360	480	510	600	680
Carcass weight (psf)*	0.90	1.04	1.35	1.80	1.56	2.25	2.08
Carcass thickness (in.)	.120	.150	.190	.260	.235	.330	.320
Weight per 1/32 in. of cover (psf)	.192	.192	.192	.192	.192	.192	.192
Minimum recommended							
pulley diameter (in.)	12	16	18	24	24	30	30
Relative tear factor							
base index 6MP50 = 100	250	300	300	500	425	600	500
Impact rating							
base index 6MP50 = 100	360	425	450	600	550	750	600

*Metric equivalents: psf × 4.882 428 = kg/m²; in. × 25.4 = mm.

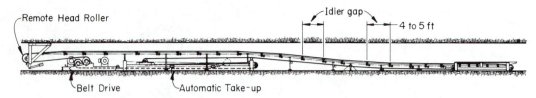

FIG. 2. Belt must be strong enough to carry load between idlers. Metric equivalent for idler gap: 1.34 m.

(36 in.) wide with main belts in the 1066.8 to 1219.2 mm (42 or 48 in.) range. The basic components of belt conveyors are the belt drive and head pulley, the belt takeup, the tail pulley, the troughing and return idlers, the support frame, and the belt itself. In addition, federal law requires that a water line with outlets every 91.4 m (300 ft) be installed parallel to all belts for their entire length, that water or foam fire control systems be provided at all conveyor drives, and that an alarm system capable of stopping the belts be provided.

The most common problem, or characteristic of a problem, with a haulage system is the presence of a spill. Consistent spillage of coal and rock anywhere along a belt line indicates there may be a problem with the system. To effect solutions to any spillage, one must be able to recognize the type of spillage and to identify its cause.

Analyzing Belt Spillage

Belt spillage can be classified in three general categories: (1) belt line spills, (2) transfer point spills, and (3) drive, feeder, and surge bin spills. The classifications point out the differences in either location or probable cause of each spill for the construction supervisor or engineer to focus in on during an analysis.

Belt Line Spills: Belt line spills are defined by the first general category because of the following characteristics:

1. They are located along a belt line in the absence of any belt line structure such as a transfer point, chute, feeder, drive, and so on.
2. They are caused by (a) transitional mining elements such as falling roof or ribs, oversized chunks of coal or rock riding on the belt, (b) alignment problems with the belt chains, stands, ropes, structure, or rollers, or (c) elevation problems causing the material to "roll" on the belt and breach the trough formed by the belt structure.

Any one of these potential belt line spills can be repeated time and time again in any given haulage system, as well as being a contributing factor to the cause of another spill (i.e., roof fall knocks a belt line out of alignment).

To analyze whether a haulage system is suffering from belt line spillage, a construction foreman, mine foreman, or engineer must travel the belt line over a definite multishift period and observe any pattern of spillage that exists. A pattern occurs when repeated spillage is present in the same spots. If a pattern is evident, then steps toward defining the cause may be taken by reviewing the specifics of the spill point and asking the following questions:

1) Is there evidence of any recent roof or rib fall?
2) How steep an elevation is the belt line at this point?
3) Is the belt structure at this spot aligned with the rest of the belt line?
4) Is the belt running out of trough or off the return idler at this point?
5) Does the unit(s) normally feeding this belt have a feeder-breaker at the loading end to aid in reducing the chunks, and if so, is there trouble with the feeder-breaker settings?

Once the belt line spill can be characterized, corrective action can be taken to eliminate the problem.

Transfer Point Spills: Transfer point spills are characterized mainly by their location: at the transfer point. These spills can be caused by one or more of the following factors:

Vertical misalignment, consisting of poor positioning of the dumping belt over the receiving belt.

Horizontal misalignment, shifting the discharge stream centerline away from the calculated impact spot. This misalignment can be caused by poor installation of either the dumping or receiving belt.

Hang-ups of oversized coal or rock at the

transfer point, which effectively block the material stream from changing belts.

Leakage, either through the sideboards, wipers, rollers, or chute structure. This leakage can be from fines or coarse material which builds up over a number of shifts.

At each transfer point in the haulage system, an observer can track the spillage pattern to isolate one or more of the causes for the spill. Of course, the hardest cause to pinpoint is the horizontal alignment problem (unless it is blatant!), but this can be traced by eliminating the other possibilities.

Drive, Feeder, and Surge Bin Spills: Spills at static, long-term structures such as drives, feeders, and surge bins are less frequent, but usually are more severe than the previous two spillage areas. Large static structures, when installed correctly, should suffer only minimal spillage due to structure leakage, coal fines, and float dust. If major spills have or are occurring, then a problem with installation or structure alignment may be present. Close visual examination of the physical structure may be necessary to determine what, if any, alignment or construction problems are causing the spillage.

Once the cause of any given belt spill is determined, design efforts are directed toward eliminating the problem from a long-term solution standpoint. Past experience in dealing with belt haulage spills has brought many practical solutions to mine operators for use on a day-to-day basis. My next discussion will review some of these alternative design solutions for controlling spillage, as well as some fundamental characteristics of modern underground belt haulage.

Alternative Installation Aspects for Drives, Takeups, and Transfer Points

When considering the installation of a haulage system, an engineer or foreman must choose among many alternative designs and installations. For construction engineering purposes, these alternatives reduce to considering among the following:

1) Suspension from the roof or anchoring in the floor.
2) Available bang-boards, sideboards, chutes, wipers, and cleaners for belt haulage.
3) Dust and fire suppression, sensors, and deluge system choices for underground haulage.

Each installation is based upon site specific data, so only a general discussion of the possible alternatives is available for the purposes of this text.

Suspension vs. Floor Anchorage: Suspension from the roof or anchorage in the floor relates directly to the installation decisions for the head roller, belt drive, and takeup components.

The head roller component of any haulage system can be installed as a part of the belt drive attached (as shown in Fig. 3), or as a remote setup (Fig. 4). If it is attached, then the support arms are an extension of the belt drive frame. When the head roller is a remote setup, it can be suspended from the roof or anchored in the floor. Most operations today suspend the roller from the roof for various reasons. Table 3 summarizes the advantages and disadvantages for each of the alternative head roller setups. The most important reason remote head rollers are suspended from the roof is to increase the dumping height of the transfer point to the required distance as per system design.

Turning to the belt drive, it is almost always anchored into the floor during installation. The ease of placement and structure stability outweigh the advantages of roof suspension in all but the most extreme cases.

The takeup system employed in underground belt installations can be either automatic or manual, depending upon the site characteristics and desires of management. A manual takeup is composed of taking up belt slack with a screw type of adjustment (Fig. 5). These types are recommended only for short setups [under 91.4 m (300 ft), center to center]. The takeup adjustment of a manual takeup should be at least 1% of the length of the belt, plus any additional length required to splice the belt.

An automatic takeup or tensioning unit uses hydraulic pressure, air, or gravity to take up belt slack and maintain belt tension (Fig. 6). Automatic units have the flexibility to tension long and large belts. Reeving arrangements for floor anchored hydraulic or air pressured units can produce travel up to 9.1 m (30 ft) as shown in Fig. 7. Gravity type setups are suspended from the belt and, while having less tension flexibility, require almost no maintenance, do not overtension the belt, and are very reliable. The recommended adjustment for automatic takeups is roughly split between conveyors over and under 91.4 m (300 ft). Longer belt lines use 0.3

FIG. 3. Head roller used as a component of the belt drive (courtesy of Hewitt-Robins Conveyor Equipment Division).

FIG. 4. Head roller shown as a remote setup (courtesy of Hewitt-Robins Conveyor Equipment Division).

BELT DRIVES, TAKEUPS, AND TRANSFER POINTS

TABLE 3. Advantages and Disadvantages of Head Roller Setups

Advantages	Disadvantages
Roof-mounted	
Increases the effective dumping height of the belt	Installation time needed to hang the frame
Increases the clearance and maneuverability around the roller	Anchor points are susceptible to vibration fatigue and failure over time
Easier cleaning and inspection activities	Low adjustment potential
Floor-mounted	
Short installation time	Cleaning is harder around roller
Increased adjustment potential	Lower dumping height inherent in the setup unless adjusted for
Easy attachment of safety systems	Subject to floor heave
Stable anchorage	Corrosion and failure potential much greater in wet conditions

m per 30.5 m (one ft per 100 ft) of conveyor length as the guide, while shorter belt lines estimate an adjustment of 0.6 m per 30.5 m (two ft per 100 ft) of conveyor length. These takeup movements vary directly with the type of belt and the length of belt lines.

Installation of automatic takeup units occurs as a supplement to the belt drive placement. Hydraulic units are anchored behind the belt drive. Gravity units are suspended either behind the head roller or behind the belt drive proper (Fig. 8).

Alternative Bang-Boards, Sideboards, Chutes, Wipers, and Cleaners: Coal is not moved by belt alone. It takes auxiliary equipment such as bang-boards, sideboards, and chutes at transfer points and wipers or cleaners along the belt line to keep spillage down to a minimum. The performance of this equipment depends a great deal on the installation design and maintenance. Several different approaches to installing these haulage aids give some flexibility to the construction foreman or engineer during design and installation.

Bang-Boards. Bang-boards are devices which deflect a material stream falling from one belt into the material path of the second belt (Fig. 9) when a 90° material transfer is required. Its position at the transfer point is critical to the reduction of spillage. Alternative approaches for bang-boards revolve around differences in anchorage, materials, and shape.

Anchorage differences for bang-boards found

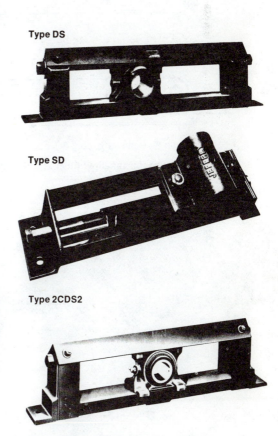

FIG. 5. Mechanical takeups used for short or temporary setups (courtesy of Jeffrey Manufacturing Division, Dresser Industries, Inc.).

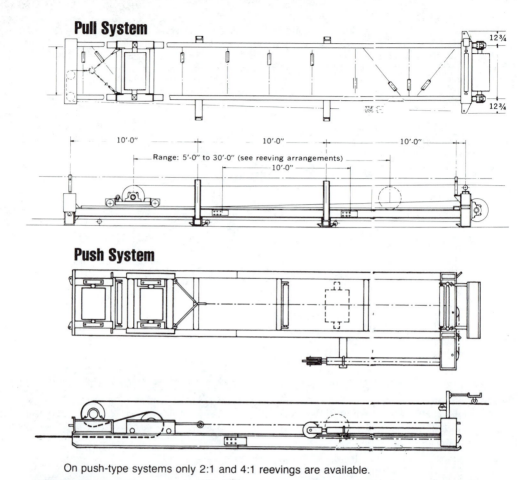

On push-type systems only 2:1 and 4:1 reevings are available.

FIG. 6(a). Horizontal takeups: push and pull systems. Metric equivalent: ft × 0.3048 = m.

FIG. 6(b). An automatic takeup for modern conveyor systems (courtesy of Hewitt-Robins Conveyor Equipment Division).

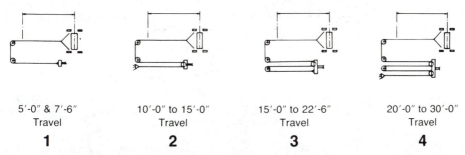

5'-0" & 7'-6" Travel	10'-0" to 15'-0" Travel	15'-0" to 22'-6" Travel	20'-0" to 30'-0" Travel
1	2	3	4

FIG. 7. Typical reeving arrangements for automatic takeups (courtesy of Hewitt-Robins Conveyor Equipment Division). Metric equivalent: ft × 0.3048 = m.

in underground installations include the roof, floor, and belt line points. Roof installations as shown in Fig. 10 require chain supports to suspend the board at the correct height and deflection angle. Guide chains or turnbuckles to adjust the correct deflection angle can also be mounted on the belt structure (Fig. 11).

Floor installations are designed on a self-supporting stand with support arms holding the board at the correct height and deflection angle (Fig. 12). These installations give less fine adjustment than the roof-mounted boards because the stands can only slide and change position. In addition, more maintenance inspections and repair work is usually needed because of the higher fatigue factor in the support arms and welds of the structure. Floor mounted bang-boards are used only in situations where the roof is too fractured or soft to absorb the high vibration stress of the bang-board.

Belt line installations, like floor-mounted boards, are steel structures with arms supporting and angling the board in the desired position (Fig. 13). These installations are mounted on rigid channel stringers and are the least desirable installation alternative from the construction engineer's standpoint. First, belt line installations have little adjustment after initial placement. Secondly, they are high vibration structures which create stress at the welds on the belt line and support arms. Finally, of the three anchorage alternatives (roof, floor, or belt line), belt line installations require the most rigorous inspection and activity schedule to maintain the performance, as well as being the most complicated to design, manufacture, and install at the construction site.

Material alternatives for bang-boards center on obtaining high abrasive resistant materials such as steel plate or ceramic lining for use as a deflective board. By far the most popular today is steel plate of 12.7 to 15.8 mm (1/2 to 5/8 in.) thick. Wooden bang-boards reinforced with scrap conveyor belt or steel sheeting are sometimes used as a temporary measure in short-lived installations. Conversely, ceramic linings for use in bang-board installation are used only on life-of-mine transfer points or at high abrasive material setups where frequent bang-board replacement is difficult and/or costly. The deciding factors for bang-board materials are convenience, application, and cost.

The final alternative to bang-boards is the *design shape*. Most bang-boards are just a steel plate cut to useable forms (Fig. 14). However, concave bangboards, dubbed ''chutes,'' also find some application in life-of-mine projects.

Sideboards. Sideboards, or skirtboards, are devices which help channel and maintain the flow of coal and rock at the belt line entrances such as transfer points, chutes, and tailpieces (Fig. 15). They are often called skirtboards because of the flexible rubber material (usually 60 durometer) which sheds abrasive particles and still remains stiff. Thicknesses range from 6.4 to 25.4 mm (1/4 to one in.) with widths from 76.2 to 304.8 mm (3 to 12 in.). The skirtboard material is shipped in 15.2 m (50 ft) rolls with a price range of $1.50 for 6.4 × 76.2 mm (1/4 × 3 in.) to $14.75 for 25.4 × 254 mm (1 × 10 in.) per 0.3 m (linear ft) (1981 dollars). Skirtboard installations are very important to prolonging conveyor belt life and for centering the dumped material on the moving conveyor belt.

Chutes. Chutes, for the most part, are an armored extension of the bang-board design. Their primary application is at life-of-mine transfer points where reliability and performance is the main concern. Chutes are commonly found in

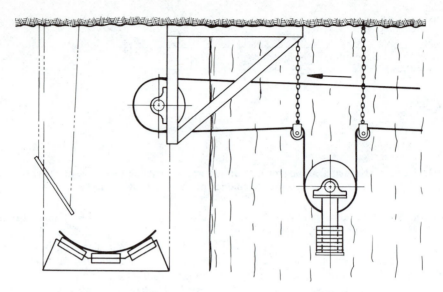

FIG. 8(a). Gravity takeup unit behind the head roller.

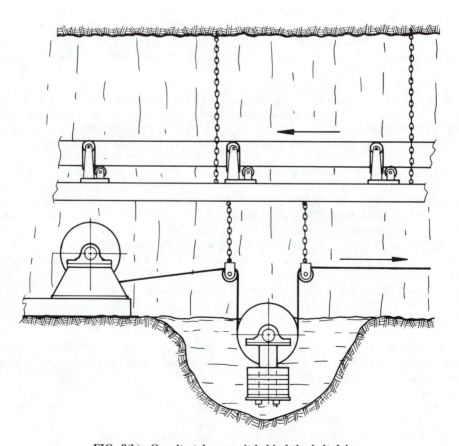

FIG. 8(b). Gravity takeup unit behind the belt drive.

FIG. 9. Bang-boards are used to deflect the coal from one belt to another. (Crickmer and Zegeer, 1981).

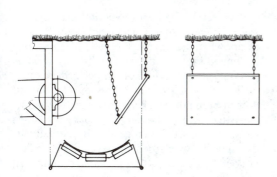

FIG. 10. Roof supported bang-board (four point).

FIG. 11. Turnbuckles attached to the structure form deflection angle.

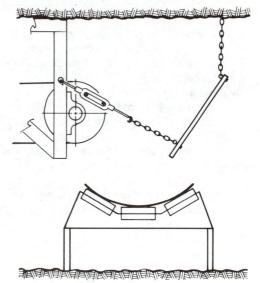

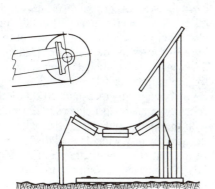

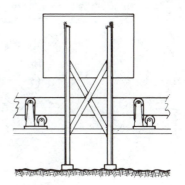

FIG. 12. Floor supported bang-board.

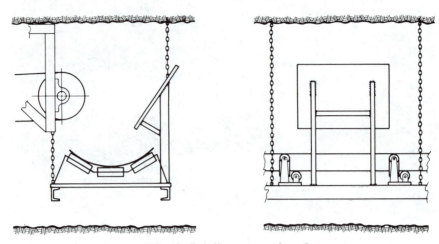

FIG. 13. Belt line support board.

piggyback installations for mainline belts (Fig. 16). Chutes can be floor or roof mounted and anchored at the desired height and position. They are made of heavy-gage steel plate and used with or without a liner.

Chutes can load material in the same direction as the belt travel, or on an angle to it. Parallel loading is, of course, the easiest to accomplish. The chute must be able to impact some forward velocity to the material (by gravity) before falling on the belt, hence the inclined design of the chute. If the velocity of the material is not great enough, it may be necessary to reduce the speed of the receiving belt. However, lowering the belt speed suggests that a wider belt may be necessary when this approach is taken. Chutes, like the one in Fig. 17, are site specific and must be carefully designed for correct performance.

Wipers and Cleaners. Wipers and cleaners fill a need common to all conveyor belt systems. Coal fines, float dust, and rock particles which stick or are partially embedded in the belt carcass must be swept (or scraped) from the surface as soon as possible to avoid damage to the belt and structure.

There are many types of belt wipers and belt cleaners available using a variety of cleaning methods. Brushes, blades, and rubber wipers are typical approaches to cleaning design. The most common types for coal mining are spring-loaded or counterweighted cleaners and the plow (*V*) wiper.

The spring-loaded and counterweighted types of wipers (Fig. 18) are designed to clean the carrying side of the belt surface at the head roller. Their performance depends upon the placement and regular maintenance of the wiper. Fig. 19 shows the design specifications of typical wipers. These models average around $500 for the spring-loaded wiper, $600 for a single counterweighted wiper, and $1000 for a double counterweighted wiper. Prices will naturally vary with the model and size of the belt.

The plow or *V* wiper is used to clean the inside of the belt at the belt tailpiece (Fig. 20). It can be spring-loaded or weighted to achieve the necessary cleaning performance. The cost of a fitted plow wiper varies between $325 and $400 de-

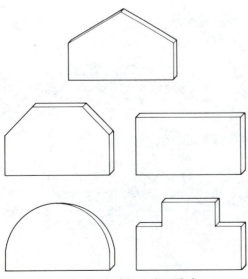

FIG. 14. Typical bang-board shapes.

FIG. 15. Typical sideboard installation (courtesy of Conveyor Components Co.).

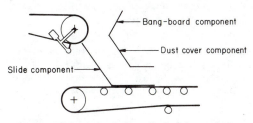

FIG. 16. Piggyback installation for a chute.

pending on the size and type of belt and wiper material.

Dust and Fire Suppression, Sensor, and Deluge Systems: As part of the 1969 Mine Health and Safety Act, minimum fire detection and prevention was required on all haulage systems. Since then, various designs for complying with the act have been placed in underground coal mines. Most systems rely on pneumatic, smoke, or heat-sensing systems which activate both an alarm and suppression system. In most cases, the sensing system runs down the middle of the belt line, although it is also frequently placed on one side of the belt.

The belt line suppression and deluge systems based on water are composed of piping, valves, and fire hose placed along the belt line (Fig. 21) at predetermined spots. Valves are placed every 91.4 m (300 ft) with fire hose located every 304.8 m (100 ft). Around belt drives, takeups, and electrical controls, deluge systems must be designed and installed to blanket the entire area with the extinguishing agent (Fig. 22). Maintaining system pressure becomes very important to such installations for achieving the required performance, especially if the system relies on water as the extinguishing agent. Foam and dry chemical systems for fire protection are also available, as shown in Figs. 23 and 24. These systems usually provide a self-contained pres-

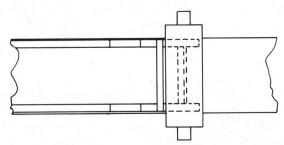

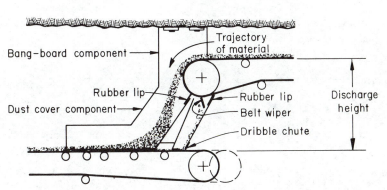

FIG. 17. Typical vertical view of an underground chute. The dust cover is optional.

Fig. 18(a). A spring-loaded wiper (courtesy of Conveyor Components Co.).

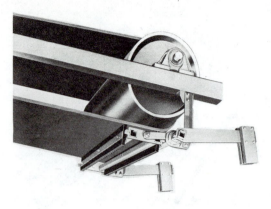

FIG. 18(b). A counterweighted wiper (courtesy of Conveyor Components Co.).

surizing component to power the extinguishing agent to the fire.

All deluge, sensor, and dust suppression systems are fairly site specific. Therefore, construction and installation of any system requires close communication among the manufacturer, design engineer, and construction foreman. The objective of any system becomes one of safe, efficient, and cost-effective performance throughout the life of the belt line.

INSTALLING BELT DRIVES, HEAD ROLLERS, AND TAKEUPS

Once the decision on systems, equipment, and design alternatives has been made, the construction engineer and/or supervisor must begin planning for its installation. Three major steps are required:

1) *Preparation,* which covers the site orientation and supply of the correct equipment.

2) *Installation,* including placing and anchoring the equipment, and hooking up the auxiliary systems.

3) *Operation and Maintenance* of the system by adjusting the tension, checking for vibration, and lubricating the drive.

These major steps can be broken down into several subtasks which are common to any belt construction project. For the purposes of this discussion, I will illustrate the installation steps needed for putting in a belt drive with an automatic takeup and a remote suspended head roller. Since construction details are site specific, I will concentrate on the general management approach and underlying engineering principles of installing the system.

To begin with, belt drives are normally located at the belt head, and may consist of either a single pulley (driven by one or two motors), two pulleys run off a single motor, called a geared tandem drive, or two pulleys driven by a synchronous (primary) motor and a squirrel-cage induction (secondary) motor. Fig. 25 shows three common drive arrangements. Typical motor horsepower for various types, sizes, and lengths of belt are given by Meader (1973) below:

Belt Type	Range of Length	Belt Widths	Range of Drive
Section	2000–2500 ft*	30,36,42, in.	50–150 hp
Gathering or butt	3000–4000	36,42,48	100–200
Main line	3000–6000	42,48,60	200–400
Slope	1000–2000	48,60,72	250 up

*Metric equivalents: ft × 0.3048 = m; in. × 25.4 = mm; and hp × 0.745 699 = kW.

BELT DRIVES, TAKEUPS, AND TRANSFER POINTS

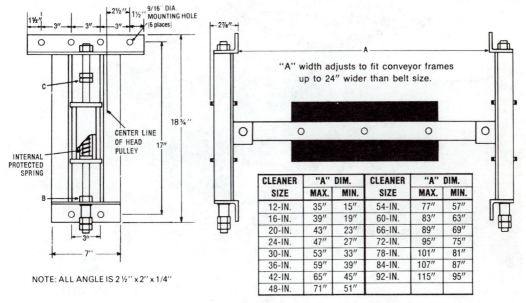

CLEANER SIZE	"A" DIM. MAX.	"A" DIM. MIN.	CLEANER SIZE	"A" DIM. MAX.	"A" DIM. MIN.
12-IN.	35"	15"	54-IN.	77"	57"
16-IN.	39"	19"	60-IN.	83"	63"
20-IN.	43"	23"	66-IN.	89"	69"
24-IN.	47"	27"	72-IN.	95"	75"
30-IN.	53"	33"	78-IN.	101"	81"
36-IN.	59"	39"	84-IN.	107"	87"
42-IN.	65"	45"	92-IN.	115"	95"
48-IN.	71"	51"			

NOTE: ALL ANGLE IS 2½" x 2" x 1/4"

FIG. 19(a). Specifications of a spring-loaded wiper (courtesy of Conveyor Components Co.). Metric equivalent: in. × 25.4 = mm.

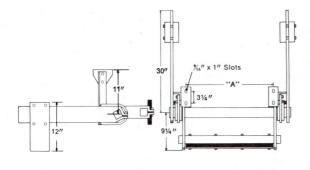

BELT WIDTH	DIMENSION A MIN.	DIMENSION A MAX.
12"	10"	32"
18"	16"	36"
20"	18"	40"
24"	22"	44"
30"	28"	50"
36"	34"	56"
42"	40"	62"
48"	46"	68"
54"	52"	74"
60"	58"	80"
72"	70"	92"

Model CW is the standard more commonly used unit. However, where there is a heavier buildup of materials on the belt, the two bladed MODEL CWD should be purchased.

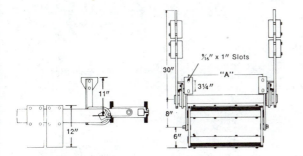

CW/CWD INSTALLATION INSTRUCTIONS

1. Mount the cleaner so that the leading edge of the first wiper is directly below the center of the head pulley and the blade arm is approximately parallel with the belt.
2. Locate the counterweight arms so they are approximately horizontal.
3. Slide counterweights on arms and position for most effective cleaning.
4. Locate stops to prevent the wiper support from contacting the belt as the wiper wears.
5. Counterweight arms can be offset, lengthened or shortened in field by customer if necessary.

FIG. 19(b). Specifications of a counterweighted wiper (courtesy of Conveyor Components Co.). Metric equivalent: in. × 25.4 = mm.

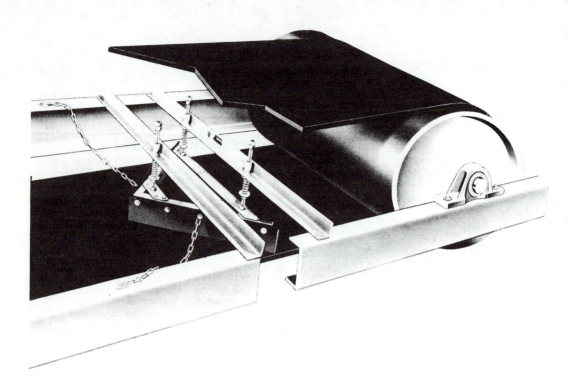

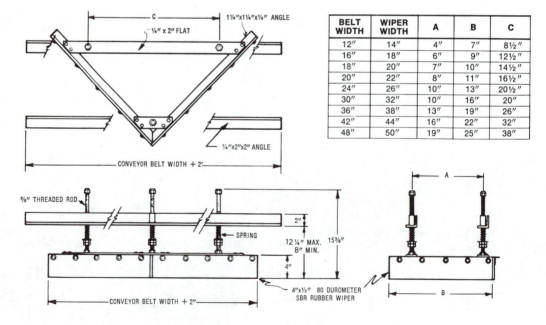

FIG. 20. A plow cleaner with specifications (courtesy of Conveyor Components Co.). Metric equivalent: in. × 25.4 = mm.

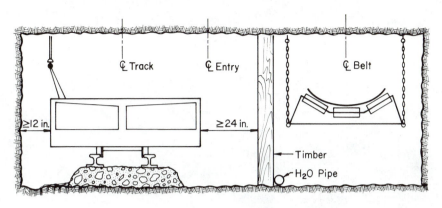

Vertical View

FIG. 21. Waterline pipe placed along the belt line.
Metric equivalent: in. × 25.4 = mm.

TYPICAL DELUGE WATER SPRAY SYSTEM APPLICATION

FIG. 22(a). Typical water deluge system (courtesy of Automatic Sprinkler Corp.).

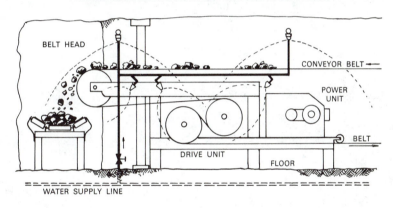

FIG. 22(b). Typical water deluge system (courtesy of Pyott-Boone Electronics).

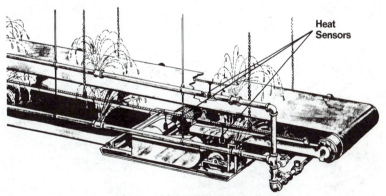

All belt drives are composed of several types of pulleys (as mentioned earlier), each with a different function. The main types are:

Drive pulley, which is attached to the motor (or speed reducer, etc.) and provides the force to turn the belt. It is almost always a lagged (fabric covered) pulley.

Head pulley is the main pulley at the discharge end of the belt. A head pulley may or may not also be a drive pulley.

Snub pulley is a smaller diameter pulley 304.8 to 355.6 mm (12 to 14 in.) used to increase the angle of wrap on a drive pulley by "snubbing up" the belt on the drive pulley in the desired position.

Tail pulley is the main pulley used at the load-

FIG. 23. Foam generator system (courtesy of Automatic Sprinkler Corp.).

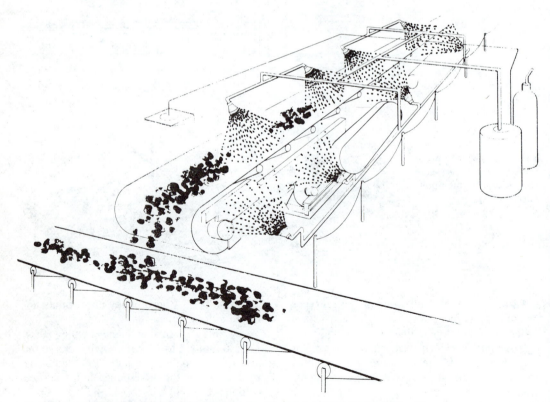

FIG. 24. Dry chemical system (courtesy of Automatic Sprinkler Corp.).

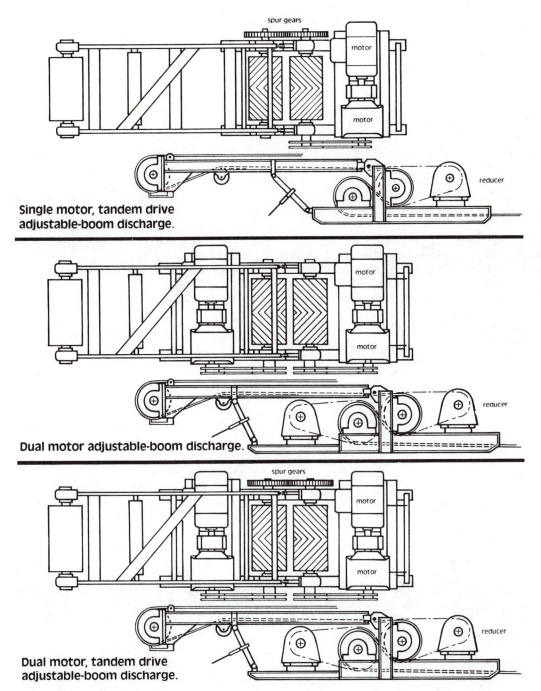

FIG. 25. Typical belt drive arrangements (courtesy of Goodman Equipment Corp.).

ing end of the belt. It can be crowned and winged for increased efficiency and performance.

Takeup pulley identifies the pulleys (one fixed and one moveable) used in an automatic belt takeup unit.

Drive or head pulleys do not have to be crowned (i.e., peaked at the center), but crowned tail, snub, or takeup pulleys facilitate belt training. However, all pulleys should be flat-faced and smooth when the belt is made of very high modulus material. Wing-type pulleys are often used at the tail or for snubs or bends. Wing pulleys aid in keeping the belt clean; belt damage due to coal spills is less severe when this type of pulley is used. However, the stress developed by belt flexing as the belt passes over small diameter wing-type pulleys, 304.8 to 355.6 mm (12 to 14 in.), is severe enough to reduce the life of the mechanical joints in the belt.

The belt takeup is a natural extension of the belt drive construction. It is used to apply the correct tension to the belt. Most newly installed belt takeups are automatic for reliability, durability, and ease of operation in an underground setup. However, it is important that all takeups be adjusted to apply the correct tension to the belt. Insufficient belt tension can cause belt slip at the drive and excessive sag of the loaded belt between carrying idlers (which can result in spillage, shifting and movement of the coal, slightly increased power requirements, and increased belt stress). Excessive belt tension can result in decreased splice life, increased belt elongation, and increased stress on terminal pulleys and their bearings. Therefore, automatic takeups are located as close to the drive as possible to minimize belt slip. Manual takeups are usually located at the end of the conveyor, where they are convenient to adjust and least costly to install.

A key parameter in a belt drive design is the drive factor, which defines the combined efficiency of the belt drive motor and speed reducer, and is the basis of the overall horsepower required by the support system. In order to decrease the drive factor, and hence the amount of horsepower to the motor (and the tension required in the belt's slack side), mine operators sometimes use lagged, as opposed to plain, drive pulleys. Lagged pulleys are pulleys covered with fabric or rubber material which increases the friction between the belt and pulley. Lagged pulleys become particularly effective in wet conditions where the belt has a tendency to slip.

The drive factor also can be decreased by increasing the angle of belt wrap around the drive pulley. Snub pulleys are often used in conjunction with single-pulley drives to increase the effective angle of wrap. Of course, it is possible to achieve a larger angle of wrap with dual-pulley drives than with single pulley drives, which is one of the reasons dual-pulley drives are popular. (The popularity of this type of drive also owes to its compactness and low overall height, making it ideal for use in underground coal mines.) However, poorly designed geared-tandem drives can be problematic, because unusual stresses in the belting and drive equipment can develop. The problems which can result include: low life for mechanical belt splices, early belt failure at a point of belt injury, low life of the tandem pulley gears, pulley structural failures, and excessive wearing of lagged pulleys.

Preparation

The discussion on site preparation contains three general points. The first point considers the necessary site orientation for the drive, head roller, and takeup. The second point looks at supplying the site with the correct materials and equipment, and the third point discusses the scheduling of required labor and installation time needed to complete the entire project. Again, only a general discussion will be followed due to the specific nature of belt line installation.

Site Orientation: Site orientation is concerned with (1) proper alignment of the belt drive and takeup at the desired location, and (2) the positioning of the suspended head roller. Because the manufacturer delivers the drive as a single structure, the minimum number of orientation spads needed by the construction foreman is approximately five. Depending upon the size of the belt and type of drive, the specific spad positions will shift, but their purpose stays the same.

The location of a belt drive and takeup in the path of the belt line boils down to a choice of entry or intersection placement. Generally, the advantages of access and maneuverability favor the intersection location over placement in the entry. In addition, if the belt line is designed to run down the middle of the entry, a production sight spad can be used in the alignment of the belt drive and takeup.

The five site spads needed to orient the belt drive include three alignment spads and two po-

BELT DRIVES, TAKEUPS, AND TRANSFER POINTS

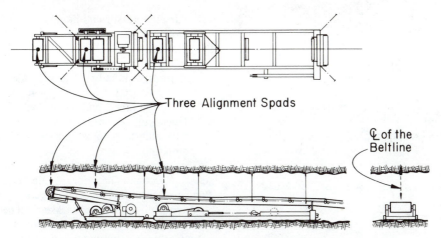

FIG. 26. Alignment spads on the belt drive drums.

sition spads. The three alignment spads orient the belt drive and takeup to be parallel to the belt line while the position spads fix the location of the belt drive with respect to the mine construction site and anchorage points.

The alignment spads are placed on the belt line centerline so their relative position will be on the center of the driving drums (Fig. 26). The production sight spad, mentioned earlier, can be used for one such spad if desired. In addition, optional spads can be placed in line with the drive alignment spads to further help position the takeup unit (Fig. 27).

The position spads are placed on either side of the leading alignment spad. They are positioned over the pillow blocks of the driving drum on the centerline of the drum shaft (Fig. 28). This eliminates any frame inconsistencies within the drive structure itself which would tend to throw the belt out of alignment.

Other spads for orientation can and are frequently used. Spads to align the speed reducer, drive frame, and anchor holes are all possibilities at the site. Mine management should decide which spads can help in the construction of the belt line.

Spads needed for the suspended head roller include only the centerline spad for the belt line and a position spad for the pulley shaft (Fig. 29). These can easily be put up during the other spad work by the surveyors.

Supplying the Site: One of the better innovations in underground coal mine construction is the scoop-tractor. This useful machine is a very big key in the timely supply of the site with the proper equipment and materials for instal-

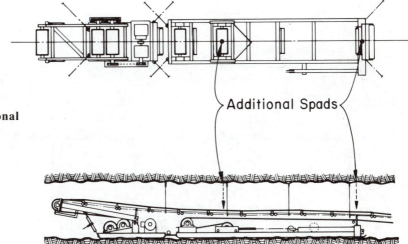

FIG. 27. Additional alignment spads.

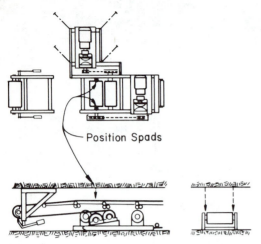

FIG. 28. Position spads located on the belt drive pillow blocks.

lation. If a scoop is not available, then a shuttle car or track motor (with block and tackle) is used.

The equipment taken to the site includes the belt drive, takeup, and controls, plus miscellaneous support materials and equipment. A partial list of support materials includes:

    The auxiliary system (deluge and sensor) components.
    Rotary air drill and steel for drilling 0.9 to 1.2 m (3 to 4 ft) holes.
    Air compressor and hose.
    Bolts or flag pins and turnbuckles for anchorage.
    9.5 mm (3/8 in.) rope for threading the drive.
    A track jack and roof jack.
    203.2 × 203.2 mm × 2.4 m (8 × 8 in. × 8 ft) timbers and several header boards.
    Guarding for the drive and takeup.
    Miscellaneous tools and supplies for construction.

The delivery of the materials and equipment can be accomplished during idle shifts in advance of the scheduled construction date. All the delivered items should be stacked neatly and orderly close to the position of the belt drive. The final task in preparation is that of construction scheduling.

Construction Scheduling: Construction scheduling is the task of planning for the belt drive installation. It includes estimating the labor and time required for each installation task.

The first step in construction scheduling is to break down the installation tasks into logical units. For the belt drive and head roller, the tasks can be broken down as follows:

Belt Drive
    Aligning the drive and takeup structure
    Anchoring the structure
    Placing the auxiliary systems

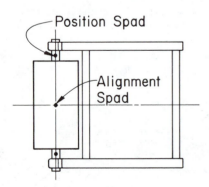

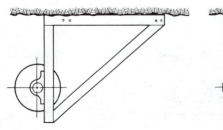
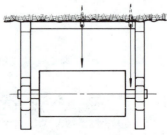

FIG. 29. Spads for head rollers.

TABLE 4. Labor and Time Estimates for Belt Drive and Takeup Tasks*

Task	No. of Men	No. of Shifts		Total Manshifts
Aligning the structure	2	1/2		1
Anchoring the structure	3	1		3
Auxiliary systems	2	3		6
Guards	1	1/2		1/2
Troubleshooting	2	3/4		1 1/4
			Total	12

*Does not include supervision or training requirements.

    Placing the machinery guards
    Troubleshooting the setup
Head Roller
    Marking and drilling the hanger holes
    Hanging the head roller
    Leveling and tightening the structure
    Placing the auxiliary systems
    Inspection of the setup

Once the tasks are outlined, estimates can be made for scheduling the installation.

For the purposes of the discussion, Tables 4 and 5 present a general estimate of labor and time for installing a belt drive and suspended head roller, respectively. These figures can be taken one step further for cost estimating purposes if and when required. Chap. 10 explores this topic in further detail. Once the estimates have been made, the scheduling of the installation work can proceed in order to finish before production demands are placed on it.

Installing the Belt Drive and Takeup

Once the crew has been assembled and the time scheduled, the installation work can begin. Again, the installation steps are: aligning the drive and takeup structure, anchoring the structure, placing the auxiliary systems, placing the machinery guards, and troubleshooting the setup. It is important to remember each step can be treated as an individual task with different teams of workers performing each one.

Aligning the Drive and Takeup Structure: This task requires at least two workers and the use of a scoop tractor. The objective is to position the belt drive and takeup according to the site orientation spads.

The first step is to roughly level the belt drive area with the scoop or by shovel. The immediate area should be dry, with good drainage. If extremely soft bottoms are encountered, then dumping crushed stone and pouring at least 76.2 mm (3 in.) of concrete as a pad may help level the site. Otherwise, 76.2 to 101.6 mm (3 to 4 in.) of track ballast should be sufficient for support purposes.

The drive is carefully loaded into the scoop from a frontal position so that the drive structure will slide or be slightly lifted off the floor. The scoop maneuvers the drive to the desired location and places it in the approximate position under the spads. Considerable positioning using the scoop bucket and blade actions may be necessary to line up the drive within a few mm (in.) of the spad position.

While the scoop is loading the takeup unit, one worker with the roof and track jacks will begin fine tuning the drive position. Here, by using the jacks, the drive can be inched into the correct position.

The takeup is usually bolted into the final position behind the drive (it may not originally, but with modifications it will). The takeup is centered under the alignment.

TABLE 5. Labor and Time Estimates for Suspending the Head Roller*

Task	No. of Men	No. of Shifts		Total Manshifts
Marking and drilling	2	1/2		1
Hanging and leveling the head roller	3	1		3
Auxiliary systems	1	1/2		1/2
Inspection	1	1/2		1/2
			Total	5

*Does not include supervision or training requirements.

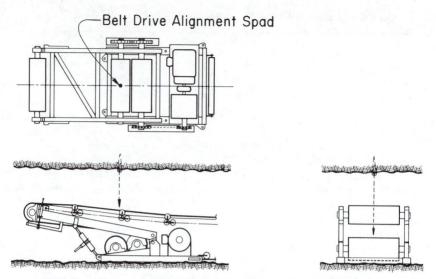

FIG. 30. Alignment spad for belt drive aligned with belt line center.

At the end of this task, the drive and takeup should be positioned exactly so that:

1) The belt drive alignment spads mark the centerline of the belt drive structure and belt line (Fig. 30).

2) The position spads mark the centerline of the drive drum shaft and mark the pillow block positions.

3) The takeup pulley center is aligned with the belt line centerline.

4) Both the drive and takeup are level in all directions.

Considerable fine adjustments may be necessary to line up all the orientation spads. However, it is extremely important to gain this position and maintain it throughout the anchoring task.

Anchoring the Drive and Takeup: There are two alternatives with anchoring the structure. The first utilizes the flag pin and turnbuckle method called *variable anchorage,* and the second uses roof bolts, known as *fixed anchorage.* Both have advantages and disadvantages for anchoring.

The *variable anchorage* method requires eight points for proper anchorage of the drive. Two drill holes, approximately 0.6 m (2 ft) from all four corners, are marked in a pattern similar to Fig. 31. A rotary drill powered by an air compressor drills these holes approximately 0.9 m (3 ft) deep at a slight angle (less than 15°) away from the structure frame (Fig. 32). Flag pins, which are 31.8 × 762 mm (1¼ × 30 in.) solid steel pins with an eye flange (deemed the "flag") welded on one end, are hammered into the holes until the eye flange meets the floor. Hook-ended turnbuckles are opened and connected to the pin and belt drive. This procedure is repeated for each of the anchorage holes at all corners of the drive.

The advantages of variable anchorage include the adjustment potential offered by the turnbuckle arrangement and the ability to drill the anchor holes in unencumbered places. In addition, the flag pins provide a measure of anchorage along the entire length of the pin shaft.

A disadvantage of variable anchorage is the tendency of the structure to drift off position from vibration and human error over a period of time. In addition, wet bottom conditions can cause a lapse of anchorage over time.

Fixed anchorage requires only six anchor points; three on each long side of the structure (Fig. 33). Roof bolts are used as single-point anchorage systems normal to the plane of the belt drive. Here holes are drilled as close to vertical as possible under the structure bolt holes. Bolts and modified 4 × 4 roof plates are used to anchor the drive and takeup.

The advantages of fixed anchorage depend largely on the competency of the bottom material. However, its main advantage is the use of bolts over pins and turnbuckles and the tendency for the drive to be solidly anchored.

The disadvantages of fixed anchorage lie in the fact that little or no adjustment is possible after the bolts are tightened.

BELT DRIVES, TAKEUPS, AND TRANSFER POINTS

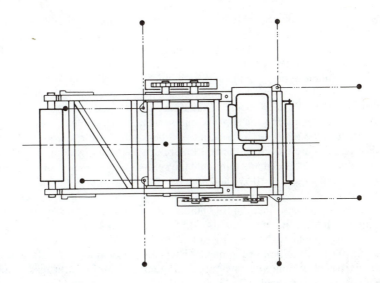

FIG. 31. Anchorage position for variable anchorage, 8 points.

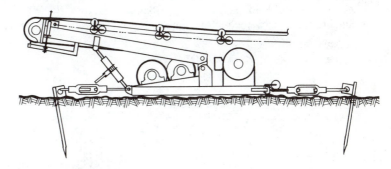

The takeup unit can be anchored with either the variable or fixed point method. Because one end is usually securely bolted to the drive frame, it is necessary to firmly support the other end to make sure the takeup frame is level with the belt drive. The procedure for anchoring the takeup is the same as that for the drive.

Placing the Auxiliary Systems: This task encompasses the work of hooking up the electrical, alarm, suppression, and/or deluge systems of the drive.

The electrical hookup from the in-line power splitter to the belt drive motors is a fairly straightforward connection. Standard three conductor cables are hung from the switchhouse to the motors and controls on insulated hangers. A certified electrician should make the connection to ensure proper hookup.

The alarm, suppression, and deluge systems for the drive, takeup, and motor controls center on providing coverage to all possible fire hazard areas. If water is the primary extinguishing agent,

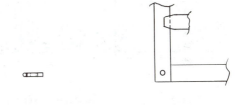

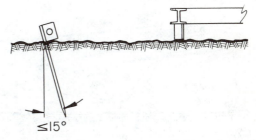

FIG. 32. Anchorage pin at a slight angle away from structure.

then the piping, nozzles, valves, and sprinkler heads making up the system should be drawn out on paper first. Once these positions are designed, then the system can be easily transferred to the installation site. Foam or dry chemical systems usually have predetermined positions which must be followed during installation.

The final item concerns the slip and spill switch placement in the belt line. These control rollers and/or paddles are placed in the belt line return to monitor the belt speed and eliminate the spillage due to slippage at the drive and pileups from big chunks in the belt line. These switches are connected to the belt line control cable.

Placing the Machinery Guards: Machinery guards are mandated by law for all exposed moving parts of mining equipment. The belt drive is no exception. Fencing placed around the drive and takeup should be substantial, clearly marked, and moveable for easy cleaning. Chain link fencing which is framed at the correct positions (height, length, etc.) is excellent for this purpose. The guarding, once fitted for the site, is removed and placed to one side until the belt drive is ready for operation.

Troubleshooting the Setup: Once the proceeding tasks have been accomplished, an inspection and troubleshooting of the setup is desirable. This task includes reviewing the alignment and anchorage of the drive, the auxiliary system placement and hookup, as well as the cleanup of the site of trash, tools, and extra equipment.

An important task of the troubleshooting step is ensuring the correct performance of the drive motors and controls, as well as the hydraulic takeup and controls. An experienced worker or engineer should check each operating point carefully for leaks, imbalances, blockages, and vibrations. Correct settings for both the belt drive motors (RPM) and hydraulic takeup (takeup pressure for tension) should be checked.

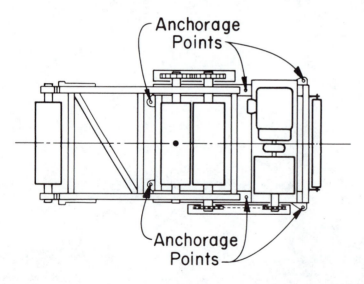

FIG. 33. Fixed anchorage points (6 points).

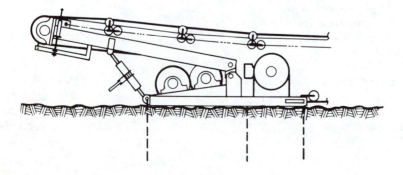

One final note concerns the threading of a new drive. It is desirable to thread the drive last when putting a belt together on the first setup of a belt line. This means starting in front of the drive, pulling the belt around the head, down the troughing side, then through the tail, through the takeup and finally through the drive. Enough slack should be pulled through the takeup for the belt to go completely through the drive. This is done so that no tension is placed on the belt being threaded in the drive. By threading the drive last, one eliminates the work of pulling the entire belt through the drive and then threading the belt line, which compounds the effort needed to pull the belt due to the frictional force of the belt being pulled around the drums.

Installing the Suspended Head Roller

Installing a suspended head roller can be a difficult task when geologic and mine conditions (roof rock competency, seam height, and mining clearances) are not right. However, planning and prior inspection during site orientation should minimize any adverse local conditions. Sometimes, excavation of roof material is required for obtaining the transfer height. If this is necessary, the work is scheduled for an idle shift. The receiving belt line is opened and removed (along with the structure) from the site. The roof material is graded and cleaned up in a manner similar to overcast grading discussed in Chap. 3. The receiving belt is then reconnected and trained to normal operating conditions.

As mentioned earlier, the installation steps for hanging a head roller include: (1) marking and drilling the holes, (2) hanging the head roller, (3) placing the auxiliary systems, and (4) inspecting the setup. Although the head roller is suspended as a remote structure, it is critical to the success of the belt line that one consider the head roller as a key component to efficient materials handling.

Marking and Drilling the Holes: The first step after the site orientation task is marking and drilling the hanger holes for the head roller frame. The key to this task is the use of a template.

A template, usually made of 25.4 × 152.4 mm (1 × 6 in.) framing lumber, is designed and constructed to resemble the roof contact surface of the head roller frame (Fig. 34). Bolt holes are marked and later drilled in the exact position of the holes found on the frame. The centerlines of the belt and head roller shaft are marked also. A template may be reused many times.

Once the template is made, it is placed on the mine roof and held in position by a roof jack. Caution should be exercised to align the template with the centerline of the belt line and position the roller shaft centerline with the appropriate orientation spad. Spray paint or chalk is used to transfer the hole locations from the template to the roof.

The bolt holes are drilled with a standard jackleg or stoper drill. Either can be powered by an air compressor brought to the site. The holes are usually drilled at least 0.9 m (3 ft) deep to ensure good anchorage. Local conditions should dictate the depth needed when scaly or fractured roof is present. Extra utility holes for a chain jack or other construction device may be necessary with high mine roofs.

Hanging the Head Roller: After the holes are drilled, hanging the roller is accomplished with the aid of the scoop, chain jacks, and roof jacks. The head roller is placed in the bucket and lifted to within 0.6 m (2 ft) of the roof (or as high as possible) and maneuvered into the approximate position. Further height can be gained by jacking the head frame to the approximate location. Bolts with header plates are inserted through the head roller frame into the drilled roof holes and tightened only so the head roller can be suspended and the scoop removed. Two or more roof jacks are used to level the head roller and place it against the roof. Header boards, washer plates, or other wedging materials may be necessary between the frame and any undulations in the mine roof to provide good roof contact and keep the head roller level.

Once the head frame is flush and leveled against the roof, the bolts are securely tightened to hold the head roller in position.

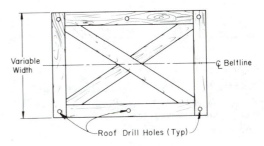

FIG. 34. Head roller template.

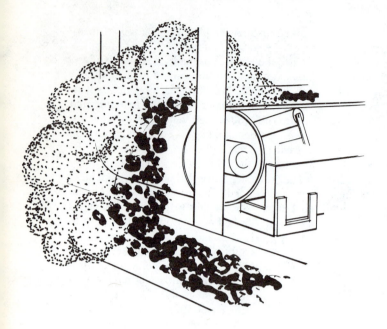

FIG. 35. Fire system at the head roller (courtesy of Automatic Sprinkler Corp.).

Placing the Auxiliary Systems: Auxiliary systems for the head roller are a similar version of the systems in place at the belt drive. There is no electrical hookup other than the belt line fire sensing and/or suppression systems, belt line control cable and the slip/spill switches which are powered from the belt line control cable. The water, foam, or dry chemical fire extinguishing system must cover all fire hazard areas around the head roller and transfer point (Fig. 35). Site specific details are necessary for each installation and so will not be reviewed here.

Inspecting the Setup: Once the head roller is in position, and the auxiliary systems hooked up, an inspection of the entire setup should be completed. This inspection is necessary to assure quality control and avoid any little operational problems. The alignment and position of the head roller should be checked with the centerline of the belt as well as the normal discharge distance to the receiving belt centerline. Any irregularities in construction or design should be straightened out during this task and before belt operation.

Operation and Maintenance

Once the belt line is installed and the entire setup put into production purposes, one of the most important tasks begins. This task concerns the operating and maintenance procedures for the belt drive, takeup, and head roller.

There are three general items which should be mentioned. The first is a regular schedule of inspection for vibration and wear effects on the drive, takeup, or head roller. This includes inspecting for loose anchorage, frictional heat buildup from excessive wearing, as well as fatigue in the structural members.

The second item should be a consistent effort to lubricate the motors, idlers, and pillow block bearings on all moving equipment. All too often an irregular program of maintenance will lapse into disuse at the belt drive point since this equipment is a relatively low maintenance item.

The final item is one of conscientiously cleaning the float dust and spillage from the drive, takeup, and head roller areas. This is especially true around the head roller area where the wipers clean the belt. A regular program of inspection, cleaning, and rock dusting the area at least once an operating shift is recommended.

INSTALLING THE TRANSFER POINTS

The second most common static construction item of haulage systems is installation of the transfer point. Since a discussion of the head roller component has preceded this point, the focus of this section is on the completion of the transfer point.

There are three basic discussions in this section: (1) design parameters affecting transfer point performance, (2) building the transfer point, and (3) operation and maintenance of the transfer point. Each discussion deals with the general characteristics of the transfer point without applying site specific functions.

Two basic options are used in coal mining for transferring materials to another belt, namely: loading the belt parallel to belt travel (end dumping) and loading normal to belt travel (side dumping).

End dumping is probably the easiest type of dumping available to the construction engineer. The flow of material from the transfer point can be designed so the forward velocity of the falling material is nearly equal to the velocity of the receiving belt. The falling material can be directed to the center of the belt in a symmetrical pattern based mainly on the angle of repose. In addition, loading of the receiving belt is fairly insensitive to the fluctuations of the feed rate. Finally, the transfer point height need not be as high with this type of dumping, since material turbulence on the receiving belt will be minimized.

On the other hand, side dumping can be a complex problem. Any angle can be used, but 90° transfer points are the most common. With angles increasing to 90°, the height of the transfer point must also increase due to the material turbulence on the receiving belt. In addition, a bang-board or chute must be placed very carefully to lead the material to the receiving belt at the center of the trough, as shown in Fig. 36. Sideboards will have to be higher and longer on side dumping chutes because of material acceleration characteristics. Finally, there is more wear on the belt, and in most cases, impact idlers are used to aid in supporting these transfer areas. Transfer points with angles greater than 90° should be avoided, since their design becomes very complex.

Design Parameters Affecting Transfer Point Performance

There are some specific parameters of belt haulage which affect the ultimate performance of a given transfer point. These parameters will cause changes in the basic transfer point design in terms of approach, materials, and maintenance.

There are four specific parameters which affect performance more than the rest. These are:

1) *Belt capacities* of both dumping and receiving belts. The design question comes down to surge capabilities of the receiving belt and the expected loading of the receiving belt. Capacities also tie the belt width and troughing designs together.

2) *Belt speeds,* relative speeds of both the dumping and receiving belts. The proper control of the speeds can minimize any overloading at the transfer point and greatly reduce the potential spillage.

3) *Belt loading,* which ties directly to the capacities and speeds of the two belts. Loading can be estimated in terms of normal and surge levels. The design of the transfer point must account for both levels, and include a reasonable safety factor.

4) *Material mix* completes the parameter considerations by focusing on the material characteristics. This, in turn, focuses on the durability and strength of the transfer point materials needed to handle the new coal and rock mixture.

Much of these design parameters will be resolved during the initial design of the mine haulage system. However, local conditions can and often do change the original designs. The unexpected occurrence of rock faults, or a production unit change (i.e., miner to longwall)

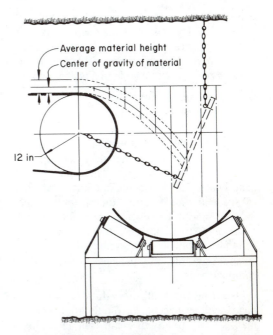

FIG. 36. Profile of material stream as it leaves the head roller (modified from Conveyor Equipment Manufacturers Association and CBI Publishing Co.). Metric equivalent: in. × 25.4 = mm.

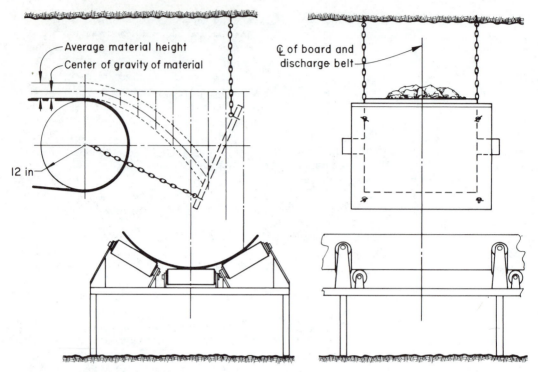

FIG. 37. Bang-board position for deflecting material stream into the center of receiving belt.

may result in a reevaluation of the transfer point capabilities. By looking at and quantifying these parameters, an engineer or construction foreman can improve the performance potential of the transfer point structure under the new condition.

Building the Transfer Point

Much of the transfer point construction has already been discussed under the section on placement of the head roller. This section looks at installing the three other necessary components to the transfer point. These are: (1) the bang-board, (2) the sideboard (skirtboard), and (3) the wiper and controls on the return. Each must be placed in such a manner as to provide efficient performance and minimize any spillage. The following is a discussion of the most common construction approaches to each of these components.

Installing the Bang-Board: A typical bang-board will be suspended from the roof with two main straps, usually of chain. The chain holes in the roof are located so the board will hang on the far side of the receiving belt and approximately 50.8 to 101.6 mm (2 to 4 linear in.) in from the edge. The length of the chain allows approximately 203.2 to 254 mm (8 to 10 in.) of clearance between the belt and the bang-board edge. The center of the board is lined up with the centerline of the dumping belt, as shown in Fig. 37.

A second pair of chains (or in many cases, turnbuckles), attached to the side of the board provide the deflection angle to the falling material. The angle is adjusted so the bulk of the material will fall into the center of the receiving belt (Fig. 38). In some cases, these deflection angle chains are dropped in lieu of supports from the rear of the bang-board. These supports are welded onto the back at a predetermined angle and at a given length to form a brace between the roof and the board. This type of design, however, is subject to more maintenance and gives less adjustment than the use of chains and turnbuckles. While it is an acceptable and common design, it generally is not recommended.

Installing the Skirtboard: Skirtboards, as mentioned earlier, are used to channel the material on the belt and provide a surge control. For transfer points, the skirtboard is usually 25.4 × 152.4 mm (1 × 6 in.) and attached by bolts to the bottom of the bang-board. The exposed

BELT DRIVES, TAKEUPS, AND TRANSFER POINTS

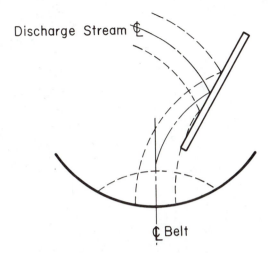

FIG. 38. Bang-board deflects material into centerline of belt.

skirtboard aids in containing the material in the center of the receiving belt while reducing the spillage potential due to a material surge at that point. The skirtboard installation is made so approximately 25.4 to 50.8 mm (1 to 2 in.) of space is left between the belt and skirtboard bottom. The space reduces the wear of the skirtboard material and allows freer operation of the receiving belt.

Installing the Wiper and Controls on the Return: The wiper on the return side is installed to prevent spilled material from reaching the tail pulley. It is located inby the transfer point on the upper side of the belt return, as shown in Fig. 39. As the figure illustrates, the plow (or V) wiper design is used to aid in channeling any spillage to the sides of the belt. The plow point is placed opposing the direction of the traveling belt. If the belt line is a rope type, then modifications for attaching it must be made (e.g., attached to a belt stand).

Slip switch rollers are also installed on the return side of the receiving belt. These controls aid in reducing spills by shutting off the dumping belt when the receiving belt is down. These switches operate on a number of principles, plus they are very reliable, low maintenance, and low cost instruments which can save a great deal of potential trouble at a transfer point.

Operation and Maintenance

Once a transfer point has been put into production, the construction foreman must continue to monitor and keep up the performance of the transfer point.

There are some routine points of maintenance which should be scheduled periodically. These include:

1) *Replacing worn sideboards, wipers, or bang-boards.* These items will wear periodically on the basis of the transfer point use. Wipers and sideboards have a typical life of nine to twelve months, while a bang-board will last between one to two years.

2) *Checking the sensor and suppression systems,* which include the fire systems and the production sequencing system. Electrical slip switch connections, piping connections, valves, sprinklers, detectors, and the pressure systems to power them are all susceptible to dust and age. Checking once every three months, or more frequently, is recommended.

3) *Eliminating material buildup,* including the spillage from the head roller and return side wipers. Shoveling all the buildup and cleaning the entire area is the easiest approach. Rock dusting the area after each cleanup will aid in eliminating a fire hazard. Cleaning once a shift is recommended for adequate performance.

4) *Maintaining effective machinery guards.* Over time, many installations with poor guarding will provide no safety protection. Checks every six months to observe the effectiveness of the installation are recommended. Corrective action should begin immediately after a safety hazard is observed.

5) *Lubricating the equipment,* which includes the head roller bearings, the belt idler bearings at the transfer point, and any light oiling of the fire or production sensing equipment. Manufacturers should recommend a lubricating schedule.

On a regular basis, these general items should maintain the performance of the transfer point and minimize the spillage buildup at any given point in time to institute corrective measures.

This concludes the discussion on the construction and installation of static haulage structures. The general procedures on belt drive, takeups, head rollers, and transfer points have been presented. Because details are site specific, each mine installation will vary to take advantage of local conditions. The possibilities for design variations, therefore, are quite extensive.

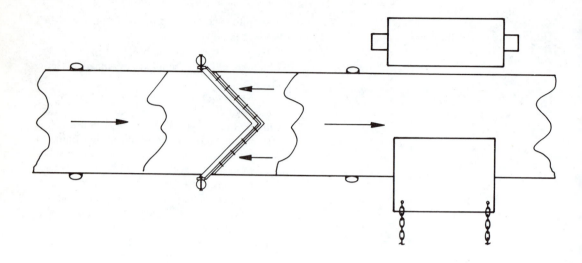

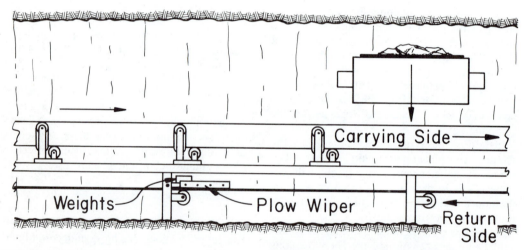

FIG. 39. *V* wiper position on the return belt inby the transfer point. This aids in keeping any dribbling material from building up at the tailpiece.

REFERENCES AND BIBLIOGRAPHY

Automatic Sprinkler Corp. of America, *Automatic Sprinkler Coal Mine Fire Protection Systems*, Cleveland, OH.

Conveyor Components Co., 1981, *Conveyor Components*, Catalogue C-217, Croswell, MI.

Conveyor Equipment Manufacturers Association, 1976, *Belt Conveyors for Bulk Materials*, CBI Publishing Co., Boston, MA.

Crickmer, D. F., and Zegeer, D. A., eds., 1981, *Elements of Practical Coal Mining*, 2nd ed., AIME, New York.

Cummins, A. B., and Given, I. A., 1973, *SME Mining Engineering Handbook*, AIME, New York.

Goodman Equipment Corp., 1979, *Ropebelt® Belt Conveyor Systems*, Catalogue GCC-255, Murfreesboro, TN.

Hewitt-Robins Inc., 1980, *Hewitt-Robins Idlers*, Catalogue 1200-80, Passaic, NJ.

Hewitt-Robins Inc., 1979, *Wire Rope Conveyors*, Catalogue 1104-79, Passaic, NJ.

Jeffrey Manufacturing Division, Dresser Industries, Inc., 1980, *Jeffrey Conveyor Accessories*, Catalogue 1180, Belton, SC.

Pyott-Boone Electronics, Technical Data Bulletin 79-1101, Tazewell, VA.

6

Belt Line Construction

This chapter is concerned with the construction of the dynamic component of belt haulage systems. The dynamic system includes the belt line and tailpiece extension and retraction movements in a production system where mine management is continually placing the belt in the most effective operating position. Correct installation of the belt and tailpiece during an extension is very important for many performance reasons. Two of the most important reasons are to eliminate spillage and safely operate the belt line.

This chapter is divided into three major discussions. The first discussion looks at analyzing the parameters of effective belt line construction. This includes a review of the parameters and management tools needed for proper scheduling of a move-up.

The second major discussion topic centers on efficient procedures for a belt line extension. Topics such as site preparation, tailpiece breakdown and setup, stringing the belt line and structure, plus operating and maintenance tasks needed after the belt line is installed and running, are covered.

The third discussion presents a list of conveyor and conveyor component manufacturers who provide services and products to the coal mining industry.

Again, any discussion on underground haulage systems must generalize much of the site specific design approaches to be practical. This text will mention, where applicable, some of the alternatives available to construction engineers in designing belt line systems.

ANALYZING THE PARAMETERS OF EFFECTIVE BELT LINE CONSTRUCTION

To properly attack the problems associated with poor belt extensions, one must identify and solve the causes of these problems. Not all possible causes are associated with the installation work, but it has been a general rule that a lot of nagging and unproductive aspects of an extended belt line can be eliminated by good sound construction.

This discussion, therefore, focuses on the construction parameters which have the major influence on materials handling. Specifically, these include: rope, structure, and belt integrity, belt line alignment, and loading and dumping interaction. Each of these parameters encompasses several rather subtle materials handling concepts instead of one single idea. Each shall be discussed briefly.

Rope, Structure, and Belt Integrity

When a new belt line is to be extended, there must be a conscious effort to blend the new set of belt line characteristics with the existing ones. The tensions of the rope and belt, as well as the position of the structure should be equal throughout the length of the belt line. For example, too little tension in the ropes on a belt line will create a sagging between support (anchor) stands, as illustrated in Fig. 1. If this condition is taken too far, then a properly tensioned belt will ride off the structure at each low spot (i.e., float when not loaded). Improperly ten-

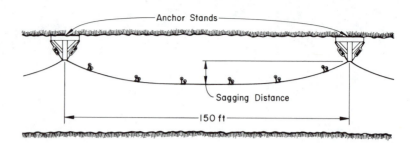

FIG. 1. Rope tension is too low to prevent sagging between stands. Metric equivalent: ft × 0.3048 = m.

sioned ropes cannot function correctly during operation, increasing the likelihood of belt spillage, slippage, and belt line imbalance (Fig. 2). If a belt line is supported by solid channels, then the tension becomes a minor factor. The popularity of channel belt lines for mainline and other permanent mine setups is partially due to this reason.

Belt tension in a routine extension also plays an important role. Over-tightened belts will float over the length of the belt line at each low or level spot. Under-tensioned belts will slip at the drive, stall out, and create tremendous spillage problems. Belt tension problems can be reduced by the proper adjustment of the belt takeup unit, whether manual or automatic. However, these units can occasionally lose power, pressure, or fail if not checked.

Structure integrity refers to the alignment of the troughing and return idlers between the anchor stands. A predesigned spacing by the construction engineer finds the best compromise between the necessary support for the belt and too much support structure (which can be costly). A typical design spacing found in modern mines is 1.2 to 1.5 m (4 to 5 ft) between troughing idlers and 3.0 m (10 ft) for return idlers, (Fig. 3). Anchor stands are placed 45.7 m (150 ft) apart, making one stand to stand setup consist of 30 troughing idlers and 15 return idlers. During an extension, the structure must be properly placed and tightened. It is important to maintain the spacing and alignment of each idler to keep the desired performance. Out-of-placement idlers are a consistent cause of belt misalignment because they force the belt center of gravity to move toward one side or the other.

Belt Line Alignment

The second parameter influencing materials handling is alignment of the belt line after the extension is completed.

It goes without saying that sight spads are placed approximately every 15.2 m (50 ft) to help in keeping the belt straight. If, for some reason, the belt line does not follow the predetermined centerline, a belt "kink" will occur. Any kink can and will make the belt center ride out of the trough and cause a spill. The construction foreman and engineer must strive to eliminate any and all belt kinks during installation.

Loading and Dumping Interaction

A belt line is used only as a carrier of materials. To get coal and rock on and off the belt line for movement, a loading and dumping system must be employed. The loading system, in this case, is commonly a feeder to tailpiece setup (Fig. 4) while the dumping is the discharge from a head roller into a chute, bang-board, or bin. I have already discussed the dumping installation, so the loading system is the focus of this discussion.

Specifically, the points needing to be stressed are the placement of the tailpiece and feeder. Proper anchorage and alignment during installation will keep spillage in this area to a minimum. In addition, sideboards, wipers, and discharge rates should be regulated by the construction engineer during the extension. By doing this, he will ensure that the coal will be chan-

FIG. 2. An imbalance on the belt ropes creates a spill situation.

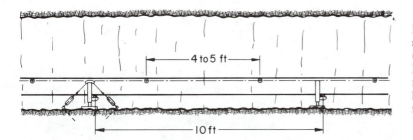

FIG. 3. Typical idler spacing on belt setups (adapted from Hewitt-Robins Conveyor Equipment Division Catalogue). Metric equivalent: ft × 0.3048 = m.

neled onto the belt and be moved according to predesigned specifications.

Materials, Labor, and Equipment for Construction

In addition to understanding the influences of belt line installation practices on materials handling, it is important to understand the essential components associated with underground belt haulage. This section looks at the materials, labor, and equipment of belt line construction which are typical of today's coal mining setups.

Materials: The materials of belt line construction refer to the variety of belt line designs and manufactured systems. They can be broken down into the frame and structure components.

Frame. Frame components are a choice between rigid or rope frame belt lines, which are hung either from the roof or supported on the floor, or both (Fig. 5). Today, most mine conveyor belts are supported by the wire-rope frame support system rather than solid frame structures because of their ease of installation, low cost, and low maintenance. The wire-rope system consists of two steel ropes, usually 15.9 or 12.7 mm (5/8 or 1/2 in.), running parallel to the conveyor belt (one on each side). They are suspended either from the roof by chains, or supported on the floor by anchor stands. The troughing and return idler structure is clamped

FIG. 4. The common loading system in mining is a feeder to tailpiece setup (courtesy of Stamler, Inc.).

FIG. 5. Belt frames can either be roof hung or floor mounted or both (courtesy of Hewitt-Robins Conveyor Equipment Division).

at right angles to the two ropes by adjustable "J" hooks.

The wire-rope system aids cleaning, extension, retraction, dismantling, and reassembly of belt lines, and is cheaper to install than the rigid frame structures. However, some problems are unique to wire-rope supports. As the rope corrodes, the troughing and return idler clamps may weaken, causing the idlers to slip out of alignment. (It is important to try to keep the idlers aligned, or the belt edge will wear causing the belt to not train properly.) Also, when two or more troughing idlers are mounted between the floor stands or roof hangers and rope tension is inadequate, the loaded belt is drawn down into the idler roll juncture points of the higher troughing idlers and pinched due to the weight of the material. As a result, longitudinal lines of belt ply adhesion loss or breaks on the belt carcass can develop. Finally, the advantage of adaptability is lessened when poor installation causes repeated alignment problems, mismatched splices, and shifting idlers due to loose structure clamps.

The above problems can be lessened or avoided by using a semirigid belt support structure, where the wire rope is replaced by steel channel stringers. The stringers are usually hung by chain from the roof because the roof is often more stable than the floor. Because the idlers are bolted rather than clamped to the stringers, it is easier to keep them aligned, and thus good belt training can be maintained. Rigid frame has a longer life than rope and a higher reliability, so, in spite of the higher cost of the semirigid support system, it is used for mainline permanent belts in place of the wire-rope system. Finally, they are almost exclusively used in slope haulage and preparation plants.

The rope frame of a belt conveyor also presents some unique anchorage problems. Because of its flexibility, wire-rope must be anchored (either in the roof or floor as in Fig. 6) in order for tension to be applied to it. This requires the use of anchor stands, called sail boats because of their shape. (Fig. 7). Chains and hangers hung from the roof or rigid stands on the floor between the sail boats help support the rope when the structure is applied and clamped. The floor stands, anchor stands, and chains used for rope frame conveyors can be adjusted for height and seam undulations, as shown in Fig. 8.

Structure: The structure component of belt materials includes the troughing and return idlers. These components provide proper support and protection for the conveyor belt and the coal being transported on it.

As discussed earlier, run of mine (ROM) coals are generally classified by CEMA as D35T. It is important, therefore, to select the structure components which are suitable for handling this kind of material. To this end, CEMA rates and

ROOF ANCHOR-SHORT BASE
demountable center post type

FLOOR ANCHOR

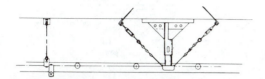
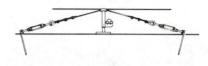

FIG. 6. Wire-rope with roof anchor and floor anchor (courtesy of Hewitt-Robins Conveyor Equipment Division).

classifies idlers on the basis of their bearing life, as given by the following estimating equation (simplified for ROM coal):

Actual Idler Load, kg (lb) = $(W_b + W_m) S \times 1.42$

W_b = Weight of belt, kg/m (lb/ft)
W_m = Weight of material, usually 74.4 kg/m (50 lb/ft)
S = Idler spacing, m (ft)

In most cases, the idlers used for coal fall in the CEMA B classifications, with some long or heavy-duty mine surface installations going to CEMA C standards. Tables 1 and 2 show the minimum load ratings for these idlers.

So what are the general types of idlers popular in the coal industry? Generally there are four types: (1) three-roller in-line idler; (2) three-roller offset idler (either single or boxed frame); (3) catenary idler, and (4) return idler.

The in-line idler (Fig. 9) usually comes in troughing angles of 20° or 35°, although deep troughing is not recommended for ROM coal. Idler diameters of 101.6 to 127 mm (4 to 5 in.) are available, depending upon the belt thickness and system design. The gap between the rollers of three-roller in-line idlers should be kept to a minimum, so that a heavily loaded flexible belt will not settle into the gaps. It has been suggested that gap size for coal conveyors not exceed 6.4 mm (1/4 in.).

Offset three-roller idlers are commonly available in troughing angles of 20°, 27½°, and 35° with rolls of 101.6 or 127 mm (4 or 5 in.) available. The two varieties are the single frame and box frame types, as shown in Fig. 10. The boxed frame allows minimum spacing between rolls to protect the belt, while single frame idlers are less costly and easier to install.

It is important that the center roller of the three-roller offset type idler be positioned so that the loaded belt does not drape down over the ends of the roller. Otherwise, wear of the belt bottom cover along two longitudinal lines will occur. If the center roller is the same length as the two concentrating rollers, the edges of the center roller should be well rounded.

There are two variations of the catenary type idler. One type consists of rubber or plastic discs, or cylinders molded on a steel cable. The steel cable, which is strung between the belt's longitudinal structural members, is free to rotate. The other type, which is not a true catenary,

ROOF ANCHOR, SHORT BASE
with non-demountable center post

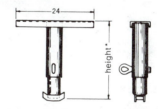

ROOF ANCHOR, SAILBOAT TYPE
with demountable center post

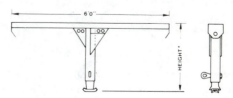

FIG. 7. Roof anchor stands (courtesy of Hewitt-Robins Conveyor Equipment Division). Metric equivalents: in. × 25.4 = mm; ft × 0.3048 = m.

CONSTRUCTION ENGINEERING IN UNDERGROUND COAL MINES

TABLE 1. Load Ratings for CEMA B Idlers, kg (lb)*

Belt Width (in.)	Trough Angle			Return
	20°	35°	45°	
18	410	410	410	220
24	410	410	410	190
30	410	410	410	165
36	410	410	396	155
42	390	363	351	140
48	380	353	342	130

*Metric equivalents: lb × 0.453 592 4 = kg; in. × 25.4 = mm.

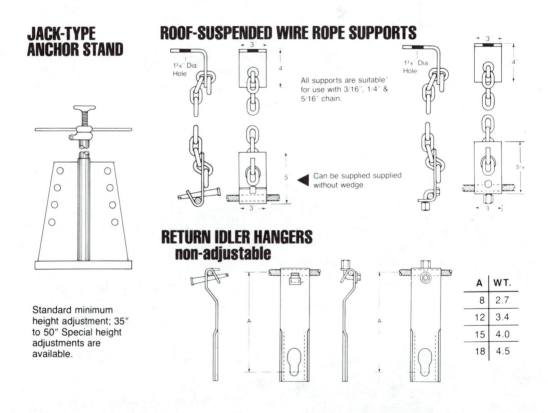
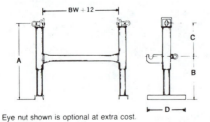
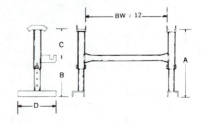

FIG. 8. Floor stands, anchor stands, and chains used for rope frame conveyors can be adjusted (courtesy of Hewitt-Robins Conveyor Equipment Division). Metric equivalent: in. × 25.4 = mm.

TABLE 2. Load Ratings for CEMA C Idlers, kg (lb)*

Belt Width (in.)	Trough Angle			Return
	20°	35°	45°	
18	900	900	900	475
24	900	900	900	325
30	900	900	900	250
36	900	837	810	200
42	850	791	765	150
48	800	744	720	125
54	750	698	675	—
60	700	650	630	—

*Metric equivalents: lb × 0.453 592 4 = kg; in. × 25.4 = mm.

consists of three in-line conventional idler rollers connected to each other by hinges.

Return idlers, shown in Fig. 11, are used to support the belt on the return side. Rubber discs are sometimes used to help keep the belt clean. In addition, belt training is very important on the return run, where most damage to the belt edge occurs. To aid in proper training, the return run idlers can be crowned and self-aligning idlers spaced approximately 61 m (200 ft) apart can be installed. Self-aligning troughing idlers can also be used.

An idler troughing angle of 20° was used exclusively in the past, but as belts have become more pliable, the $27^1/_2°$ and 35° configurations with their greater capacity have become more popular. The catenary type idler allows the mine operator greater flexibility in choosing the troughing angle, since the angle is determined by the distance between the conveyor's longitudinal structural members. When the larger troughing angle is used, transfer points must be raised to provide a surge capacity. In addition, design of vertical curves and transitions from the last idler to a terminal pulley take on added importance.

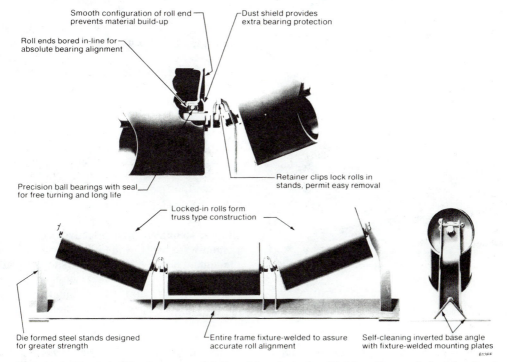

FIG. 9. In-line belt idler (courtesy of Jeffrey Manufacturing Division, Dresser Industries, Inc.).

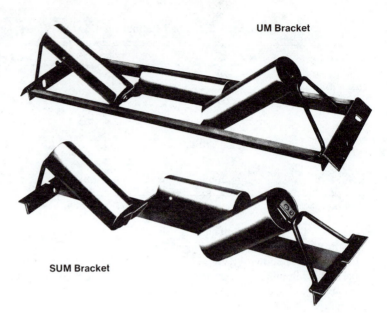

FIG. 10. Single frame and box frame wire-rope idlers (courtesy of Jeffrey Manufacturing Division, Dresser Industries, Inc.).

Wire Rope Return idlers are available with return hangers of varying drops for both wedge type and J-bolt type mounting.

Disc Return idlers have rubber (or Urethane) discs which permit the belt to flex. This usually loosens any material adhering to the belt and permits it to fall free.

Return Training idlers work on the same principle as Troughing/Training idlers; when the belt touches either guide roll, the idler swivels and corrects the belt alignment.

FIG. 11. Three types of return idlers (courtesy of Jeffrey Manufacturing Division, Dresser Industries, Inc.).

Curves and terminal transitions must be designed to minimize belt stress and wear. Idlers on a vertical curve should be positioned so that each idler is on the circumference of the curve. Furthermore, the idlers should be spaced at no more than half the spacing used in level areas. The distance from the last standard troughing idler to a terminal pulley should be based on (1) the idler troughing angle, (2) the belt's modulus of elasticity, (3) the pulley mounted level, and (4) the belt tension at that point. An idler of a lesser troughing angle can be used between the last standard idler and the pulley to help support the belt through the transition.

Spacing of troughing idlers should be based on the weight of the coal and the minimum belt tension. Normally, the troughing idlers are spaced 0.9 to 1.5 m (3 to 5 ft) apart, while return idler spacing is usually twice troughing idler spacing. Since these parameters are fundamental to conveyor design, the prospective equipment manufacturers should be contacted for further help in design.

Labor Requirements: Effective belt line extensions less than 91.4 m (300 linear ft) require a minimum number of six workers to complete in one eight-hour shift. For extensions greater than 91.4 m (300 linear ft), the number rises by a factor of 2.5 persons per additional 30.5 m (100 linear ft).

The labor force should be broken into two groups when completing all the tasks needed for a belt line extension. One group contains four

BELT LINE CONSTRUCTION

TABLE 3. Labor Requirements for a Belt Extension

Task	No. of Men	No. of Man-hours	Total Hours
Breaking down and moving the tailpiece	2	1	2
Installing anchor stands and ropes	4	2	8
Breaking the belt and splicing new piece	2	1/2	1
Reanchoring the tailpiece	2	4	8
Placing the belt, structure, and auxiliary systems	4	4	16
Training the belt, troubleshooting	6	2	12
			47

workers, the other has two. All the tasks can be defined as the following: preparation (preliminary); breaking down and moving the tailpiece; installing the anchor stands and ropes; breaking the belt and splicing the required piece in; reanchoring the tailpiece; placing the belt, structure, and auxiliary safety systems; and training the belt and troubleshooting the system.

Each of these tasks can be further broken down in terms of the required time to keep it within the 48 man-hour level. To this end, Table 3 shows the estimated time requirements, without the preparation steps of material placement, drilling of hanger holes, and site orientation. Of course, Table 3 shows a major portion of time for training and troubleshooting, but this provides a buffer for any problems which may arise with other tasks. Fewer laborers can be used (i.e., four workers), but the efficiency of the work drops and the time buffer is greatly reduced.

The labor for the preparation task has not been identified because of its variable nature. However, an estimate of 24 to 30 man-hours is a general figure for a 91.4 m (300 ft) extension. This preparation task will be explored further in later discussions.

Tool Requirements for an Extension: There are basic tools needed for extending and installing any belt line. These tools are dependent upon design parameters such as tailpiece anchorage method, belt splicing method, the choice of roof or floor mounted belts, and belt line framing method (i.e., wire rope or rigid frame). One of the essential tools common to all efficient extension operations is the horizontal belt winder. Here, the additional piece of belt can be spliced and spooled out to the existing belt line on demand. It helps to eliminate belt kinks and threading problems, because in most cases, the winder and belt can be placed into a scoop and easily located on the path of the belt line.

A second essential tool is the equipment used to move the tailpiece. Under good mining conditions, a powerful feeder can move the tailpiece to the new location. If poor conditions are encountered, a scoop, shuttle, or track motor (with block and tackle) must do the job. It is important to have this equipment available when required to move the tailpiece, since reanchoring is a time-consuming task.

A final major tool category concerns the splicing of the new piece of belt (Fig. 12). Belts can be spliced by a variety of mechanical methods (hooks, bolts, etc.). The common element, however, remains the preparation of the belt ends for splicing. Here splicing tools such as a belt knife (Fig. 13), and fastener tool (Fig. 14) are used. Ideally, the belt to be spliced into the new belt will be sized and the ends made up before the actual splicing. However, this is not always the case, so having the tools and splicing materials handy will avoid unnecessary trouble.

Other tools include the assorted hand tools to loosen and tighten bolts, turnbuckles, and pipe couplings. Power tools such as a rotary air drill are necessary to reanchor the tailpiece. Other miscellaneous tools (including a belt tensioning device) are used as the individual setup requires.

INSTALLING A BELT LINE EXTENSION

The following is a discussion on the general procedures for extending a belt. The example will focus on a roof-mounted, wire-rope conveyor typical of panel development. Movements of the work force will be discussed in the approximate order necessary to complete the task.

The four major task groupings include:

1) *Preparation,* which covers site orientation, drilling the hanger and anchor sailbase holes, delivering the materials, stringing the structure, and making up the ends of a new piece of belt.

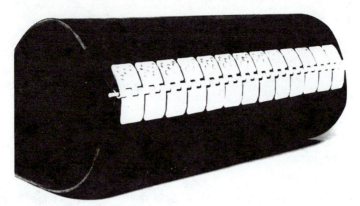

FIG. 12. Typical mechanical belt splice (courtesy of Flexco, Flexible Steel Lacing Co.).

2) *Belt line assembly,* which includes stringing the ropes, threading the belt, assembling the structure, and connecting the auxiliary safety systems.

3) *Tailpiece assembly,* which covers the breakdown and reanchoring of the tailpiece, as well as attaching the belt to the tailpiece.

4) *Operation and maintenance,* which looks at the activities of troubleshooting and training the belt, lubrication, cleanup, and building belt line accessories (crossovers, tailpiece bunker, etc.).

Although the belt line assembly and tailpiece assembly are done concurrently, a separate discussion is desirable for clarity.

Preparation

As it has been with other underground construction projects, preparing the site for an extension is very important. It requires planning and organization by mine management to effectively coordinate the preparation of site orientation, drilling the hanger sail plate holes, delivering the extension materials, stringing the structure, and making up the new piece of belt.

Adequate preparation will help smooth all of the later tasks of the extension.

Site orientation is the task of providing sight spads for the construction engineer or foreman to follow during installation. The spads to be placed are (1) centerline spads, (2) belt line offset spads, and (3) tailpiece position spads. The centerline spads are usually placed every 12.2 to 15.2 m (40 to 50 ft) and mark the path of the belt line center (Fig. 15). These sights can only be taken directly from the mine orientation grid if the belt line goes down the middle of the entry. Otherwise, new centerline spads are needed (Fig. 16).

Offset spads are used to mark the position of the wire ropes, hangers, and sailbase holes for the belt line frame (Fig. 17). These spads are measured from the centerline spad at exactly

FIG. 13. A belt knife used to cleanly slice the belt end for splicing (Anon., 1982).

FIG. 14. A belt fastener tool used to place and secure a mechanical splice (courtesy of Flexco, Flexible Steel Lacing Co.).

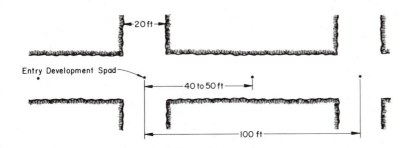

FIG. 15. Centerline spads for belt are placed every 12.2 to 15.2 m (40 to 50 ft). Note belt centerline in this case is the same as the entry centerline. Therefore, the entry development spads can be used.

one-half the width of the designed belt line. These spads will naturally vary with the width of the belt. Sight strings can then be strung from the last set of anchor stands to the new spads. Hanger holes are marked every 4.6 m (15 ft), with anchor sailbase holes marked every 45.7 m (150 ft), or the designed distance. It is important to position the hanger and sailbase holes with the appropriate offset spads to keep the belt line in alignment.

Tailpiece position spads mark the position and alignment of the tailpiece in the new location. A centerline spad is located over the spot where the tail pulley shaft crosses it, and a second centerline spad locates the front edge of the tailpiece proper. Another spad is offset from the rear centerline spad and located on an edge of the tailpiece as an aid in aligning the tailpiece. The position of each of these spads is shown in Fig. 18.

Drilling the hanger and sailbase holes can be accomplished by using a roof bolting machine, stoper, or jackleg drill. The holes are drilled into competent rock with a depth of at least 762 mm (30 in.). Chain hangers and the sailbases are then bolted to the roof. The sailbases are then leveled against the roof, both front to back and side to side. During the drilling, chain hanger plates on roof bolts are inserted into the hanger holes and tightened.

The next step is prompt delivery of all the belt line materials needed for the move. For this example, a 91.4 m (300 ft) extension includes the following materials:

91.4 m (300 ft) of water pipe, sensing and control cable, and other auxiliary system components (valves, outlets, etc.)
189 m (620 ft) of 12.7 mm (1/2 in.) rope [four 47.2 m (155 ft) pieces]
two sets of roof-hung anchor sailbases
40 hanger chains, with one hook end approximately 1.2 m (4 ft) long
60 troughing idlers
30 return idlers
eight 19 mm (3/4 in.) turnbuckles with rope thimbles
180 J-bolts
four anchor flag pins with turnbuckles
184.4 m (605 ft) of conveyor belt
miscellaneous rope clamps, bolts, and supplies

A supply tram or scoop tractor can deliver the materials as close to the extension site as possible.

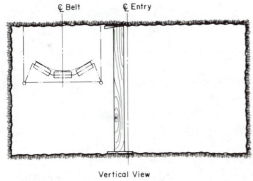

FIG. 16. Centerline of the belt offset from the entry centerline.

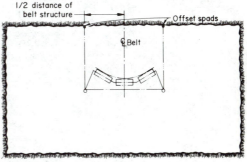

FIG. 17. Offset spads are used to mark the hanger and sailbase holes of the belt line.

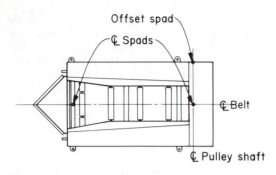

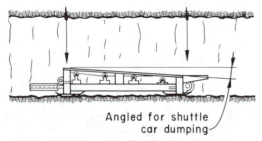

FIG. 18. Sight spads for tailpiece placement.

Once the materials are delivered, stringing the belt structure along the intended site is the next step. The troughing idlers and return idlers are placed along the entry in their approximate position, as shown in Fig. 19. The anchor sails are placed near the bases for quick mounting. The flag pins are delivered as a backup in the event the pins currently being used cannot be reused. The overall stringing operation tries to place all the materials in the best position for installation.

The final task consists of making up the ends of the new piece of belt. This requires putting splices on one or both ends of the 182.9 m (600 ft) belt to be added as the extension. Mechanical splices come as rivets (shown earlier), staples, or bolted fasteners, while vulcanized splices for belts provide an alternative method. For panel belts of relatively short distance and low useful life, mechanical fasteners are satisfactory. In mainline belts, however, vulcanizing belt splices may be more cost effective, in addition to being more durable. An important point to remember on splicing is to have a good belt edge to attach the fastener to. This means no rips or ragged ends, and most importantly, the end is at right angles to the sides of the belt (Fig. 20). If, in some instances, belts of unequal widths are spliced together, the centerlines of both belts should match, leaving the outside edges unequal.

Splicing is often the cause of belt trouble. Poor makings of either end of a belt can create a "kink" which is continually throwing the belt out of alignment.

Once these preparatory tasks are complete, mine management can schedule the belt extension. The previous preparation tasks have made it possible to easily complete the move-up in one shift (barring unforseen circumstances).

Belt Line Assembly

Belt line assembly is done with a crew of four workers, some hopefully, with prior experience. The tasks included in this step are: threading the belt, stringing and tensioning the ropes, assembling the structure, and connecting the auxiliary safety systems. Although these tasks are presented as a separate installation step, their timely completion depends partially on the other belt crew breaking down and moving the tailpiece.

Threading the Belt: The first task for the belt assembly crew is the threading of the new belt. This task is coordinated with the tailpiece crew, since the tailpiece will pull the new belt from the winder along the belt line.

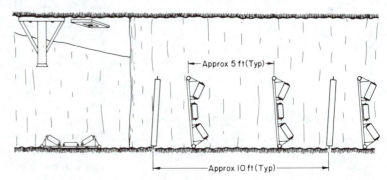

FIG. 19. The stringing of belt line structure before the extension. Metric equivalent: ft × 0.3048 = m.

BELT LINE CONSTRUCTION

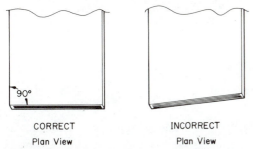

FIG. 20. Belt edge must be at right angles for belt splicing.

The first step is for the tailpiece crew to break down the tailpiece in terms of anchorage and belt line attachments. This will be discussed in the next major section of this chapter.

Concurrent to the work of the tailpiece crew, the belt line crew spots a splice in the existing belt where it is convenient to break it open and add the new piece, and then releases the belt takeup unit (if automatic).

The belt line crew anchors the return belt with clamps on the side of the splices away from the tailpiece. The connecting pin is released from the splice, freeing the ends for splicing. Hopefully, enough belt has been measured to allow a few feet of slack for splicing after the belt has been pulled.

The new piece of belting is then spliced to the tailpiece side of the splice. It is important to be sure the carrying side of the new belt is spliced on the lower side of the return (the carrying side has a thicker cover), as shown in Fig. 21.

The tailpiece crew will then pull the tailpiece with a scoop, feeder, shuttle car, or track motor, and in doing so, will thread the belt from the winder (when the winder spools if off) through the tailpiece. The belt will lay on the mine bottom in the approximate position needed for later work.

Stringing and Tensioning the Ropes: The next task for the belt line crew is stringing the wire rope frame and tensioning the ropes to their designed load. To do this, the following steps are required: mounting and adjusting the anchor sails, anchoring one end of each rope and pulling the slack out from the other, anchoring the free end of each rope, hanging the chains and supporting the rope, and final tensioning of the belt. It is best for operating performance if the four man crew is broken into two groups of two workers, placed on each side of the belt line.

Mounting and adjusting the anchor sails on the sail plates is the first task. The sails are bolted securely to the plates previously bolted to the roof during the preparation task. Two sets of anchor sails are mounted for a 91.4 m (300 ft) extension [assuming 45.7 m (150 ft) spans]. Adjusting the sails defines the distance from the roof that the ropes will be strung. The center post of the anchors can be adjusted so the belt line ropes stay relatively level to previous extensions and parallel to seam dimensions (Fig. 22).

The second task is anchoring the rope at one end and pulling the slack out of the rope from the free end. To anchor one end, the rope is wrapped around a thimble and secured with a rope clamp. One end of a turnbuckle is inserted into the eye of the anchor sail, while the other is attached to the loop of rope. The rope is then positioned under the post end to establish the position of the rope. The completed anchorage system is conceptualized in Fig. 6. The free end of the rope is looped and clamped loosely, allowing the rope to slide around the thimble when

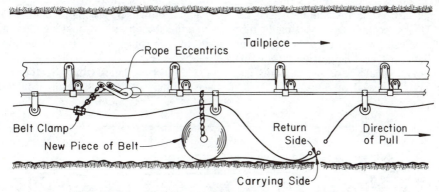

FIG. 21. Correctly splicing the new belt into the return of the existing belt.

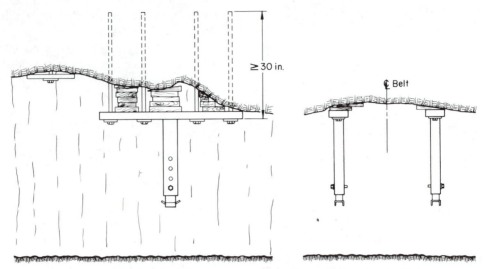

FIG. 22. Blocking and leveling sailbase anchors. Metric equivalent: in. × 25.4 = mm.

jacked. The looped end is hung on a turnbuckle which in turn is hung from the eye of the anchor sail. A chain jack and two rope eccentrics are used to take the slack out of the rope. A tensioning device can be used to achieve the desired tension.

To anchor the free end, the rope clamps around the thimble are tightened to hold the belt. Reversing the position of the clamps so pinching forces are exerted in opposite directions is recommended.

The next step is hanging the support chains from the chain hanger plates bolted to the roof. The hookless chain ends are dropped into the *key* slot which is cut (or stamped) into the hanger. The chain is then looped under the rope and is either bolted or hooked back up to itself. In doing this, a snug supporting fit between the rope and chain is achieved (Fig. 23).

The final task is the fine tuning of the belt tension. This is accomplished by using the turnbuckle for adjustment (they should be open during installation). Once the tension is satisfied, the turnbuckles should be secured by stiff wire running from each end of the turnbuckle and crossing in the middle to form a figure 8. This helps keep the turnbuckle from loosening due to vibration creep.

Once the ropes are up and secured, the structure can be placed. This task is enhanced by the preparation task which strung the structure in the approximate locations along the extension entry.

Assembling the Structure: With the ropes in place, the troughing and return idlers can now be placed. This is done by sliding the troughing idlers through the middle of the belt and lifting the carrying side up. The troughing idlers are secured to the rope with J-bolts. A measuring stick can be used to obtain the proper lateral spacing. Troughing idlers are placed on the entire length of the belt line extension. The return idlers are then placed under the return side of the belt and secured to the rope (either with hangers or hangers and short chains). One point to remember: if offset troughing idlers are used, the center roller should be placed toward the tailpiece (Fig. 24).

Connecting the Auxiliary Systems: The final task of the belt line crew before troubleshooting

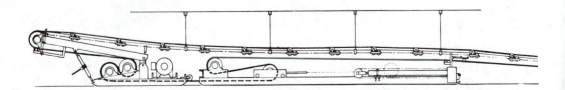

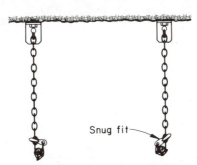

FIG. 23. Chains wrapped around belt ropes producing a snug fit.

the running belt is the installation of the auxiliary systems. This includes connecting the water pipe along the length of the extension, flushing it out, and connecting the required valves. The sensing and fire alarm systems need to be strung along the belt line to ensure adequate safety. Insulating hooks placed on the chains can be used to support these lines. In addition, many mines run communication, control, and other lines along the belt line for convenience. These must be extended to the new location if time permits.

Tailpiece Assembly

Up until now, only the work of the belt line crew has been discussed. Here the tailpiece crew and their tasks are explored. The tasks ahead of this crew are breaking down the tailpiece, pulling the tailpiece, aligning the tailpiece, drilling and anchoring the tailpiece, and attaching the tailpiece to the belt line. All of these tasks are accomplished by two workers, plus occasional additional help.

Breaking Down the Tailpiece: The initial effort of the tailpiece crew goes into breaking down the tailpiece, which means getting it ready to move to the new location. If a feeder is present, then it must also be moved (luckily, most feeders are self-tramming).

The first steps for the crew consist of loosening the turnbuckles of the wire ropes between the tailpiece and last set of anchor sails, usually less than 4.6 m (15 ft). With the tension off the ropes, the clamps can be loosened and the ropes detached from the anchor stands (Fig. 25). Concurrent to this, the turnbuckles between the anchor pins and the tailpiece are loosened and detached from the tailpiece. The anchor pins are pulled up and saved for the new setup.

Another task during breakdown is the recovery of any slip and spill switches, leveling timbers, shovels, picks, jacks, and other accessories needed at the new location.

The final task is getting the tailpiece ready to move and moving the belt. This means getting a chain or short rope and attaching the feeder, scoop, or shuttle car to the tailpiece frame (using a Y connection). If a track motor is used, a block and tackle setup is used to pull the tailpiece into the new position (Fig. 26).

Pulling the Tailpiece: The tailpiece is then pulled slowly to the new position. Along with the tailpiece, the new belt is pulled into position by the crew as a coordinated effort with the belt line extension crew splicing in the new belt. The tailpiece is moved to the new location and pulled into rough alignment under the orientation spads. It should be noted that when soft bottoms are present, the tailpiece will have a tendency to dig into the floor. The piling of gob in front of the tailpiece can effectively block the efforts of the crew to pull it forward. Therefore, a construction supervisor should try and raise

FIG. 24. Offset troughing idlers should point to the tailpiece (courtesy of Hewitt-Robins Conveyor Equipment Division).

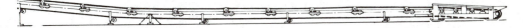

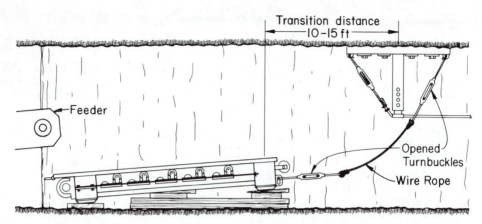

FIG. 25. Loosened turnbuckles at the transition point to move tailpiece. Metric equivalent: ft × 0.3048 = m.

the tailpiece front to eliminate this problem. The raised bucket of a scoop or conveyor boom on a shuttle car can aid quite nicely in most cases (Fig. 27).

Aligning the Tailpiece: The next step is positioning the tailpiece under the sight spads. Roof and track jacks are used to give finer adjustment of the tailpiece. Ideally, the tailpiece is in a position similar to the last setup, so the same wire ropes can be used and anchoring will not require any makeshift work.

In aligning the tailpiece, an effort should be made to level the tailpiece from side to side and slightly elevate the front of the tailpiece for better transition to the roof-hung belt line (Fig. 28).

Drilling and Anchoring the Tailpiece: Once the tailpiece is in position, anchorage holes for the pins are drilled. As mentioned previously in Chap. 5, the anchor pins are usually 31.8 mm (1¼ in.) solid steel pins with an *eye* welded to one end and 609.6 to 762 mm (24 to 30 in.) long. A rotary air drill hooked to an air compressor (for power) is used to drill a hole 0.9 m (3 ft) deep. Four holes are drilled, one at each corner, in such a position that the turnbuckle arrangement will exert a force against the pull of the belt line, as shown in Fig. 29. The holes are angled slightly away from the tailpiece to aid in anchoring the structure.

The next step is to open up the turnbuckles and attach them to the anchor pins in the holes and the tailpiece. Care must be taken to ensure the tailpiece stays in alignment during the tightening of the turnbuckles.

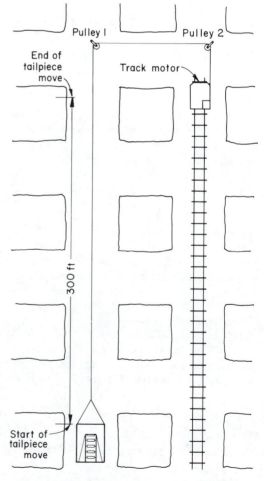

FIG. 26. Block and tackle setup for belt move. Metric equivalent: ft × 0.3048 = m.

BELT LINE CONSTRUCTION

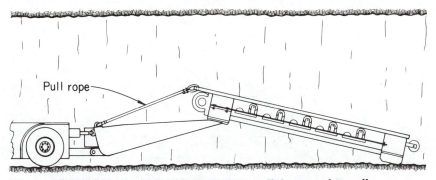

FIG. 27. Scoop tractor lifts the tailpiece off the ground to pull.

Attaching the Tailpiece to the Belt Line: The final task of the tailpiece assembly step is reanchoring the tailpiece to the belt line. If designed and positioned correctly, the wire ropes used to attach the belt to the tailpiece in the previous setup can be reused. The looped end of the rope loosened for breakdown is reattached to the turnbuckle and anchor stand. A chain jack is used to pull the slack and put tension in the short rope. The rope clamps are then retightened to hold the tension.

In many cases in an underground setup, no troughing structure is placed in this transition area between the tailpiece and first anchor stand. In fact, the return structure usually consists of the slip switch or training roller only. One of the reasons for this absence of structure on the carrying side is the angle required to lift the belt to the rope height. Often, the idlers cannot be placed on the ropes because the ropes are too wide or the conveyor belt does not have enough slack in the carrying side. However, the potential damage to the belt in this 4.6 to 6.1 m (15 to 20 ft) span does not appear serious unless there is a high percentage of rock in the ROM material mix.

Operation and Maintenance

At this point, the belt is ready for various operation and maintenance measures. These include (1) training the belt, (2) lubricating the belt and tailpiece idlers, (3) extension cleanup, and (4) building the belt line accessories.

Training the belt includes several steps of troubleshooting and inspection activities. First, the takeup is activated and the slack is taken out of the belt. A foreman should check the takeup pressure to ensure the proper tension. Once the belt is running, it usually must be trained to run straight. However, the better the belt line installation, the less training of the belt is required.

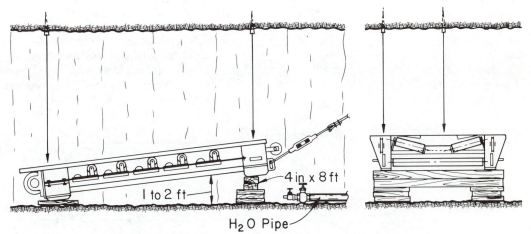

FIG. 28. Aligning the tailpiece in the transition (side to side). Metric equivalents: in. × 25.4 = mm; ft × 0.3048 = m.

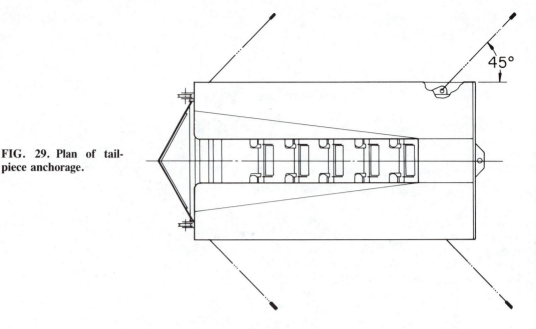

FIG. 29. Plan of tailpiece anchorage.

If training idlers have been installed, then manual training is minimal. These idlers pivot from a central point on the frame. The misaligned belt traverses these idlers and is urged back to the center of the belt line.

If manual training of the troughing idlers is done, then one starts from the tailpiece and walks one side of the belt to the belt head. Where the belt is going out of alignment, a foreman or worker must manually pivot several troughing idlers to train the belt back in line. The important point to remember is not to try and make the entire adjustment with one idler. A dozen idlers should be adjusted slightly to make the same adjustment (Fig. 30). Skewing a training idler in either direction results in the following adjustments:

1) If you skew one side *toward* the belt head, the belt will climb the opposite side.

2) If you skew one side *away* from the belt head, the same side of the belt will climb up the troughing side.

Almost anyone can be trained to adjust a belt line, however, too much training is as bad as too little. An effort should be made to balance the belt training desires of management and the training needs of the belt line.

The second operation and maintenance measure is lubrication of the belt idlers and tailpiece idlers. The roller bearings are lubricated as per the manufacturer's suggestions. These preventative maintenance procedures should be strictly followed to increase the useful life of the belt. A regular schedule should be set up and followed closely.

Another operation and maintenance measure is cleaning up the extension area after the move. Here, excess tools, materials, and trash which can cause stumbling hazards are picked up and removed. The entire area should be rock dusted

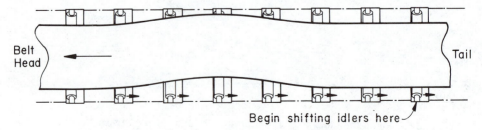

FIG. 30. Training the belt by slightly adjusting many idlers.

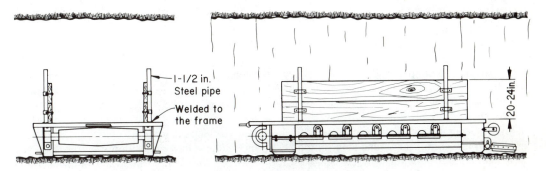

FIG. 31. Tailpiece bunker. Metric equivalent: in. × 25.4 = mm.

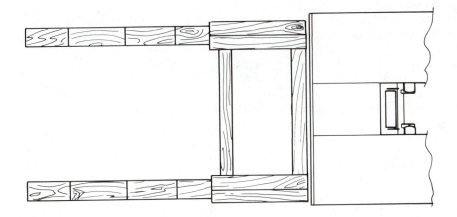

Plan View

Vertical View

FIG. 32. Feeder ramp built of timbers and header boards.

heavily to comply with federal laws.

The final measures concern the belt line accessories of crossovers, tailpiece bunkers, and setting of the feeder on the tailpiece.

Crossovers are simply bridges across a belt line. It is extremely unsafe for anyone to attempt to cross a moving belt line. However, many employees will do just this if a crossover is not in sight. Generally, one crossover every 213.3 m (700 ft) is recommended, and they should be well marked and built for easy crossing (i.e., wide enough and tall enough for the average worker). Unfortunately, human factors engineering is not always considered when a crossover is designed. Crossovers are always used more when placed in areas where frequent crossings are expected (by doors, pumps, transformers, and so on).

A second accessory is the tailpiece surge bunker. While it is referred to by several names, it generally consists of sides built on the tailpiece to cut down on spillage. Often rocks get jammed as they fall from the feeder, causing a pileup until a spill switch turns off the belt. Sides built on the tailpiece 1.2 to 1.5 m (4 to 5 ft) high over the length of the frame will allow an operator to handle smaller sized jams without spillage (Fig. 31). Locating the spill switch inside the bin sides can readily contain any major spill. This design is especially useful for 0.9 and 1.1 m (36 and 42 in.) production panel belts where large surges of coal can be expected.

The final accessory topic is the placement of the feeder on the tailpiece. This requires a certain skill in equipment maneuvering, since the centerline of the feeder should correspond to the centerline of the belt. In most cases, a ramp is built to allow the feeder to walk up and be in position (Fig. 32). The centerline spads of the tailpiece should serve as guides in positioning the feeder. Once in position, the transition area between the feeder throat and tailpiece frame should be blocked to reduce spillage and dragging of fines under the feeder (Fig. 33).

REFERENCES AND BIBLIOGRAPHY

Anon., 1982, "Rivet Fastener Belt Splices Reduce Conveyor Downtime," *Coal Mining and Processing*, Vol. 19, No. 11, Nov., p. 59.

Conveyor Equipment Manufacturers Association, 1976, *Belt Conveyors for Bulk Materials*, CBI Publishing Co., Boston, MA.

Crickmer, D. F., and Zegeer, D. A., eds., 1981, *Elements of Practical Coal Mining*, 2nd ed., AIME, New York.

Cummins, A. B., and Given, I. A., 1973, *SME Mining Engineering Handbook*, AIME, New York.

Flexco, Flexible Steel Lacing Co., Bulletin SR-100, Downers Grove, IL.

Hewitt-Robins Inc., 1980, *Hewitt-Robins Idlers*, Catalogue 1200-80, Passaic, NJ.

Hewitt-Robins Inc., 1979, *Wire Rope Conveyors*, Catalogue 1104-79, Passaic, NJ.

Jeffrey Manufacturing Division, Dresser Industries, Inc., 1980, *Jeffrey Belt Idlers*, Catalogue 1156-R, Belton, SC.

FIG. 33. Feeder-tailpiece junction blocked with rock dust bags to contain any spillage (courtesy of Hewitt-Robins Conveyor Equipment Division).

EQUIPMENT MANUFACTURERS

There are many different choices to be made among belt components. A majority of manufacturers can supply a wide array of technical and service advice for installing and maintaining an efficient belt line.

Table 4 is a listing of all manufacturers of belt equipment discussed in Chaps. 5 and 6. This list is taken from the annual *Buyers Guide*, published by McGraw-Hill Mining Services.

TABLE 4. Equipment Manufacturers, by Product Type

Air Compressors

1. portable, self-propelled, mine
2. portable, self-propelled, surface
3. stationary
4. rotary

Acme Machinery Co.
Airtek, Inc. (3, 4)
Alemite & Instrument Div., Stewart-Warner Corp. (3)
Allis-Chalmers (3, 4)
Atlantic Track & Turnout Co. (2)
Atlas Copco, Inc. (1, 3, 4)
Atlas Copco MCT AB (1, 2, 3, 4)
Berry Davidson Co. (3, 4)
Chicago Pneumatic Equipment Co. (3, 4)
CompAir Construction & Mining Ltd.
Cypher Co., The (1, 2)
Davey Compressor Co. (1, 2, 3, 4)
Dover Conveyor & Equipment Co., Inc. (3, 4)
Dresser Industries, Inc. (3, 4)
Dresser Industries, Inc., Drilling Equipment (3, 4)
Fuller Co., A Gatx Co. (3, 4)
GS Industries, Inc. (2, 3)
Gardner-Denver/Cooper Ind., Mining & Construction Group (1, 2, 3, 4)
Gardner-Denver, Industrial Machinery Div. (3, 4)
Guyan Machinery Co. (3, 4)
Ingersoll-Rand Co. (3, 4)
Iowa Mold Tooling Co., Inc.
Jaeger Machine Co. (3)
Jeffrey Mining Machinery Div., Drilling Equipment (1, 2, 3, 4)
Joy Mfg. Co. (3, 4)
Le Roi Div., Dresser Industries, Inc. (3, 4)
Logan Corp. (1, 2, 3)
Long-Airdox Co. A Div. of the Marmon Group, Inc. (3)
M.A.N. Maschinenfabrik Augsburg-Nurnberg, Div. GHH Sterkrade (3, 4)
Midway Equipment, Inc. (1, 2, 3, 4)
Nash Engineering Co. (3, 4)
Permco Inc. (4)
Quincy Compressor Div., Colt Industries (3, 4)
Rish Equipment Co. (1, 3, 4)
Schramm Inc. (2, 3, 4)
Sepor, Inc. (2, 3, 4)
Siemens Corp. (3, 4)
Strojexport, pzo (3)
Sullair Corp. (1, 2, 3, 4)
Tamrock Drills (3)
Tidewater Supply Co. (1, 2, 3, 4)
Wilson, R. M., Co. (2)
Worthington Pump Inc.

Bearings

1. ball
2. bronze, babbitt
3. carbon
4. journal
5. needle
6. roller
7. sleeve
8. thrust
9. retainers

A.C.R. Equipment Co. Inc., Parts Div. (2)
Abex Corp. (2, 4, 7, 8)
Abex Corp., Friction Products Group (7, 8)
Abex Corp., Railroad Products Group (2, 4, 7, 8)
Ampco Metal Div., Ampco-Pittsburgh Corp. (2)
Banner Bearings (1, 2, 4, 5, 6, 7, 8)
Bearing Service Co. (1, 5, 6, 8)
Bearings, Inc. (1, 2, 4, 5, 6, 7, 8)
Big Sandy Electric & Supply Co., Inc. (1, 6)
Browning Mfg. Div., Emerson Electric Co. (1, 6)
Bruening Bearings, Inc. (1, 2, 4, 5, 6, 7, 8)
Burrell Construction & Supply Co.
Continental Conveyor & Equipment Co., Inc. (6, 8)
Dixie Bearings, Inc. (1, 2, 4, 5, 6, 7, 8)
Dresser Industries, Inc. (4)
ELMAC Corp. (2, 8)
FAG Bearings Corp. (1, 5, 6, 8)
FMC Corp., Bearing Div. (1, 2, 6, 7)
Fafnir Bearing, Div. of Textron (1, 4, 6)
Fairmont Supply Co. (1, 2, 5, 6, 7, 8, 9)
Federal-Mogul Corp. (1, 2, 4, 5, 6, 7, 8)
Guyan Machinery Co. (1, 2, 5, 6, 7, 8)
Hanco International Div. of Hannon Electric Co. (2)
Helwig Carbon Products, Inc. (3)
Huwood-Irwin Co. (1, 6)
Ingersoll-Rand Co. (1, 5, 6, 8)
Jeffrey Mfg. Div., Dresser Industries Inc. (1, 2, 5, 6, 7, 8)
Keene Corp. (1, 5, 6)
Leman Machine Co. (2)
Logan Corp. (2, 6, 7)
Marathon Coal Bit Co., Inc. (1, 5, 6, 7, 8)
McNally Pittsburg Mfg. Corp. (6)
Morse Chain, Div. of Borg-Warner Corp. (1, 6)
NL Bearings/NL Industries (2, 4, 7, 8)
NL Industries, Inc. (2, 6, 7, 8)
National Mine Service Co. (1, 6)
Ohio Carbon Co. (3, 7, 8)
Persingers Inc.
Phelps Dodge Industries, Inc. (2)
Poly-Hi/Menasha Corp. (6, 7)

TABLE 4. (continued)

Portec, Inc., Cast Products Div. (2)
Power Transmission Equipment Co. (1, 2, 4, 5, 6, 7, 8, 9)
Pure Carbon Co., Inc. (3, 7, 8)
Rexnord Inc. (1, 6, 7)
Rexnord Inc., Bearing Div. (6, 7)
SKF Industries, Inc. (1, 5, 6, 8)
Stoody Co., WRAP Div. (7)
Timken Co., The (6, 8)
Torrington Co., The Bearings Div. (1, 4, 5, 6, 8)
TRW Bearings Div. (1, 6)
Union Carbide Corp. (3)
Wescott Steel Inc. (7, 8)
West Virginia Armature Co. (2, 4)
Wilson, R. M., Co. (1, 5, 6, 8, 9)

Belt-Loading Stations, Automatic

Aggregates Equipment Inc.
Davco Corp.
Dowty Corp.
Dravo Wellman Co.
FMC Corp., Materials Handling Systems Div.
Fairfield Engineering Co.
Hartman-Fabco Inc.
Huwood-Irwin Co.
Jeffrey Mfg. Div., Dresser Industries Inc.
Jim Pyle Co.
Jold Mfg. Co., Inc.
KHD Industrieanlagen AG, Humboldt Wedag
Kennedy Van Saun Corp.
Lancaster Steel Products
Mannesmann Demag Baumaschinen Bereich Lauchhammer
Marathon Steel Co.
McNally Pittsburg Mfg. Corp.
Mintec/International, Div. of Barber-Greene
PHB Material Handling Corp.
Process Equipment Builders, Inc.
RAHCO, R. A. Hanson Co.
Rexnord Inc.
Schroeder Bros. Corp.
Stamler, W. R., Corp., The
Watt Car & Wheel Co.
Webb, Jervis B., Co.
Webster Mfg. Co.
Willis & Paul Corp., The
Wilson, R. M., Co.

Controllers

1. electric & parts
2. electric-trackswitch & derail
3. electronic
4. hydraulic
5. locomotive
6. pneumatic
7. temperature
8. signal
9. solid state

AMF Inc. (3)
Acurex Corp., Autodata Div. (3)
Airtek, Inc. (2, 3, 4, 5, 6, 7, 8, 9)
Allen-Bradley Co. (1, 7)

Alnor Instrument Co. (7)
American Meter Div., Singer Co., The (6, 7)
Anixter Mine & Industrial Specialists (1)
ARO Corp. (6)
Astrosystems, Inc. (3)
A-T-O Inc. (1, 3)
Babcock & Wilcox (3, 6, 7)
Barnes Engineering Co. (7)
Baylor Company (1, 3, 4)
Beckman Instruments, Inc. (1, 3, 6, 7)
BIF, a unit of General Signal (3, 4, 6)
Bristol-Babcock Div., Acco Industries Inc. (3, 6, 7)
British Coal International
Cashco, Inc. (6, 7)
Compton Electrical Equipment Corp. (1)
Continental Conveyor & Equipment Co., Inc. (1, 9)
Control Products, Inc. (1, 3)
Controlled Systems Inc. (3, 5)
Cross Electronics Inc. (1, 3, 7, 8, 9)
Cutler-Hammer Products, Eaton Corp. (1, 3, 9)
Dana Industrial, Dana Corp. (3, 9)
Davis, John & Son (Derby) Ltd. (7)
Downard Hydraulics, Inc. (4)
Electrical Automation (1)
ELMAC Corp. (1, 3, 9)
Ensign Electric Div., Harvey Hubbell Inc. (1)
Fairmont Supply Co. (1)
Fisher Controls Co. (3, 6, 9)
Flight Systems, Inc. (1, 9)
Foxboro Co., The (1, 3, 6, 7, 9)
GTE Products Corp., Electrical Equipment (1, 3)
General Electric Co. (1, 3)
General Equipment & Mfg. Co., Inc. (1, 2, 3, 6)
Goodman Equipment Corp. (5)
Guyan Machinery Co. (9)
HB Electrical Mfg. Co., Inc. (1)
Honeywell Inc., Process Control Div. (3, 5, 7, 9)
Jeffrey Mining Machinery Div., Dresser Industries Inc. (1, 5)
Leeds & Northrup, a unit of General Signal (3, 7)
Line Power Manufacturing Corp. (1, 9)
Louis Allis Div., Litton Industrial Products, Inc. (3)
Mag-Con, Inc. (7)
Mining Controls, Inc. (1, 3, 9)
Montgomery Elevator Co. (9)
Moog Inc., Industrial Div. (3, 4)
Morse Chain, Div. of Borg-Warner Corp. (3)
National Mine Service Co. (3, 5, 9)
North American Mfg. Co. (4)
OCENCO, Inc.
Ohio Thermal, Inc. (7)
Owens Mfg., Inc. (1)
Pace Transducer Co., Div. of C.J. Enterprises (4, 6)
Preiser/Mineco Div., Preiser Scientific Inc. (3, 6, 7)
Process Equipment Builders, Inc. (3, 4, 6)
Reliance Electric (1, 3, 7, 8, 9)
Robicon Corp. (1, 3, 7)
S & S Corp. (1, 9)
Safetran Systems Corp., Mining & Urban Transit, Signal & Control Div. (2, 8)
Sevcon, Div. Technical Operations, Inc. (1, 3, 5, 9)
Siemens-Allis Inc. (1, 3)

TABLE 4. (continued)

Square D Co. (1, 3, 7, 9)
Taylor Instrument Co. (3, 6, 7)
U.S. Electrical Motors Div. Emerson Electric Co. (1, 3, 9)
Vanan Associates (7)
Wallacetown Engineering Co. Ltd. (1)
Webb, Jervis B., Co. (3)
West Virginia Armature Co. (1, 5)
Westinghouse Electric Corp. (1, 3, 6, 7)
Wichita Clutch Co., Industrial Group, Dana Corp. (6)
Wiegand, Edwin L., Div., Emerson Elec. Co. (7)

Controls

1. cable-type
2. inductive-carrier-remote
3. liquid-level
4. solids-level
5. static
6. remote, automatic, R.R.
7. conveyor
8. remote, underground machines
9. slurry monitoring

Airtek, Inc. (6, 8)
Allen-Bradley Co. (3)
American Meter Div., Singer Co., The (2, 3, 5, 6)
Armco Autometrics (9)
Automation Products, Inc. (3, 4)
Babcock & Wilcox (3)
Baylor Company (7)
Bertea Industrial Products (8)
Bindicator (3, 4)
Bristol-Babcock Div., Acco Industries Inc. (3)
British Coal International
CWI Distributing Co. (3)
Communication & Control Eng. Co. Ltd. (3, 5, 7)
Compton Electrical Equipment Corp. (1, 3, 4, 7)
Continental Conveyor & Equipment Co., Inc. (7)
Control Products, Inc. (1, 7)
Controlled Systems Inc. (5, 7)
Conveyor Components Co. (1, 4, 7)
Cross Electronics Inc. (1, 5, 8)
Crouse-Hinds Co. (7)
Cutler-Hammer Products, Eaton Corp. (4, 5, 6, 7)
Davis, John & Son (Derby) Ltd. (6, 7, 8)
Delavan Electronics, Inc. (3, 4)
Delavan Corp. (3, 4)
Diversified Electronics, Inc.
Downard Hydraulics, Inc. (8)
Eaton Corp., World Headquarters (7)
Eaton Corp., Industrial Drives Operations (7)
Eickhoff America Corp. (8)
Electric Machinery Mfg. Co. (3)
Electrical Automation (3, 4, 7)
ELMAC Corp. (7)
Endress & Hauser, Inc. (3, 4)
FMC Corp., Material Handling Equipment Div. (4)
Fairfield Engineering Co. (6, 7)
Fisher Controls Co. (3, 4)
Foxboro Co., The (3, 9)
Fuller Co., A Gatx Co. (4)
GTE Products Corp., Electrical Equipment (5, 7)
General Electric Co. (3, 5, 7)
General Electric Co., Gen. Purpose Control Dept. (7)
General Electric Co., Transportation Systems Business Div. (6)
General Equipment & Mfg. Co., Inc. (6, 7)
General Resource Corp. (7)
Gould Inc., Distribution & Controls Div., R.B. Denison (3, 4, 7)
Grad-Line, Inc. (1, 6, 8)
Gundlach, T. J., Machine Div., Rexnord Inc.
Guyan Machinery Co. (3)
Hawker Siddeley Dynamics Engineering Limited (6, 7, 8)
Honeywell Inc., Process Control Div. (3)
Huwood-Irwin Co. (7)
Huwood Limited (6, 7)
Jabco, Inc. (6, 7)
Jeffrey Mining Machinery Div., Dresser Industries Inc. (7)
Joy Mfg. Co., Denver Equipment Div. (3)
Kay-Ray Inc. (3, 4)
Leeds & Northrup, a unit of General Signal (3, 9)
Line Power Manufacturing Corp. (1, 5, 7)
Louis Allis Div., Litton Industrial Products, Inc. (5)
3 M Co. (5)
Mag-Con, Inc. (3, 4, 7)
Martin-Decker (4)
Metal Craft Inc. (7)
Micro Switch, A Div. of Honeywell
Mid-West Conveyor Co. (7)
Milltronics Inc. (3, 4)
Mining Controls, Inc. (1, 7)
Monitor Manufacturing (4)
Moog Inc., Industrial Div. (1, 6, 8)
Muncie Power Products (1)
National Electric Coil Div. of McGraw-Edison Co. (3, 6)
National Mine Service Co. (8)
Ohmart Corp. (3, 4)
P&R Systems Corp. (8)
Pace Transducer Co., Div. of C.J. Enterprises (3)
Preiser/Mineco Div., Preiser Scientific Inc. (3, 4, 7)
Reliance Electric (5, 7)
Revere Corp. of America (3)
Rexnord Inc.
Robicon Corp. (5, 7)
Safetran Systems Corp., Mining & Urban Transit, Signal & Control Div. (6)
Siemens-Allis Inc. (5)
Square D Co. (3, 7)
Stevens International Inc. (3, 4)
Taylor Instrument Co. (3)
Texas Nuclear (3, 4)
Union Switch & Signal Div., Amer. Standard Inc. (6)
Wallacetown Engineering Co. Ltd. (6, 7, 8)
Weatherhead Div., Dana Corp. (1)
Webb, Jervis B., Co. (7)
WESMAR Industrial Systems Div. (3, 4)
West Virginia Armature Co. (7)

Conveyor-Belt Parts, Services

1. clamps
2. cleaners
3. cleats
4. cold vulcanizing

TABLE 4. (continued)

5. cutters
6. drive pulleys
7. fasteners, splicing materials
8. idler pulleys
9. loading stations, mine, automatic
10. repair kits
11. repair material
12. repair service
13. splicing, shop & field
14. tighteners
15. trippers
16. vulcanizers
17. winders
18. control switches
19. steel cord reinforcement
20. guides

Aggregates Equipment Inc. (7, 8)
Albright Mfg. Co., Inc. (7)
Anderson Mavor (USA) Ltd. (8)
BTR Trading, Inc. (7, 10, 11)
Babcock Hydro-Pneumatics Ltd. (15)
Banner Bearings (6, 8)
Barber-Greene Co. (2, 6, 8, 15)
Bearings, Inc. (2, 6, 8, 14)
Bekaert Steel Wire Corp. (19)
Big Sandy Electric & Supply Co., Inc. (6, 7)
Bonded Scale & Machine Co. (2, 6, 7, 8)
Browning Mfg. Div., Emerson Electric Co. (6, 8)
Bruening Bearings, Inc. (2, 6, 8, 14)
C-E Raymond (8, 15)
C-E Raymond, Combustion Engineering, Inc. (8, 15)
Cincinnati Rubber Mfg. Co. Inc. (3, 10, 11)
Clouth Gummiwerke Aktiengesellschaft (11)
Compton Electrical Equipment Corp. (18)
Concrete Equipment Co., Inc. (6, 8)
Continental Conveyor & Equipment Co., Inc. (2, 6, 8, 15)
Control Products, Inc. (18)
Conveyor Components Co. (2, 18)
Crouse-Hinds Co. (18)
Dick Inc., R. J. (6, 8)
Dixie Bearings, Inc. (2, 6, 8, 14)
Dodge Div., Reliance Electric Co. (6, 8)
Dowty Corp. (2, 6, 8, 9)
Dresser Industries, Inc. (2, 6, 8, 15)
Durex Products, Inc. (2, 8, 10)
Eaton Corp., World Headquarters (6, 8)
Eaton Corp., Industrial Drives Operations (6, 8)
ELMAC Corp. (2, 5, 6, 7, 8, 18)
FMC Corp., Material Handling Equipment Div. (6, 8)
Fafnir Bearing, Div. of Textron (8)
Fairmont Supply Co. (6, 7, 8, 14, 18)
Fenner America Ltd. (2, 5, 7, 13)
Fenner International Ltd. (2, 5, 7, 13)
Fletcher Sutcliffe Wild, Ltd. (6, 8, 9)
Flexible Steel Lacing Co. (1, 2, 3, 5, 7, 11, 14)
Foamcraft, Inc. (2, 8)
GEC Mechanical Handling Ltd. (6, 8, 15)
General Electric Co. (18)
General Equipment & Mfg. Co., Inc. (18)
General Splice Corp. (1, 2, 4, 5, 7, 10, 11, 13, 16, 20)
Goodman Equipment Corp. (6, 8)

Goodrich, B.F., Co., Engineered Products Group (2, 6, 7, 8, 11, 13, 15, 16)
Goodyear Tire & Rubber Co. (7, 12, 13)
Gould Inc., Distribution & Controls Div., R.B. Denison (18)
Greengate Industrial Polymers Ltd. (13)
Guidler Co., The (20)
Guyan Machinery Co. (6, 9, 18)
Hammermills, Inc., Sub. of Pettibone Corp. (8)
Hartman-Fabco Inc. (1, 2, 6, 7, 8)
Hayden-Nilos Conflow Ltd. (2, 7)
Hewitt-Robins Conveyor Equipment Div. Litton Systems, Inc. (2, 8)
Holz Rubber Co., Inc., a Randtron Sub. (3, 6, 11)
Huwood-Irwin Co. (2, 6, 8, 15)
Huwood Limited (8)
Imperial Oil & Grease Co. (2)
Interroll Corp. (8)
Jabco, Inc. (18)
Jeffrey Mfg. Div., Dresser Industries Inc. (1, 2, 6, 8, 9, 14, 15, 20)
Joy Mfg. Co., Electrical Products (8)
Kennedy-McMaster Inc. (18)
Kolberg Mfg. Corp. (2, 6, 8)
Lancaster Steel Products (2, 6, 8, 9, 14, 15)
Leman Machine Co. (6, 12)
Linatex Corp. of America (2, 11)
Lippmann-Milwaukee, Inc. (2, 6)
Logan Corp. (1, 2, 6, 7, 8, 10, 15, 16)
Long-Airdox Co. A Div. of the Marmon Group, Inc. (7, 8, 9, 12, 13, 17)
Mag-Con, Inc. (18)
Manson Services, Inc. (1, 3, 4, 5, 6, 7, 8, 10, 11, 12, 13, 16, 17)
Marsh, E. F., Engineering Co. (6, 8)
Martin Engrg. Co. (2, 14)
Maschinenfabrik Buckau R. Wolf Aktiengesellschaft (15)
Material Control, Inc. (2)
MATO (7)
McNally Pittsburg Mfg. Corp. (6, 9, 15)
Metal Craft Inc. (6, 8, 11, 12)
Micro Switch, A Div. of Honeywell (18)
Midwest Industrial Supply Inc. (2)
Mine Equipment Co. (2)
Mining Controls, Inc. (18)
Molded Dimensions Inc. (2)
National Mine Service Co. (5, 7, 13, 16)
Owens Mfg., Inc. (6, 8, 14, 18)
PHB Material Handling Corp. (1, 6, 8, 9, 11, 12, 15)
Pipe Systems, Inc. (8)
Portec, Inc., Pioneer Div. (2, 6, 8, 14, 15)
Power Transmission Equipment Co. (1, 3, 6, 7, 8)
Preiser/Mineco Div., Preiser Scientific Inc. (2)
Pyott-Boone Electronics (18)
Ramsey Engineering, Co. (18)
Rema-Tech (1, 3, 4, 10, 11, 12, 13)
Rexnord Inc. (2, 6, 8, 9, 14, 15)
Rexnord Inc., Corp. Relations Dept. (6, 8)
Rexnord Inc., Process Machinery Div. (6, 8)
Rish Equipment Co.
Rung, D.G., Industries, Inc. (10, 11, 16)

TABLE 4. (continued)

Safetran Systems Corp., Mining & Urban Transit, Signal & Control Div. (18)
Scandura Inc. (3, 7, 10)
Schaefer Brush Mfg. Co. (2)
Sepor, Inc. (2)
Shaw-Almex Industries Ltd. (16)
Shingle, L.H., Co. (1, 7, 10, 11, 13)
Square D Co. (18)
Stephens-Adamson Inc. (2, 6, 8, 15)
Strojexport, pzo
Stryker Machine Products Co. (6, 8)
Templeton, Kenly & Co. (14)
Titan Mfg. Co. Pty. Ltd. (6, 8)
Transall Div., Dick-Precismeca, Inc. (2, 8)
Trelleborg AB (2)
Trelleborg, Inc. (2)
Unilok Belting Co., Div. of Georgia Duck and Cordage Mill (7)
Uniroyal, Inc.
U.S. Steel Corp.
Van Gorp Corp., Sub Emerson Elec. Co. (6, 8)
Vibco Inc. (2)
Vulcan Materials Co., Conveyor Belt Dept. (12, 13, 16, 17)
Wajax Industries Ltd. (1, 5, 7)
Wallacetown Engineering Co. Ltd. (18)
Webb, Jervis B., Co. (6, 8, 9, 14, 15, 18)
Webster Mfg. Co. (8, 9)
West Virginia Armature Co. (6, 8, 11, 12, 18)
West Virginia Belt Sales & Repairs Inc. (1, 2, 5, 6, 7, 8, 10, 11, 12, 13, 14)
Willis & Paul Corp., The (2, 15)
Wilson, R. M., Co. (1, 2, 3, 6, 7, 8, 9, 13)
Workman Developments, Inc. (2, 8)

Conveyor Belting

Aggregates Equipment Inc.
BTR Trading, Inc.
Bonded Scale & Machine Co.
Boston Industrial Products Div., American Biltrite Inc.
British Coal International
CRP Industries, Inc.
Celanese Fibers Marketing Co.
Central States Industries Inc.
Cincinnati Rubber Mfg. Co. Inc.
Clouth Gummiwerke Aktiengesellschaft
Concrete Equipment Co., Inc.
Cypher Co., The
Daniels, C. R., Inc.
Dick Inc., R. J.
Dowty Corp.
Dresser Industries, Inc.
Eagle Crusher Co., Inc.
Eaton Corp., Industrial Drives Operations
ELMAC Corp.
Fairchild Inc.
Fairmont Supply Co.
Fenner America Ltd.
Fenner International Ltd.
Flexowall Corp.
Goodall Rubber Co.

Goodrich, B.F., Co., Engineered Products Group
Goodyear Industrial Products
Goodyear Tire & Rubber Co.
Greengate Industrial Polymers Ltd.
Hartman-Fabco Inc.
Huwood-Irwin Co.
Indiana Steel & Fabricating Co.
Lancaster Steel Products
Logan Corp.
Long-Airdox Co. A Div. of the Marmon Group, Inc.
Manheim Mfg. & Belting
Manson Services, Inc.
National Mine Service Co.
Persingers Inc.
Power Transmission Equipment Co.
Rost, H. & Co.
Scandura Inc.
Semperit of America, Inc.
Sepor, Inc.
Shingle, L.H., Co.
TBA Industrial Products Ltd.
Tidewater Supply Co.
Trelleborg AB
Trelleborg, Inc.
Unilok Belting Co., Div. of Georgia Duck and Cordage Mill
Uniroyal, Inc.
U.S. Steel Corp.
Vulcan Materials Co., Conveyor Belt Dept.
Wajax Industries Ltd.
West Virginia Belt Sales & Repairs Inc.
Williams Patent Crusher & Pulv. Co.
Yokohama Rubber Co. Ltd., c/o Marubeni America Corp.

Conveyor-Pulley Lagging

Aggregates Equipment Inc.
Big Sandy Electric & Supply Co., Inc.
Bonded Scale & Machine Co.
Cincinnati Rubber Mfg. Co. Inc.
Concrete Equipment Co., Inc.
Dick Inc., R. J.
Dowty Corp.
FMC Corp., Material Handling Equipment Div.
Fairmont Supply Co.
General Splice Corp.
Goodall Rubber Co.
Goodrich, B.F., Co., Engineered Products Group
Goodyear Tire & Rubber Co.
Guyan Machinery Co.
Hartman-Fabco Inc.
Holz Rubber Co., Inc., a Randtron Sub.
Indiana Steel & Fabricating Co.
Lancaster Steel Products
Leman Machine Co.
Linatex Corp. of America
Logan Corp.
Manson Services, Inc.
Marsh, E. F., Engineering Co.
Metal Craft Inc.
Power Transmission Equipment Co.
Rema-Tech

TABLE 4. (continued)

Scandura Inc.
Trelleborg AB
Van Gorp Corp., Sub Emerson Elec. Co.
Vulcan Materials Co., Conveyor Belt Dept.
West Virginia Belt Sales & Repairs Inc.
Wilson, R. M., Co.

Conveyor Skirt Board

Aggregates Equipment Inc.
BTR Trading, Inc.
Bonded Scale & Machine Co.
Boston Industrial Products Div., American Biltrite Inc.
C-E Raymond
Cincinnati Rubber Mfg. Co. Inc.
Concrete Equipment Co., Inc.
Continental Conveyor & Equipment Co., Inc.
Conveyor Components Co.
Durex Products, Inc.
Fairmont Supply Co.
Foamcraft, Inc.
GEC Mechanical Handling Ltd.
Gates Rubber Co., The
Goodrich, B.F., Co., Engineered Products Group
Goodyear Tire & Rubber Co.
Guyan Machinery Co.
Hartman-Fabco Inc.
Holz Rubber Co., Inc., a Randtron Sub.
Indiana Steel & Fabricating Co.
Kanawha Mfg. Co.
Kolberg Mfg. Corp.
Lancaster Steel Products
Linatex Corp. of America
Marsh, E. F., Engineering Co.
Martin Engrg. Co.
Mine Equipment Co.
Molded Dimensions Inc.
Persingers Inc.
Poly-Hi/Menasha Corp.
Portec, Inc., Pioneer Div.
Schaefer Brush Mfg. Co.
Tidewater Supply Co.
Townley Engineering & Mfg. Co., Inc.
Trelleborg AB
Trelleborg, Inc.
Webster Mfg. Co.
West Virginia Belt Sales & Repairs Inc.
Willis & Paul Corp., The
Wilson, R. M., Co.
Workman Developments, Inc.

Conveyors

  1. apron
  2. armored longwall
  3. belt
  4. belt, extensible
  5. belt-feeding
  6. bucket
  7. bucket-wheel
  8. cable-belt
  9. chain & chain & flight
10. decline
11. dewatering
12. elevating
13. elevating, mine-transfer, car loading
14. mine bridge
15. mine, flexible-chain
16. chain, mobile-head
17. portable
18. rope & button
19. screw
20. sectional
21. shaking, vibrating
22. spiral lowering
23. stockpiling & recovery
24. stacker

AEC Inc. (2, 3, 9)
Acco Industries Inc., American Chain Div. (9)
Acco Industries, Inc., Mining Sales Div. (2, 6, 7, 9, 16, 23)
Aggregates Equipment Inc. (3, 6, 9, 17, 19, 21, 23)
Allis-Chalmers (3, 6, 12, 23)
American Alloy Steel, Inc.
Anchor Conveyors Div., Standard Alliance Indus., Inc. (1, 3, 6, 9, 12)
Anderson Mavor (USA) Ltd. (2, 3, 4)
Applied Mining Equipment & Services Co. (9, 12, 17)
A-T-O Inc.
Auto Weigh Inc. (3, 5)
Babcock Hydro-Pneumatics Ltd. (1, 3, 9, 11, 21, 23)
Babcock-Moxey Ltd. (3, 7, 23, 24)
Badger Construction Co., Div. of Mellon-Stuart Co. (3, 23)
Barber-Greene Co. (3, 10, 12, 17, 18, 23)
Bearings, Inc. (3, 6, 9)
Bolton R. B. (Mining Engineers) Ltd. (9, 15)
Bonded Scale & Machine Co. (1, 3, 5, 9, 12)
Bristol Steel & Iron Works, Inc. (3, 23)
British Coal International
British Jeffrey Diamond, Div. of Dresser Europe S.A. (U.K. Branch) (2, 9)
Bruening Bearings, Inc. (3, 6, 9)
C-E Raymond (3, 5, 8, 9, 10, 12, 23)
C-E Raymond, Combustion Engineering, Inc. (3, 6, 9, 12)
CM Longwall Mining Chain Div., Columbus McKinnon Corp. (2, 9, 15)
Cable Belt Conveyors Inc. (3, 8)
Cable Belt Ltd. (8)
Calweld, Div. of Smith International, Inc. (3, 17, 19)
Cambelt Intl. Corp. (3, 5, 10, 12, 13, 14, 17, 20, 23, 24)
Campbell Chain Co. (2, 9, 15)
Can-Tex Industries (3, 17)
Canton Stoker Corp. (19, 21)
Capital City Industrial Supply Co. (3)
Capital Conveyor of Columbus (1, 3, 5, 6, 9, 19, 23)
Card Corp. (3, 4, 8)
Carman Industries, Inc. (11, 21)
Cincinnati Mine Machinery Co. (9)
Columbia Steel Casting Co., Inc. (1)
Concrete Equipment Co., Inc. (3, 12, 17, 19)
Connellsville Corp. (1, 6, 9, 11, 21)
Continental Conveyor & Equipment Co., Inc. (3, 5, 7, 10, 13, 17, 20, 23, 24)

TABLE 4. (continued)

Crown Iron Works Co. (19)
Daniels Company, The (9)
Davco Corp. (3, 5, 23)
Davy McKee (Minerals & Metals) Ltd. (23)
Dayton Automatic Stoker Co. (19)
Deister Machine Co., Inc. (21)
Dixie Bearings, Inc. (3, 6, 9)
Dosco Corp. (4, 14)
Dover Conveyor & Equipment Co., Inc. (1, 3, 4, 5, 6, 8, 9, 12, 13, 14, 17, 19, 23)
Dowty Corp. (2, 3, 4, 5, 9, 14)
Dravo Corp. (6, 7, 14, 18, 23)
Dresser Industries, Inc. (1, 3, 5, 6, 8, 9, 11, 12, 13, 14, 15, 16, 17, 19, 20, 21, 22, 23)
Eagle Crusher Co., Inc. (17, 24)
Eickhoff America Corp. (2, 3, 4, 9, 15)
ELMAC Corp. (3, 20)
Eriez Magnetics (21)
ESCO Corp. (9)
FMC Corp., Material Handling Equipment Div. (12, 19, 21)
FMC Corp., Materials Handling Systems Div. (1, 3, 9, 10, 12, 14, 23)
Fairchild Inc. (3, 13, 14, 23)
Fairfield Engineering Co. (1, 3, 4, 5, 6, 9, 10, 11, 12, 17, 19, 23)
Fairmont Supply Co. (6, 9, 12, 13, 15, 19)
Fate Intl. Ceramic & Processing Equip., Div. of Plymouth Locomotive Works, Inc. (3)
Feeco International, Inc. (3, 9, 10, 12, 17, 19, 23)
Fenner International Ltd. (3, 12)
Fletcher Sutcliffe Wild, Ltd. (3, 23)
Flexowall Corp. (3, 5, 11, 12, 13)
Fuller Co., A Gatx Co. (9)
GEC Mechanical Handling Ltd. (1, 3, 19, 21)
General Kinematics Corp. (21)
General Resource Corp. (19, 21)
Goodman Equipment Corp. (3, 4, 21)
Goodrich, B.F., Co., Engineered Products Group (3, 5, 13, 14, 23)
Gruendler Crusher & Pulverizer Co. (3, 12)
Gundlach, T. J., Machine Div., Rexnord Inc. (3, 17, 20, 21, 24)
Guyan Machinery Co. (9)
Hammermills, Inc., Sub. of Pettibone Corp. (3, 17, 23)
Hartman-Fabco Inc. (3, 5, 12, 17, 21, 23, 24)
Hazemag, Dr. E. Andreas GmbH & Co. (1)
Hemscheidt America Corp. (2)
Herold Mfg. Co., Inc. (1, 2, 9, 15, 17, 21, 22)
Hewitt-Robins Conveyor Equipment Div. Litton Systems, Inc. (3, 4, 8)
Heyl & Patterson, Inc. (7, 23)
Holmes Bros. Inc. (3, 6, 22)
Huwood-Irwin Co. (2, 3, 4, 5, 9, 13, 15)
Huwood Limited (2, 3, 4, 9)
Hycaloader Co. (3, 17)
Indiana Steel & Fabricating Co.
Industrial Resources, Inc. (1, 3, 5, 6, 9, 12, 19, 20, 21, 22, 23)
Interroll Corp. (3)
Iowa Manufacturing Co. (3, 13, 17, 23)
Irvin-McKelvy Co., The (3, 9, 21, 22, 23)
Janes Manufacturing Inc. (1, 9, 11, 12)
Jeffrey Mfg. Div., Dresser Industries Inc. (1, 3, 4, 5, 6, 9, 10, 12, 19, 20, 21, 22, 23, 24)
Jeffrey Mining Machinery Div., Dresser Industries Inc. (9, 14, 15, 16, 17)
Jenkins of Retford Ltd. (3, 9, 23)
Joy Mfg. Co. (2, 4, 14)
KHD Industrieanlagen AG, Humboldt Wedag (3, 6, 8, 9, 10, 11, 12, 13, 19, 21, 23)
Kaiser Engineers, Inc. (23)
Kanawha Mfg. Co. (1, 3, 4, 6, 9, 12, 18)
Keene Corp. (6)
Koch Transporttechnik GmbH (1, 3, 6, 7, 9, 13, 19, 23, 24)
Kolberg Mfg. Corp. (3, 5, 12, 13, 17, 23, 24)
Koppers Co., Inc.
Krupp Industrie-und Stahlbau (7)
Lancaster Steel Products (1, 3, 4, 5, 6, 9, 13, 17, 19, 20, 21, 23)
Lee, A. L., Corp. (9)
Lippmann-Milwaukee, Inc. (1, 3, 6, 17, 20, 23)
Lively Mfg. & Equipment Co. (1, 3, 5, 23)
Logan Corp. (1, 9, 12, 17, 19)
Logan Mine Supply Co., Inc. (5, 8)
Long-Airdox Co. A Div. of the Marmon Group, Inc. (2, 3, 4, 5, 9, 10, 12, 13, 14, 16, 20, 23)
Long-Airdox Construction Co., a Member of the Marmon Group (23)
Magco Ltd. (21)
M.A.N. Maschinenfabrik Augsburg-Nurnberg, Div. GHH Sterkrade (1, 3)
Mannesmann Demag Baumaschinen Bereich Lauchhammer (3, 4, 5, 7, 10, 12, 14, 23)
Manufacturers Equipment Co., The (1, 3, 5, 6, 9, 12, 13, 19, 20)
Marathon Mfg. Co. (3, 23)
Marathon Steel Co. (3, 4, 5, 12, 17, 23)
Marion Power Shovel Div., Dresser Industries, Inc. (7, 17, 23, 24)
Marsh, E. F., Engineering Co. (1, 3, 5, 6, 10, 12, 13, 17, 20, 23)
Maschinenfabrik Buckau R. Wolf Aktiengesellschaft (3, 5, 7, 23)
McNally Pittsburg Mfg. Corp. (1, 3, 19, 22, 23)
Metal Craft Inc. (3, 8)
Midwestern Industries, Inc., Screen Heating Transformers Div. (11)
Mid-West Conveyor Co. (3, 4, 5, 6, 7, 8, 10, 13, 20, 23, 24)
Mine Systems Inc. (3, 10, 12, 17, 23)
Mining Equipment Mfg. Corp. (3)
Mining Machine Parts (9)
Mining Progress, Inc. (2, 9)
Mining Supplies, Ltd. (2, 9, 15, 17)
Mintec/International, Div. of Barber-Greene (3, 5, 7, 17, 23)
Montgomery Elevator Co. (12, 13)
Myers Whaley Co. (1, 3)
National Air Vibrator Co. (21)
National Iron Co. (1)
National Mine Service Co. (2, 3, 9)
Nicholson Engineered Systems (1, 3, 5, 6, 8, 9, 10, 12, 19, 23)
Ore Reclamation Co. (3, 9)

TABLE 4. (continued)

Owens Mfg., Inc. (3, 4, 5, 10)
PHB Material Handling Corp.
Peerless Conveyor & Mfg. Co., Inc. (3, 5, 23)
Pente Industries Inc. (1, 3, 5, 6, 9, 10, 12, 17, 19, 21, 23, 24)
Persingers Inc.
Portec, Inc., Pioneer Div. (1, 3, 5, 12, 17, 20, 23)
Power Transmission Div., Dresser Industries (1, 9, 12, 13)
Power Transmission Equipment Co. (1, 3, 6, 7, 8, 9, 11, 12, 20, 21)
Process Equipment Builders, Inc. (3, 5, 6, 12, 13, 17, 19, 23)
RAHCO, R. A. Hanson Co.
Renold Inc. (1, 11, 21)
Rexnord Inc. (1, 3, 4, 5, 6, 9, 10, 11, 12, 13, 17, 20, 21, 22, 23)
Rexnord Inc., Corp. Relations Dept. (3)
Rexnord Inc., Process Machinery Div. (3, 5, 6, 17)
Rexnord Inc., Vibrating Equipment Div. (11, 12, 21, 22)
Richmond Mfg. Co. (12)
Rish Equipment Co. (3, 17, 23)
Robins Engineers & Constructors (Hewitt-Robins) (3, 4, 5, 7, 10, 12, 14, 20, 23)
Sala International (3)
Salem Tool Co., The (12)
Savage, W. J. Co. (3)
Sepor, Inc. (3, 6, 12, 17)
Serpentrix Conveyor Corp. (3, 5, 10, 11, 12, 13, 15, 17, 20, 22, 23)
Simplicity Engineering (21)
Specialty Services, Inc. (3)
Sprout-Waldron Div., Koppers Co., Inc. (3, 6, 12, 19)
Stamler, W. R., Corp., The (5, 12, 13)
Standard Metal Mfg. Co. (3, 5, 6, 12)
Stephens-Adamson Inc. (1, 3, 13, 21, 23)
Sturtevant Mill Co. (19)
Telsmith Div., Barber-Greene Co. (3, 5, 17, 23)
Thyssen Mining Equipment (2, 9)
Titan Mfg. Co. Pty. Ltd. (2, 3)
Trelleborg, AB (3)
Trelleborg, Inc. (3)
Underground Mining Machinery Ltd. (2)
Unilok Belting Co., Div. of Georgia Duck and Cordage Mill (3)
Union Chain Div. (1, 6, 9, 12, 13, 23)
Uniroyal, Inc.
Universal Engineering Corp. (1, 3, 17, 23)
Universal Industries (3, 6, 12)
Universal Road Machinery Co. (3, 6)
Vibco Inc. (21)
Vibranetics, Inc. (12, 21)
Vibra Screw Inc. (5, 19, 21)
V/O Machinoexport (2, 3, 9)
Voest-Alpine Intl. Corp. (3)
Wajax Industries Ltd. (2, 3, 4, 9, 15, 20, 23)
Webb, Jervis B., Co. (1, 3, 4, 5, 6, 8, 9, 10, 12, 17, 18, 19, 20, 21, 23)
Webster Mfg. Co. (1, 3, 5, 6, 9, 10, 12, 13, 19, 20, 21, 23)
Weserhuette Inc. (1, 3, 4, 7, 12, 14, 17, 23, 24)
West Virginia Armature Co. (3, 4, 8, 14)
West Virginia Belt Sales & Repairs Inc. (1, 3, 12)

Westfalia Lunen (2)
Williams Patent Crusher & Pulv. Co. (1, 3)
Willis & Paul Corp., The (1, 3, 5, 6, 9, 10, 12, 19, 20, 23)
Wilmot Engineering Co. (9, 12)
Wilson, R. M., Co. (1, 3, 5, 6, 9, 10, 11, 12, 13, 17, 18, 19, 20, 21, 22, 23, 24)

Drills, Coal, Underground

1. hand-held, electric
2. hand-held, hydraulic
3. hand-held, pneumatic
4. post-mounted
5. self-propelled
6. methane degasification

Acker Drill Co., Inc. (4)
Atlas Copco MCT AB (5)
British Coal International
CompAir Construction & Mining Ltd. (3)
Deron Corp. (2, 4)
Dowty Corp. (1, 2, 3, 4, 5)
Dresser Industries, Inc.
Dresser Industries, Inc., Drilling Equipment (4)
ELMAC Corp. (4)
English Drilling Equipment Co. Ltd. (4, 6)
FMC Corp., Mining Equipment Div. (5)
Fletcher, J. H., & Co. (5, 6)
Gardner-Denver/Cooper Ind., Mining & Construction Group (3, 4)
Hausherr, Rudolf und Sohne (5, 6)
Huwood-Irwin Co. (1, 2, 3)
Ingersoll-Rand Co. (3, 5)
Jaeger Machine Co. (3)
Jeffrey Mining Machinery Div., Drilling Equipment (4)
Joy Mfg. Co. (5)
Kent Air Tool Co. (3)
Lebco, Inc., Illinois Div. (2)
Linden-Alimak, Inc. (3)
Milwaukee Electric Tool Corp. (1)
Mining Developments Ltd. (5)
Mining Equipment Mfg. Corp. (5)
Mining Machine Parts (2)
Morgantown Machine & Hydraulics, Inc., Div. Natl. Mine Service Co. (2, 4)
National Mine Service Co. (2)
S & S Corp. (2)
Schroeder Bros. Corp. (2, 5)
Strojexport, pzo
Sullair Corp. (3)
Thor Power Tool Co. (1, 3)
Torque Tension Ltd. (2, 4, 5, 6)
Turmag (G.B.) Ltd. (3, 4, 5, 6)
Victor Products (Wallsend) Ltd. (1, 2, 3, 4, 5)
V/O Machinoexport (2, 3)
Wallacetown Engineering Co. Ltd. (1)

Drills, Roof

1. hanger installation
2. self-propelled
3. stopers
4. rotary

TABLE 4. (continued)

AEC Inc. (2, 4)
Acme Machinery Co. (1, 2, 3, 4)
Atlas Copco MCT AB (3)
CompAir Construction & Mining Ltd. (3)
Deron Corp. (4)
Dowty Corp. (2)
Dresser Industries, Inc. (3)
Dresser Industries, Inc., Drilling Equipment (3)
Eimco Mining Machinery, Envirotech Corp.
FMC Corp., Mining Equipment Div. (2)
Fairchild Inc. (2, 4)
Fairmont Supply Co. (1)
Fletcher, J. H., & Co. (2, 4)
Gardner-Denver/Cooper Ind., Mining & Construction Group (2, 3)
Hausherr, Rudolf und Sohne (4)
Ingersoll-Rand Co. (2, 3)
Jeffrey Mining Machinery Div., Drilling Equipment (3)
Joy Mfg. Co. (2)
Kent Air Tool Co. (3)
Lebco, Inc., Illinois Div.
Lee-Norse Co. (2, 4)
Linden-Alimak, Inc. (2, 3)
Long-Airdox Co. A Div. of the Marmon Group, Inc.
Mining Developments Ltd. (2, 4)
Ohio Brass Co., a Sub. of Harvey Hubbell Inc. (1)
Titan Mfg. Co. Pty. Ltd. (2)
Torque Tension Ltd. (2, 3, 4)
Tread Corp.
Turmag (G.B.) Ltd. (4)
Victor Products (Wallsend) Ltd. (4)
V/O Machinoexport

Drives

1. adjustable & selective speed
2. belt
3. chain
4. flange-mounted
5. fluid, hydraulic
6. gear, worm-gear
7. shaft-mounted
8. V-belt
9. variable-speed
10. variable speed, hydraulic
11. eddy-current
12. screw-conveyor

Abex Corp. (5, 9, 10)
Abex Corp., Denison Div. (10)
Allen-Bradley Co. (1, 9)
American Standard, Industrial Products Div. (1, 2, 5)
Banner Bearings (1, 2, 3, 6, 8, 9)
Baylor Company (1, 5, 7, 9, 10)
Bearings, Inc. (1, 2, 3, 4, 6, 7, 8, 9)
Big Sandy Electric & Supply Co., Inc. (1, 2, 3, 4, 5, 6, 7, 8, 9)
Bonded Scale & Machine Co. (2, 3, 6, 7)
Browning Mfg. Div., Emerson Electric Co. (1, 2, 3, 6, 7, 8, 9)
Bruening Bearings, Inc. (1, 2, 3, 4, 6, 7, 8, 9)
Compton Electrical Equipment Corp. (1, 11)
Cone Drive Div., Excello Corp. (4, 6, 7)
Continental Conveyor & Equipment Co., Inc. (7)
Controlled Systems Inc. (1, 2, 9)
Cutler-Hammer Products, Eaton Corp. (1, 9)
Dana Industrial, Dana Corp. (1, 5, 9, 10)
Dayco Corp. (1, 2, 8, 9)
Deere & Co. (10)
Dixie Bearings, Inc., (1, 2, 3, 4, 6, 7, 8, 9)
Dodge Div., Reliance Electric Co. (2, 3, 4, 7, 8, 12)
Dominion Engineering Works Ltd. (6, 7)
Dowty Corp. (2, 4, 9)
Dresser Industries, Inc. (3, 6, 7, 9)
Dynex/Rivett Inc. (5, 10)
Eaton Corp., World Headquarters (1, 2, 4, 5, 6, 7, 8, 9, 10, 11)
Eaton Corp., Industrial Drives Operations (1, 2, 4, 6, 7, 8, 9, 11)
Eickhoff America Corp. (3, 4)
Electric Machinery Mfg. Co. (1, 9, 11)
FMC Corp., Drive Div. (1, 5, 6, 7, 9)
Fairchild Inc. (2)
Fairmont Supply Co. (1, 2, 3, 4, 6, 7, 8, 9)
Falk Corp., The, Sub. of Sundstrand Corp. (1, 4, 5, 6, 7, 9, 10, 12)
Federal Supply & Equipment Co., Inc. (5)
Fenner America Ltd. (8)
Fenner International Ltd. (1, 2, 3, 6, 7, 8, 9)
Flender Corp. (1, 4, 6, 7)
Fluidrive Engineering Co. Ltd. (5, 10)
Funk Mfg. Div. of Cooper Industries Inc. (6)
GEC Mechanical Handling Ltd. (10)
GTE Products Corp., Electrical Equipment (1, 9)
Gates Rubber Co., The (1, 8)
General Electric Co. (1, 2, 3, 4, 6, 7, 8, 9, 11)
General Electric Co., DC Motor & Generator Dept. (1)
Goodman Equipment Corp. (2)
Goodrich, B.F., Co., Engineered Products Group (8)
Guyan Machinery Co. (1, 2, 3, 6, 7, 8, 9)
Hansen Transmissions Inc. (1, 6, 7)
Harnischfeger Corp. (1)
Horsburgh & Scott Co. (6)
Huwood-Irwin Co. (2, 3, 6)
Huwood Limited (1, 2, 3, 4)
Illinois Gear/Wallace Murray Corp. (6)
Janes Manufacturing Inc. (3)
Jeffrey Mining Machinery Div., Dresser Industries Inc. (3)
Jold Mfg. Co., Inc. (2)
Kanawha Mfg. Co.
Koppers Co., Inc. (1, 7, 9)
Koppers Co. Inc., Engineered Metal Products Group (1)
Leads & Northrup, a unit of General Signal (9)
Lima Electric Co., Inc. (1)
Logan Corp. (2, 3, 6, 7, 8, 9)
Louis Allis Div., Litton Industrial Products, Inc. (1, 2, 3, 4, 6, 8, 9, 11)
Lucas Fluid Power (5, 10)
Metal Craft Inc. (2)
Mining Progress, Inc. (1, 4, 5, 7)
Mining Supplies, Ltd. (3, 4)
Moog Inc., Industrial Div. (10)
Morse Chain, Div. of Borg-Warner Corp. (1, 2, 3, 6, 7, 9)
National Iron Co. (7)

TABLE 4. (continued)

Peerless Pump (5, 9, 10)
Persingers Inc.
Philadelphia Gear Corp. (1, 6, 7, 9, 10)
Power Transmission Div., Dresser Industries (1, 3, 4, 6, 7, 9)
Power Transmission Equipment Co. (1, 2, 3, 4, 5, 6, 7, 8, 9, 10, 11)
Reliance Electric (1, 2, 4, 6, 8, 9, 11)
Renold Inc. (1, 6, 7, 9, 10)
Rexnord Inc. (3, 5)
Richmond Machine Co., Amer. Reducer Div. (7, 12)
Robbins & Myers, Inc. (1, 4, 6, 7, 9, 10)
Robicon Corp. (9)
Rockwell International, Automotive Operations (6)
Scharf Co. (5, 10)
Siemens-Allis Inc. (1)
Siemens Corp. (9, 10)
Sperry Vickers Div., Sperry Corp. (1, 5, 10)
Sperry Vickers, Tulsa Div. (4, 6, 7)
Sterling Power Systems, Inc., A Sub. of The Lionel Corp. (4, 5, 9)
Titan Mfg. Co. Pty. Ltd. (2)
Twin Disc, Inc. (1, 5)
Uniroyal, Inc. (2, 8, 9, 10)
U.S. Electrical Motors Div. Emerson Electric Co. (1, 2, 4, 6, 7, 9, 11)
Webb, Jervis B., Co.
West Virginia Armature Co. (2)
West Virginia Belt Sales & Repairs Inc. (2)
Westfalia Lunen (2, 6)
Westinghouse Electric Corp. (1, 6, 9)
Wichita Clutch Co., Industrial Group, Dana Corp. (9)
Wilmot Engineering Co. (8)
Wood's, T.B., Sons Co. (1, 2, 7, 8, 9, 10)
XTEK, Inc. (6)

Feeder-Breakers, Underground

Continental Conveyor & Equipment Co., Inc.
ELMAC Corp.
FMC Corp., Mining Equipment Div.
Long-Airdox Co. A Div. of the Marmon Group, Inc.
McJunkin Corp.
Mine Equipment Co.
National Mine Service Co.
North American Galis Co.
Ohio Brass Co., a Sub. of Harvey Hubbell Inc.
Owens Mfg., Inc.
S & S Corp.
Schroeder Bros. Corp.
Stamler, W. R., Corp., The
Wallacetown Engineering Co. Ltd.
Westfalia Lunen

Feeders

1. apron
2. chain
3. chemical, chloride, lime, reagent, etc.
4. continuous-weighing
5. grizzly
6. mine-car handling
7. mine transfer to belt or car
8. oscillating
9. plate
10. reciprocating
11. rotary
12. screw
13. vibrating

A & K Railroad Materials, Inc. (5)
Aggregates Equipment Inc. (1, 5, 12, 13)
Allis-Chalmers (13)
Allis-Chalmers, Crushing & Screening Equipment (5, 13)
Auto Weigh Inc. (3, 4)
Babcock Hydro-Pneumatics Ltd. (1, 8, 9, 10, 11, 12, 13)
Barber-Greene Co. (1, 5, 9, 10, 13)
BIF, a unit of General Signal (3, 4, 12)
Bonded Scale & Machine Co. (9, 10)
Branford Vibrator Co., The, Div. of Electro Mechanics, Inc. (13)
Calgon Corp. (3)
Campbell Chain Co. (2)
Can-Tex Industries (8, 13)
Canton Stoker Corp. (10, 12, 13)
Capital Controls Co. (3)
Card Corp. (11)
Carman Industries, Inc. (3, 5, 7, 12, 13)
Carus Chemical Co. (3)
Chlorinators Inc. (3)
Clarkson Co. (3)
Columbia Steel Casting Co., Inc. (1, 5)
Connellsville Corp. (1, 2, 6, 9, 10, 13)
Crane Co. (3)
Deister Machine Co., Inc. (5, 13)
Detroit Stoker Co. (11)
Dorr Oliver Long, Ltd. (1, 2, 5, 11)
Dover Conveyor & Equipment Co., Inc. (1, 2, 8, 9, 10, 12, 13)
Dresser Industries, Inc. (1, 2, 12, 13)
Eagle Crusher Co., Inc.
ELMAC Corp. (7)
Eriez Magnetics (5, 8, 13)
ESCO Corp. (11)
FMC Corp., Material Handling Equipment Div. (3, 5, 12, 13)
FMC Corp., Materials Handling Systems Div. (1, 5, 10, 11)
Fairfield Engineering Co. (1, 2, 9, 10, 12)
Fairmont Supply Co. (1, 12, 13)
Ferro-Tech, Inc. (12)
Fredrik Mogensen AB (5, 13)
Fuller Co., A Gatx Co. (1, 5, 11)
GEC Mechanical Handling Ltd. (1, 5, 9, 11, 12, 13)
Galigher Co., The (3, 9)
General Kinematics Corp. (5, 8, 13)
General Resource Corp. (3, 11, 12, 13)
Goodrich, B.F., Co., Engineered Products Group (7, 11)
Gruendler Crusher & Pulverizer Co. (1, 5, 9, 10, 12)
Gundlach, T. J., Machine Div., Rexnord Inc. (13)
Hammermills, Inc., Sub. of Pettibone Corp. (1, 5, 9, 10, 13)
Hartman-Fabco Inc. (5, 8, 9, 10)
Hazemag, Dr. E. Andreas GmbH & Co. (1, 10)
Hewitt-Robins Div., Litton Systems, Inc. (1, 5, 8, 9, 10, 13)

TABLE 4. (continued)

Heyl & Patterson, Inc. (6, 10)
Hycaloader Co. (13)
Indiana Steel & Fabricating Co. (10, 11)
Industrial Resources, Inc. (1, 2, 7, 9, 10, 12)
Industrial Resources, A. L. Lee Corp. (2, 8)
Interstate Fabricators & Constructors
Iowa Manufacturing Co. (1. 5. 10, 13)
Irvin-McKelvy Co., The (9, 10, 11)
Janes Manufacturing Inc. (1, 2)
Jeffrey Mfg. Div., Dresser Industries Inc. (1, 2, 3, 5, 8, 10, 11, 12, 13)
Jenkins of Retford Ltd. (6)
Jim Pyle Co. (10)
Joy Mfg. Co., Denver Equipment Div. (3)
KHD Industrieanlagen AG, Humboldt Wedag (1, 2, 4, 5, 6, 7, 8, 9, 10, 12, 13)
Kanawha Mfg. Co. (1, 6, 10)
Kennedy Van Saun Corp. (10)
Kolberg Mfg. Corp. (9, 10)
Koppers Co., Inc.
Koppers Co. Inc., Mineral Processing Systems Div.
Krupp Industrie-und Stahlbau (5, 10)
K-Tron Corp. (4, 12, 13)
Lake Shore, Inc. (1)
Lancaster Steel Products (1, 5, 12, 13)
Lee, A. L., Corp. (10)
Lippmann-Milwaukee, Inc. (1, 2, 5, 9, 10, 13)
Lively Mfg. & Equipment Co. (1, 9, 10, 12, 13)
Logan Corp. (1, 2, 5, 8, 10, 12)
Long-Airdox Co. A Div. of the Marmon Group, Inc. (2, 7)
Ludlow-Saylor Div. G.S.I.
Magco Ltd. (13)
Manufacturers Equipment Co., The (1, 2, 3, 9, 10, 11, 12)
Marathon Steel Co. (10, 11)
Marsh, E. F., Engineering Co. (1, 9, 10)
McLanahan Corp. (5, 9, 10)
McNally Pittsburg Mfg. Corp. (10)
Merrick Scale (4)
Mid-West Conveyor Co. (11)
MikroPul Corp. (11)
Milltronics Inc. (4)
Mining Progress, Inc. (2)
Mintec/International, Div. of Barber-Greene (1, 10)
Nalco Chemical Co. (3)
National Air Vibrator Co. (13)
National Iron Co. (1, 10)
National Mine Service Co.
Nicholson Engineered Systems (1, 2, 9, 10, 11, 12)
Nolan Co., The (6)
Ohmart Corp. (4)
Owens Mfg., Inc. (2, 7)
PHB Material Handling Corp.
Pente Industries Inc. (1, 2, 9, 10, 11, 13)
Persingers Inc.
Pettibone Corp. (1)
Portec, Inc., Pioneer Div. (1, 5, 9, 10, 13)
Power Transmission Equipment Co. (1, 2, 12, 13)
Preiser/Mineco Div., Preiser Scientific Inc. (4, 13)
Process Equipment Builders, Inc. (7, 10, 12, 13)
Pulverizing Machinery, Div. of MikroPul Corp. (11)
RAHCO, R. A. Hanson Co.

Ramsey Engineering, Co. (4)
Renold Inc. (1, 10, 13)
Rexnord Inc. (1, 2, 3, 4, 5, 6, 7, 8, 9, 10, 11, 13)
Rexnord Inc. Process Machinery Div. (1, 5, 13)
Rexnord Inc. Vibrating Equipment Div. (5, 13)
Rish Equipment Co. (9, 13)
Sala Machine Works Ltd. (11)
Sepon Inc. (4, 9, 12)
Simplicity Engineering (5, 13)
Solids Flow Control Corp. (13)
Sprout-Waldron Div., Koppers Co., Inc. (11, 12)
Stamler, W. R., Corp., The (2, 6, 7)
Stephens-Adamson Inc. (1, 13)
Strojexport, pzo (2, 12, 13)
Telsmith Div., Barber-Greene Co. (1, 5, 9, 10, 13)
Thayer Scale Hyer Industries (4)
Universal Engineering Corp. (1, 5, 9, 10, 13)
Universal Road Machinery Co. (10)
Vibranetics, Inc. (5, 12, 13)
Vibra Screw Inc. (3, 4, 12, 13)
Wajax Industries Ltd. (5, 13)
Webb, Jervis B., Co. (1, 2, 12, 13)
Webster Mfg. Co. (1, 2, 9, 10)
Weserhuette Inc. (1, 5, 10, 13)
Westfalia Lunen (2)
Willis & Paul Corp., The (1, 2, 5, 12)
Wilson, R. M., Co. (1, 5, 8, 9, 10, 11, 13)

Fire-Protection Systems

AD-X Corp.
Alison Control Inc.
Ansul Co., The
A-T-O Inc.
Austin, J. P. Assoc.
Automatic Sprinkler Corp.
Big Sandy Electric & Supply Co., Inc.
Cementation Mining Ltd.
Comtrol Corp.
du Pont de Nemours, E. I. & Co. Inc.
Fairmont Supply Co.
Fiberglass Resources Corp.
Fire Protection Supplies Inc.
Gilbert Associates, Inc.
Hayden-Nilos Conflow Ltd.
Jabco, Inc.
Kidde Belleville, Div. of Walter Kidde & Co.
Logan Corp.
3 M Co.
Michael Walters Ind.
Mine Safety Appliances Co.
National Foam System Inc.
National Mine Service Co.
Persingers Inc.
Preiser/Mineco Div., Preiser Scientific Inc.
Pyott-Boone Electronics
Schroeder Bros. Corp.
Senior Conflow
Twisto-Wire Fire Systems, Inc.
Uniroyal, Inc.
Victaulic Co. of America
West Virginia Belt Sales & Repairs Inc.
Wilson, R. M., Co.

TABLE 4. (continued)

Foam Fire-Fighting Equipment

A-T-O Inc.
Dowell Division
Fire Protection Supplies Inc.
Kidde Belleville, Div. of Walter Kidde & Co.
3 M Co.
Mine Safety Appliances Co.
National Foam System Inc.
National Mine Service Co.

Lacing, Belt

Capital City Industrial Supply Co.
Cincinnati Rubber Mfg. Co. Inc.
Fairmont Supply Co.
Fenner America Ltd.
Flexible Steel Lacing Co.
General Splice Corp.
Goodall Rubber Co.
Power Transmission Equipment Co.
Shingle, L.H., Co.
Tidewater Supply Co.
Wilson, R. M., Co.

Pulley, Lagging

Aggregates Equipment Inc.
Big Sandy Electric & Supply Co., Inc.
Bonded Scale & Machine Co.
Cincinnati Rubber Mfg. Co. Inc.
Clouth Gummiwerke Aktiengesellschaft
Dover Conveyor & Equipment Co., Inc.
Fairmont Supply Co.
General Splice Corp.
Goodall Rubber Co.
Goodrich, B.F., Co., Engineered Products Group
Goodyear Tire & Rubber Co.
Guyan Machinery Co.
Holz Rubber Co., Inc., a Randtron Sub.
Lancaster Steel Products
Lebus International Inc.
Leman Machine Co.
Linatex Corp. of America
3 M Co.
Manson Services, Inc.
Marsh, E. F., Engineering Co.
Power Transmission Equipment Co.
Sempent of America, Inc.
Trelleborg AB
Van Gorp Corp., Sub Emerson Elec. Co.
West Virginia Belt Sales & Repairs Inc.

Pulleys

1. cast-iron
2. rubber-covered
3. semi-steel
4. steel
5. self-cleaning
6. conveyor-(see Conveyor-belt, pulleys, idler pulleys)

Aggregates Equipment Inc. (5)
Banner Bearings (1, 2, 3, 4, 5)
Bearings, Inc. (2, 4, 5)
Big Sandy Electric & Supply Co., Inc. (2, 4, 5)
Bolton R. B. (Mining Engineers) Ltd. (1, 4)
Bonded Scale & Machine Co. (2, 4, 5)
Browning Mfg. Div., Emerson Electric Co. (2, 4, 5)
Bruening Bearings, Inc. (2, 4, 5)
Concrete Equipment Co., Inc. (2, 4, 5)
Continental Conveyor & Equipment Co., Inc. (2, 4)
Crosby Group (1, 3, 4)
Dayco Corp. (1, 4)
Dick Inc., R. J. (2, 4, 5)
Dixie Bearings, Inc. (2, 4, 5)
Dodge Div., Reliance Electric Co. (2, 4, 5)
Dover Conveyor & Equipment Co., Inc. (4, 5)
Dresser Industries, Inc. (1, 4, 5)
Eaton Corp., World Headquarters (1, 4)
Eaton Corp., Industrial Drives Operations (1)
ELMAC Corp. (2, 4, 5)
FMC Corp., Material Handling Equipment Div. (4, 5)
Fairmont Supply Co. (2, 4, 5)
Fenner International Ltd. (1)
GEC Mechanical Handling Ltd. (1, 2)
Gates Rubber Co., The (4)
General Splice Corp. (2)
Goodman Equipment Corp. (2, 3, 4, 5)
Goodrich, B.F., Co., Engineered Products Group (2, 4)
Holz Rubber Co., Inc., a Randtron Sub. (2)
Huwood-Irwin Co. (1, 4)
Interroll Corp. (2, 3, 4, 5)
Jeffrey Mfg. Div., Dresser Industries Inc. (2, 4, 5)
Jenkins of Retford Ltd. (2, 4)
Kolberg Mfg. Corp. (2, 4, 5)
Lancaster Steel Products (2, 4, 5)
Leman Machine Co. (2, 4, 5)
Linatex Corp. of America (2)
Logan Corp. (2, 4, 5)
Manson Services, Inc. (2, 4, 5)
Marmon Transmotive Div., Sanford Day Products (3)
Marsh, E.F., Engineering Co. (2, 4, 5)
Mining Equipment Mfg. Corp. (2, 4)
Moore Co., Inc., The (3)
Owens Mfg., Inc. (2, 4, 5)
PHB Material Handling Corp. (1, 2, 3, 4, 5)
Peerless Hardware Mfg. Co.
Power Transmission Equipment Co. (1, 2, 3, 4, 5)
Rexnord Inc. (2)
Rexnord Inc., Corp. Relations Dept. (2)
Standard Metal Mfg. Co. (4, 5)
Stephens-Adamson Inc. (4)
Stryker Machine Products Co. (3, 4)
Uniroyal, Inc. (2)
U.S. Steel Corp.
Van Gorp Corp., Sub Emerson Elec. Co. (2, 4, 5)
Webb, Jervis B., Co.
West Virginia Armature Co. (4, 5)
West Virginia Belt Sales & Repairs Inc. (2, 4, 5)
Wilson, R. M., Co. (2, 4, 5)
Wood's, T. B., Sons Co. (1, 2, 3)

Speed Reducers

Alten Speed Reducer Div., Alten Foundry & Machine Works, Inc.

TABLE 4. (continued)

Banner Bearings
Bearings, Inc.
Big Sandy Electric & Supply Co., Inc.
Bonded Scale & Machine Co.
Browning Mfg. Div., Emerson Electric Co.
Bruening Bearings, Inc.
Cone Drive Div., Excello Corp.
Dixie Bearings, Inc.
Dodge Div., Reliance Electric Co.
Dresser Industries, Inc.
Eaton Corp., World Headquarters
Eaton Corp., Industrial Drives Operations
FMC Corp., Drive Div.
Fairmont Supply Co.
Falk Corp., The Sub. of Sundstrand Corp.
Fenner America Ltd.
Fenner International Ltd.
Flender Corp.
Foote-Jones Operations, Dresser Industries, Inc.
General Electric Co.
Guyan Machinery Co.
Hansen Transmissions Inc.
Horsburgh & Scott Co.
Huwood-Irwin Co.
Ingersoll-Rand Co.
James D.O. Gear Mfg. Co., Unit of Ex-Cell-O Corp.
Logan Corp.
Louis Allis Div., Litton Industrial Products Inc.
Mining Progress, Inc.
Morse Chain, Div. of Borg-Warner Corp.
Persingers Inc.
Philadelphia Gear Corp.
Power Transmission Div., Dresser Industries
Power Transmission Equipment Co.
Reliance Electric
Renold Inc.
Savage, W. J. Co.
Sterling Power Systems, Inc., A Sub. of The Lionel Corp.
U.S. Electrical Motors Div. Emerson Electric Co.
Westinghouse Electric Corp.
Wilson, R. M., Co.

XTEK, Inc.

Sprinklers, Automatic

A-T-O Inc.
Automatic Sprinkler Corp.
Fairmont Supply Co.
Hayden-Nilos Conflow Ltd.
Jabco, Inc.
Reggie Industries, Inc.
Wilson, R. M., Co.

Takeups, Conveyor

Aggregates Equipment Inc.
Banner Bearings
Bearings, Inc.
Bonded Scale & Machine Co.
Bruening Bearings, Inc.
Concrete Equipment Co., Inc.
Dixie Bearings, Inc.

Dodge Div., Reliance Electric Co.
Dowty Corp.
Dresser Industries, Inc.
ELMAC Corp.
Equipment Mfg. Services, Inc.
Fafnir Bearing, Div. of Textron
Fairmont Supply Co.
Fletcher Sutcliffe Wild, Ltd.
GEC Mechanical Handling Ltd.
General Splice Corp.
Goodman Equipment Corp.
Hartman-Fabco Inc.
Huwood-Irwin Co.
Jeffrey Mfg. Div., Dresser Industries Inc.
Kanawha Mfg. Co.
Lancaster Steel Products
Leman Machine Co.
Long-Airdox Co. A Div. of the Marmon Group, Inc.
Marathon Steel Co.
Marsh, E. F., Engineering Co.
McNally Pittsburg Mfg. Corp.
Metal Craft Inc.
Owens Mfg., Inc.
Persingers Inc.
Power Transmission Equipment Co.
Rexnord Inc.
Rexnord Inc., Bearing Div.
Titan Mfg. Co. Pty. Ltd.
Torrington Co., The Bearings Div.
Webb, Jervis B., Co.
West Virginia Armature Co.
West Virginia Belt Sales & Repairs Inc.
Willis & Paul Corp., The
Wilmot Engineering Co.
Wilson, R. M., Co.

Turnbuckles

Anixter Mine & Industrial Specialists
Bowman Distribution, Barnes Group, Inc.
Campbell Chain Co.
Capital City Industrial Supply Co.
Crosby Group
Dyson, Jos., & Sons Inc.
Elreco Corp., The
Fairmont Supply Co.
Logan Corp.
Ohio Brass Co., a Sub. of Harvey Hubbell Inc.
Service Supply Co., Inc.
Tidewater Supply Co.

V-Belt & Belt Fasteners

BTR Trading, Inc.
Bearings, Inc.
Browning Mfg. Div., Emerson Electric Co.
Bruening Bearings, Inc.
Dayco Corp.
Dixie Bearings, Inc.
Dodge Div., Reliance Electric Co.
Fairmont Supply Co.
Fenner America Ltd.
Fenner International Ltd.

TABLE 4. (continued)

Flexible Steel Lacing Co.
Gates Rubber Co., The
Goodall Rubber Co.
Goodrich, B.F., Co., Engineered Products Group
Goodyear Tire & Rubber Co.
Guyan Machinery Co.
Peerless Hardware Mfg. Co.
Persingers Inc.
Power Transmission Equipment Co.
Shingle, L.H., Co.
Tidewater Supply Co.
Trelleborg AB
Uniroyal, Inc.

Wire Rope

ALPS Wire Rope Corp.
Anixter Mine & Industrial Specialists
Armco Inc.
Bethlehem Steel Corp.
Bridon American Corp.
Broderick & Bascom Rope Co.
Capital City Industrial Supply Co.
Continental Conveyor & Equipment Co., Inc.
Fagersta, Inc.
Fairmont Supply Co.
Gebruder Kulenkampff
Greening Donald Co. Ltd.
Guyan Machinery Co.
Jennmar Corp.
Leschen Wire Rope Co.
Logan Corp.
Macwhyte Wire Rope Co.
Marathon Coal Bit Co., Inc.
National Mine Service Co.
Owens Mfg., Inc.
Paulsen Wire Rope Corp.
Persingers Inc.
Phelps Dodge Industries, Inc.
Rochester Corp.
Ryerson, Joseph T., & Son, Inc.
Tidewater Supply Co.
U.S. Steel Corp.
West Virginia Belt Sales & Repairs Inc.
Wilson, R. M., Co.
Wire Rope Corp. of America

Wire Rope Accessories

1. boom-guy pendants
2. clamps
3. clips
4. eyes
5. lubricants
6. shackles
7. slings
8. sockets
9. swaged assemblies
10. swivels
11. thimbles

ALPS Wire Rope Corp. (1, 3, 4, 5, 6, 7, 8, 9, 11)
American Hoist & Derrick Co. (2, 3, 9, 10, 11)
Anixter Mine & Industrial Specialists (1, 2, 3, 4, 6, 7, 8, 9, 10, 11)
Armco Inc. (7, 8, 9)
Bethlehem Steel Corp. (1, 7, 9)
Bowman Distribution, Barnes Group, Inc. (2, 5)
Bridon American Corp. (1, 2, 3, 4, 5, 6, 7, 8, 9, 10, 11)
Broderick & Bascom Rope Co. (1, 3, 6, 7, 8, 9, 11)
CM Chain Div., Columbus McKinnon Corp. (3)
Campbell Chain Co. (3, 6, 8, 10, 11)
Capital City Industrial Supply Co. (1, 2, 3, 4, 5, 6, 7, 8, 9, 10, 11)
Card Corp. (11)
Century Hulburt Inc. (5)
Century Oils Ltd. (5)
Columbia Steel Casting Co., Inc. (6, 8)
Connellsville Corp. (2, 3, 8, 11)
Continental Conveyor & Equipment Co., Inc. (2, 3)
Crosby Group (2, 3, 4, 6, 8, 10, 11)
D-A Lubricant Co., Inc. (5)
Dow Corning Corp. (5)
DuBois Chemicals Div. of Chemed Corp.
Elreco Corp., The
ESCO Corp. (4, 6, 7, 8, 9, 10, 11)
Exxon Co., U.S.A. (5)
Fairmont Supply Co. (1, 2, 3, 4, 6, 7, 8, 9, 11)
Greening Donald Co. Ltd. (1, 2, 3, 4, 6, 7, 8, 9, 10, 11)
Guyan Machinery Co. (7, 8, 9)
Holmes Bros. Inc. (3, 11)
Jet Lube Inc. (5)
Johnson Blocks Div., Hindeliter Energy Equipment Corp. (8, 10)
Keystone Div., Pennwalt Corp. (5)
Lebus International Inc.
Leschen Wire Rope Co. (1, 7, 9)
Logan Corp. (2, 3, 7, 11)
Macwhyte Wire Rope Co. (1, 2, 3, 4, 5, 6, 7, 8, 9, 10, 11)
Mobil Oil Corp. (5)
Ohio Brass Co., a Sub. of Harvey Hubbell Inc. (2, 3, 6)
Owens Mfg., Inc. (2, 3)
Paulsen Wire Rope Corp. (1, 2, 3, 4, 6, 7, 8, 9, 10, 11)
Peerless Hardware Mfg. Co. (3)
Penreco, Div. Pennzoil Co. (5)
Philadelphia Resins Corp. (8)
Rochester Corp. (1, 7, 9)
Sanford-Day/Marmon Transmotive, Div. of the Marmon Group, Inc. (2)
Sauerman Bros., Inc. (2, 8, 10)
Service Supply Co., Inc. (2, 3, 4, 11)
Sun Petroleum Products Co. (5)
Texaco Inc. (5)
Tidewater Supply Co. (2, 3, 6, 7, 8, 10, 11)
U.S. Steel Corp. (1, 2, 3, 4, 5, 6, 7, 8, 9, 10, 11)
Whitmore Mfg. Co., The (5)
Wilson, R. M., Co. (2, 3, 4, 6, 7, 8, 9, 10, 11)
Wire Rope Corp. of America (1, 7, 9)
Young Corp. (6, 10)

7

Surge Bins, Shops, and Special Projects

Until now, only the range of typical construction projects which are repeated over and over during the life of a coal mine have been discussed. This chapter looks at underground construction projects which occur only once or twice during the life of any given mine and are major construction projects in both time and cost. These projects usually require the services of an outside engineering and construction firm with hands-on experience in design installation. Because of this involvement with engineering and construction firms (E/C), the role of the construction engineer will change from one of actively designing and directing tasks to one of planning, inspection, and quality control. To reflect this role change in the discussions, the focus of this chapter will be changed to a broad overview of project participation.

Of course, the projects under consideration in this chapter could easily fill more than one book on their design and installation. In addition, major construction projects are extremely site specific, requiring months of data collection about site geology, physical conditions, and potential excavation and construction requirements. Therefore, no attempt shall be made to discuss any detailed construction aspects as was done in earlier chapters. The purpose of this chapter is to present the fundamental concerns of a staff engineer or supervisor. This engineer does not make the decisions on choosing the E/C firm, but assists and interacts with their project team to accomplish the goals set forth by higher management.

The construction topics to be reviewed fall into three main categories. The first category is highlighted by the installation of a main surge bin; the second is typified by the construction of an underground shop; and the third category encompasses special projects such as rotary dumps, crusher stations, and foremen's offices. These categories highlight the differences between degrees of mine responsibilities for project completion. The first topic is usually a turnkey project by an E/C firm for a given mine. The second topic represents the leading role of the mine in the project with a contractor as a support service or equipment supplier only. The third topic, special projects, is an example of a joint venture between a contractor (leading partner) and mine management. These roles can change with any given project, but the changes in management responsibilities for the construction engineer can be nicely illustrated for the purposes of this text.

The first step toward constructing major mine projects is to analyze the mine parameters which influence their design. From the mine management standpoint, these parameters should reflect the role the mine will play in the project construction. In other words, what is the nature of the proposed mine environment and what are the material, labor, and time requirements needed to assist or direct the chosen engineering and construction firm? Answering these questions requires some basic project planning and cost estimation work. Here, only a general treatment is necessary, since detailed discussions of these topics will follow in Chapters 8 and 10.

For those persons interested in the role of the E/C contractor, I suggest beginning by reviewing the article by Beall, "The Role of an Engineering and Construction Contractor," presented in the October 1980 issue of *Mining Engineering* magazine.

ANALYSIS FOR EFFECTIVE PERFORMANCE

The first step for any mine engineer considering the installation of a major project with an outside construction firm is a thorough evaluation of the site specific parameters. Coupled with this task would be a review of the available mine labor, materials, and equipment for the project. The planning and scheduling of necessary mine tasks before actual construction then can be made based upon the gathered information and approved project designs.

Parameters to Consider

The first data collection effort for planning purposes focuses on the parameters influencing the construction project design. There are four categories defining the minimum information needs (Fig. 1). These are (1) geologic, (2) physical, (3) mining, and (4) regulatory. Others may be used as necessary, but these four categories are flexible enough to include most information needs, yet comprehensive enough to ensure a successful project.

Geologic parameters include data on the coal seam, immediate roof and floor strata, and the inherent stability of the mining horizon before, during, and after construction. In this case, detailed geotechnical studies on these horizon characteristics by the technical mine staff are needed, unless this has already been completed on a mine-wide basis. If a large scale effort has already been made, the focus of the new data collection is on detailed data at the local geologic level. Geologic data parameters need to focus on the static and dynamic ground forces occurring around project anchorage points and mine strata (roof, floor, and coal). Such forces will require analysis (by force diagrams) to eliminate movement, vibration, and fatigue factors; increase the useful life of the project; and reduce stress buildup due to strata disturbance during excavation. In addition, preproject weakness zones (faults, joints, and so on) should be plotted on project designs in an effort to minimize any adverse influence on future project performance. Weakness zones can aid in project excavation if properly used in design plans.

Physical parameters include data on the presence of water and gas, mining heights, and mining dimensions. Mine management should classify each underground mine construction site

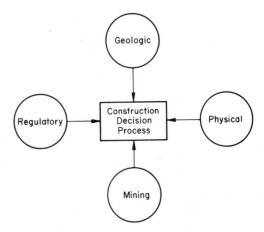

FIG. 1. Parameters influencing the decision process.

in terms of the maneuverability and accessibility needs of the project vs. the existing physical conditions. In these instances, decisions on necessary equipment (drainage pumps, muckers, etc.), construction approaches (cutting, drilling and blasting, and so on), or probable site constraints to installation (height, water, etc.) are made. Physical parameters deal mainly with local conditions, but can be expanded to include such items as the entire route of supply delivery, shifting manpower organization (i.e., what men are required and from where?), and surface and underground staging areas.

Mining parameters focus on the mining production and support systems, mine design (both present and future), mining equipment, and mine construction scheduling now in effect at the mine. The impact of the project is considered on all the parameters in terms of a benefit-cost analysis. Agreements between mine production, mine support, and construction activities will have to be reached. For example, if a continuous miner is pulled from a unit to excavate a rotary dump site, can the current mine production be maintained or lowered within acceptable limits for the two months that the machine would be used? Can the maintenance system handle the increased service loads if the miner is used to cut rock? Or, if a construction crew is pulled for two weeks to build an underground shop, will other necessary support activities suffer? These kinds of tradeoffs must be considered during the planning stages.

Regulatory parameters deal mainly with federal and state health and safety regulations as they apply to the construction project. If explosives are to be used in a mine where there were none before, then steps should be taken to make sure all safety regulations are known and followed. Discussions with MSHA and OSHA district officials and state mine officials should be included to clarify any potential problem areas. Construction workers scheduled for the project should be instructed in the safety of the required tasks, thoroughly briefed on the nature of the project, and trained for specific tasks. Legal parameters such as additional permits, mine plans, and so on should be reviewed where applicable so proper scheduling of the project can be maintained. In many cases, the regulatory parameters will require the most planning and extensive work for leading into a project.

Materials, Labor, and Equipment

Concurrent to the mine site parameter data collection task, planning and scheduling of necessary mine materials, labor, and equipment is accomplished.

The focus of this planning is twofold. First, mine management must quantify the amount of construction support it will give toward the completion of the project. Some projects will require a turnkey approach; others will not. Contract negotiations are a serious business and can be detrimental to the mine if they have little knowledge of what, if any, mine facilities will be necessary to support the contractor during the project. Supplies, regardless of who uses them, are expensive. Labor may be at a premium, and is certainly a bargaining point between the contracting parties. This is also true when purchasing large, special tools (overhead crane, etc.) for the project which may be used only once.

The second reason for this planning is to identify the extent to which the mining company will be responsible for completing any initial construction work. Such excavation, material movement, storage, and fabrication necessary before a contractor will begin the actual project construction should be specified and assigned during this planning phase. Specific items to be considered are:

1) *Materials*—any special sealing (strata), building, or prefab materials requiring special purchases.

2) *Tools*—large or single-use equipment designed or used specifically for this project. Standard tools such as cement mixers, hand tools, welders, and so on should be summarized when necessary.

3) *Labor*, which includes not only the actual construction time of the worker, but the time needed for training, orientation, and future operations and maintenance tasks.

4) *Safety equipment and instrumentation* which covers the increased health and safety equipment needed at the site during construction, as well as future safety requirements. Also, any additional monitoring instruments or sampling devices necessary for effective performance during and after construction.

At the completion of the analysis stage, mine management, and the construction engineer in particular, will have a clear idea of the impacts this construction project will have on the mine. Both current scheduling and future planning will be redefined in terms of the data gathered in this phase.

INSTALLING A SURGE BIN

One of the largest and most common construction projects found in any deep underground coal mine is a surge bin. The purpose of the surge bin is to (1) store the mine production (which occurs in cyclic patterns), and (2) regulate the ROM feed from the mine to the outside (either to a plant, loading point, or storage silo).

Main mine surge bins usually are found in the slope or shaft bottom area of a large mine, as shown in the example sketch of Fig. 2. A bin feeds the slope belt or skip hoists with the ROM material from the mine at regular rates designed to achieve a continuous flow of production.

Conceptually, a surge bin can be thought of as a big hopper which catches all the mine production. ROM material discharged from either mainline belts or track cars falls into the bin. Track systems can use either bottom dump cars or a rotary dump facility. The bin discharges its load from the bottom onto a slope belt or into skips by means of a vibrating feeder or chutes.

Most bins designed for underground coal use are the rectangular bolted plate type. In special cases, a cylindrical or suspension type may be used. All types of bins are constructed of steel and/or reinforced concrete, are generally simple in design, and require little extra in the way of materials.

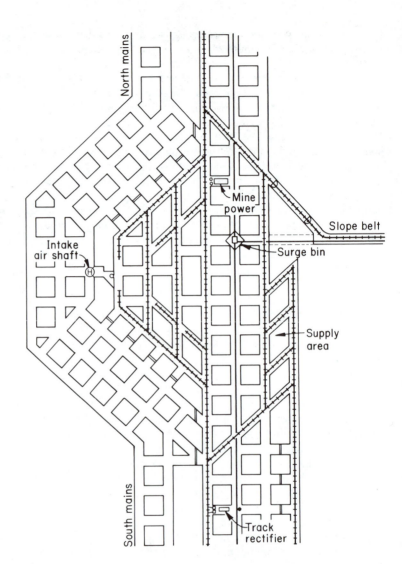

FIG. 2. Typical surge bin location on the slope bottom.

The bolted plate type of bin is the most popular in the coal industry for several reasons. First, it is generally made of lighter gaged steel than other bins, but is supported by the substructure at more points. It can be sectioned at the plant and easily erected at the site. Its rectangular shape aids in the installation process. Besides being cheaper and easier to build, bolted plate bins are durable enough to last the life of a mine under heavy service with a minimum of maintenance. Fig. 3 shows an isometric view of a typical production shaft including the bin and its accessories. The bin capacity is 907.2 t (1,000 st), with reciprocating (vibrating) feeders used to move the coal from the bin to the skips for hoisting out of the mine. This figure also shows the immense design detail required with slope and shaft construction.

Cylindrical bins are designed to place the walls of the bin in tension under loading (Fig. 4). No bending forces are present, which makes these bins better suited for large volumes without extra support structure. In fact, little framework is necessary for the bin except for the substructure. Underground cylindrical bins are designed with conical bottoms (with slope angles between 35° and 45°) to enable efficient flow of material from the bin, making the critical support areas somewhat smaller than with flat-bottom bins (Fig. 5).

Suspension bins use the weight of the coal and rock material to force the bin body to deform to an equilibrium polygon. The rim of the bin is attached to beams which act as a column and support the bin (Fig. 6). Suspension bins generally are not used because control of the material size is limited, thus unequal stresses may develop in the bin wall. Their greatest attraction in the industry is the fact that these bins are generally 40 to 50% cheaper to build than either the bolted plate or cylindrical bin.

The cost of an underground bin depends upon these parameters:

1) *Material*—based on the quality and quantity of the materials needed for the bin, bin support, foundation, and auxiliary systems.

2) *Design*—the fabrication requirements necessary to construct the bin. Complex mine conditions or extraordinary bin requirements can change the cost tremendously.

3) *Erection*—the required labor, tools, and time needed to construct the bin and bin facilities.

4) *Transportation* of the materials to the site, which includes the surface and underground staging areas plus the equipment needed to transport the materials.

Of course, the cost varies with local conditions and actual design. Individual cost estimates can roughly be made based on a 50/50 material and labor estimate plus a 15% contingency fee.

Excavation of a bin is accomplished by drilling and blasting the material down from the area between the slope bottom and mine workings, as shown in Fig. 7. This vertical distance in normal mining situations averages 22.8 to 38.1 m (75 to 125 ft) below the coal seam elevation. The volume of material to be removed must be sufficient to allow the bin, bin support, bin foundation, auxiliary functions, and construction maneuvering to occur. For example, a 635 t (700 st) cylindrical surge bin with a diameter of 9.1 m (30 ft) should have an excavation diameter of 12.2 to 15.2 m (40 to 50 ft), depending upon design (Fig. 8). This allows all piping, cables, accessways, and erection hatchways to be installed around the bin. These excavation requirements also allow any necessary site reinforcement of the interburden strata in the form of wire mesh, bolting, straps, shotcrete, or poured reinforced concrete panels to be constructed. Also, as shown earlier in Fig. 3, using a skip hoist instead of a belt will dramatically increase the amount of excavation required during shaft development.

Although the mine engineer will not take a direct role in the construction of the bin (unless it is installed later on in the mine life), he should become familiar with the design and fabrication specifications of the particular bin being erected at his mine. To do this, he must become part of the project management team at the mine which oversees the construction of these major construction projects.

The three main concerns of project management for the construction engineer can be broken down into (1) preparation, (2) assembly, and (3) operation and maintenance tasks. The following is a discussion of the functions of the construction engineer during each of these tasks from the standpoint of the mine, i.e., inspection and quality control.

Preparation

The preparation task in most cases is accomplished within one to two months of the actual construction of the bin. It is the most important task of the three for the engineer. The most common example of surge bin construction occurs during the initial slope/shaft bottom construction of a new mine. Here, the E/C contractor is most likely sinking the slope/shafts and building the main mine support facilities at the same time. Because it is generally a turnkey project, the preparation, inspection, and quality control functions reduce to the following concerns:

1) *Analyzing the mine requirements* of the bin in terms of the fully producing mine. These requirements are defined by the material characteristics and production schedules planned during mine design stages.

2) *Site orientation* for all future construction and production needs. Sight and orientation spads for the bin, bin supports, feeder controls, and the construction work needed for installation should be placed and reviewed.

3) *Grading and excavation* for the bin area and bin supports. This work may require excavating roof, floor, or both for the necessary bin capacity if the contractor does not do it.

4) *Construction scheduling* of materials, labor, and equipment needed from the mine facilities for cost estimation purposes.

The role of the construction engineer becomes one of an interfacing person between the mine

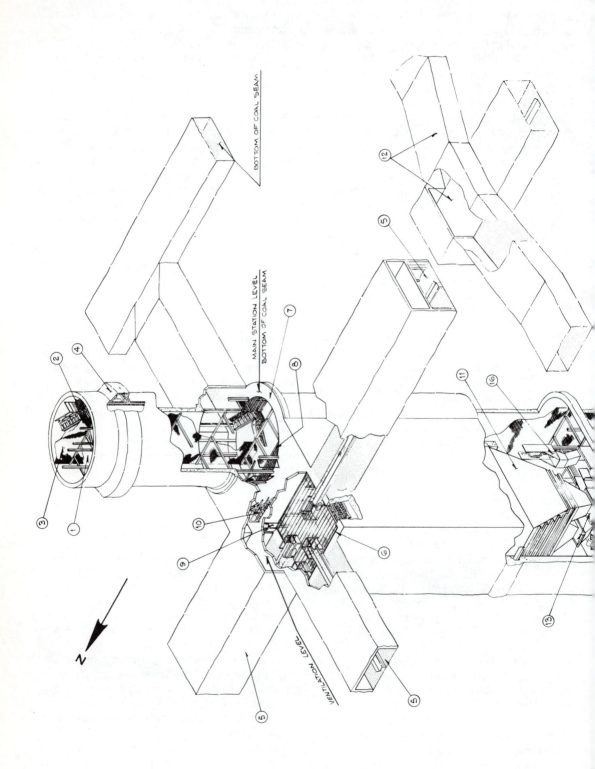

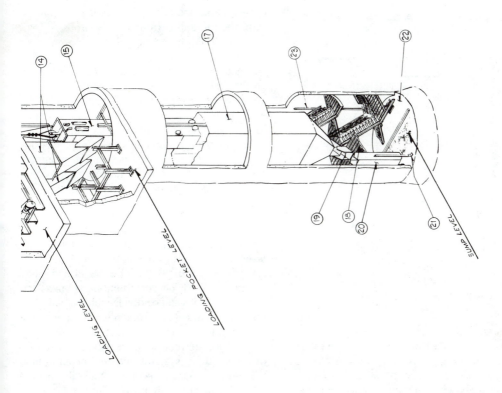

FIG. 3. A production shaft design using a bolted plate bin. Legend:
(1) skip compartment,
(2) stairs (sump to collar),
(3) expanded metal guard,
(4) water ring with discharge lines inside shaft lining to sump,
(5) conveyor entry,
(6) top of surge bin,
(7) passageway at station level,
(8) access to skip's emergency mandeck,
(9) access through air lock doors,
(10) stairs from ventilation level to station,
(11) 907.2 t (1,000 st) capacity surge bin,
(12) ventilation level ramp to mining level,
(13) reciprocating feeders,
(14) skip loaders,
(15) skip,
(16) spill skip dump scroll with chute,
(17) spill recovery bin,
(18) spill chute operating platform,
(19) spill bin chute,
(20) spill recovery skip,
(21) spill recovery skip rail guide anchoring,
(22) sump pumps location,
(23) sump discharge line to station level (courtesy of ARCO Coal Co.).

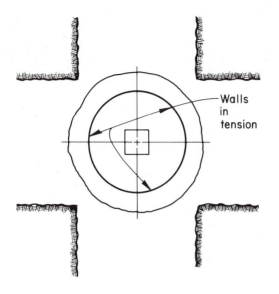

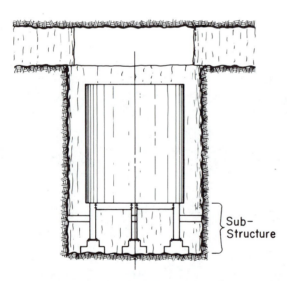

FIG. 4. A cylindrical bin placing the walls of the bin under tension.

and contractor. Generally, his task is to integrate the activities of the two parties for swift and successful project completion. In many cases, much give and take occurs in day-to-day negotiations. The construction engineer must act with confidence and management support to resolve any conflicts. He will find the preparation task very demanding in terms of project management because he holds one of the most visible management positions during these preparation activities.

Assembly

The assembly task is concerned primarily with monitoring the contractor's work. The tasks of building and supporting the bin are included, and can be broken out as the following:

1) *Laying the foundation,* which includes reinforcing the rock strata, bolting the forms, and placement of equipment and pieces to be cast into the foundation.

2) *Piecing the bin together,* requires that the foundation be poured, the steel sides placed, and bolted or welded together.

3) *Sealing the structure* of the bin with impact- or abrasive-resistant material. This lining is optional at the time of construction, but is recommended in certain cases because retrofitting this lining at a later date is difficult.

4) *Fabricating the accessories* of the bin, such as the chutes, walkways, drainage, and monitoring requirements. Sumps, a frequently attached item, are set up to catch the drainage from the bin and surrounding environs.

During this phase, the construction engineer will work closely with the contractor, providing any needed information and mine-related help on the tasks. In addition, the construction engineer must become intimately familiar with the bin, bin structure, and the accessory systems. Many potential problems can be avoided if the mine engineer understands and frequently inspects the structure construction and integrity. There is no better time for the engineer to gain this knowledge than during the construction phase.

Operation and Maintenance

Once the surge bin is operating, and the troubleshooting period of the contractor is over, the responsibilities of effective bin performance and integrity fall to the construction foreman. His inspection duties can be divided into (1) adjusting the bin monitoring instruments, (2) repairing or reinforcing any weak or damaged bin areas, and (3) performing routine maintenance inspections of the auxiliary systems and equipment around the bin. Frequent inspection, generally once a month, can and will prevent future problems from becoming damaging to mine production. If the material flow to the surge bin must be stopped, then the entire mine is stopped from production activities until the surge bin can operate again. These delays are very costly in terms of both labor and lost coal. In some cases, a

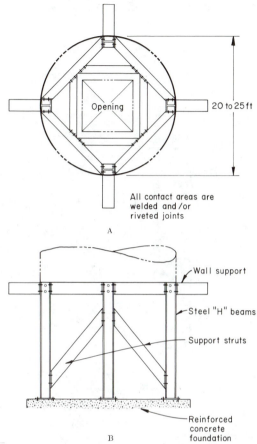

FIG. 5. Underground cylindrical bin. A: Bin bottom. B: Bin substructure. Metric equivalent: ft × 0.3048 = m.

service contract with the original bin contractor for inspections every five to eight years may be advisable.

INSTALLING AN UNDERGROUND REPAIR SHOP

This construction project discussion looks at the large mine considering the placement of major repair and fabrication facilities underground, as opposed to a surface location. The shop would have two to three major equipment bays, an overhead crane, complete warehouse, and additional environmental controls for the health and safety of the shop workers. Unlike the other major tasks, much of the construction work is accomplished by the mine labor force under mine management direction. This definition of management responsibility conforms to the second major category discussed earlier.

The construction of an underground shop also can be broken down into three tasks: (1) preparation, (2) assembly, and (3) maintenance. Each is defined by the steps required to complete the task.

Preparation

The preparation for building an underground shop is very important to the overall success of the project. This task is composed of five separate steps. These are (1) analyzing the mine repair requirements, (2) grading and cleanup of the site, (3) site orientation, (4) required labor, materials, and equipment, and (5) construction scheduling and cost estimating. All these tasks help define the construction requirements of the assembly task.

Analyzing the repair requirements focuses on matching the proposed shop capacities and designed abilities to the needs of the mine. For example, if complete rebuilding facilities are shown by past experience to be necessary underground, then space should be made available in the shop. Another example is the use of a warehouse and repair shop underground to cut delivery time and storage space from outside facilities. Many mines high on excavated benches cut into a hillside can find valuable storage space in old workings along the main travelways. In addition, specific functions such as bit sharpening, a first aid station, or electric cable repair shop can be included in the design. A typical underground shop is shown in Fig. 9.

Grading and cleanup of the chosen shop area is the first step in construction. A scoop tractor is used to clean the floor of debris and trash, while the grading for repair bays is accomplished by a continuous miner or roadheading machine. These bays are graded to a depth of 1.8 to 2.1 m (6 to 7 ft), so drainage channels and grate floor foundation can be poured. The bays are cleaned by hand, and lined with steel mesh or concrete block at a later date (Fig. 10). The entire area is thoroughly scaled. Hanger holes are drilled based on the information given during the site orientation.

During site orientation, the mine surveyors locate the required sight spads. These spads mark the locations for installation of roof- or floor-mounted machine equipment such as welders, benches, cranes, or chain jacks. Additionally, spads for hanger holes to attach piping for air, gas, or water are placed.

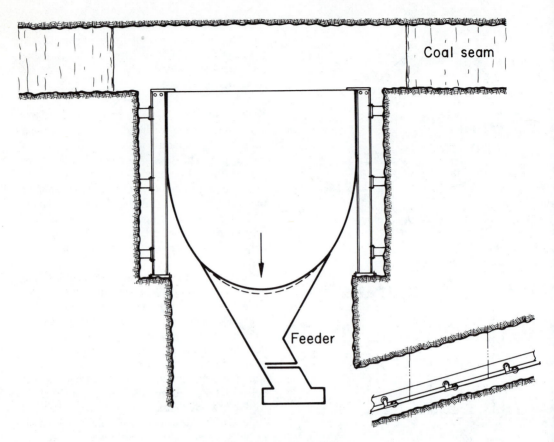

FIG. 6. Suspension bin.

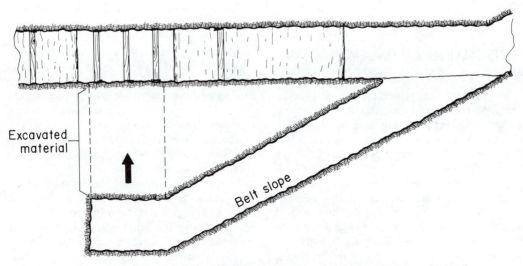

FIG. 7. Typical material to be removed for a bin when a slope belt is to be used.

The final two steps involve estimating what materials are necessary for building the shop and the required labor to complete the shop design. The materials needed for the shop include putting a *skin* on the shop to reduce degradation of the seam, roof, and floor strata from weathering and mining activities. This skin is composed of a 76.2 to 101.6 mm (3 to 4 in.) concrete floor poured throughout the entire area (except for the repair bays) and wire mesh coated with shotcrete applied to the ribs and roof to completely seal the area. Other materials, depending upon the final shop design, include concrete block for the walls of the bays, gravel for drainage under the bay floor, paint for the walls and roof, and so on.

Labor requirements are difficult to estimate

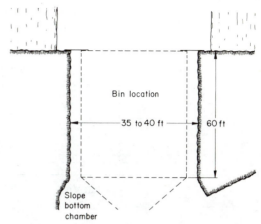

FIG. 8. Excavation requirements for a 635 t (700 st) surge bin. Metric equivalent: ft × 0.3048 = m.

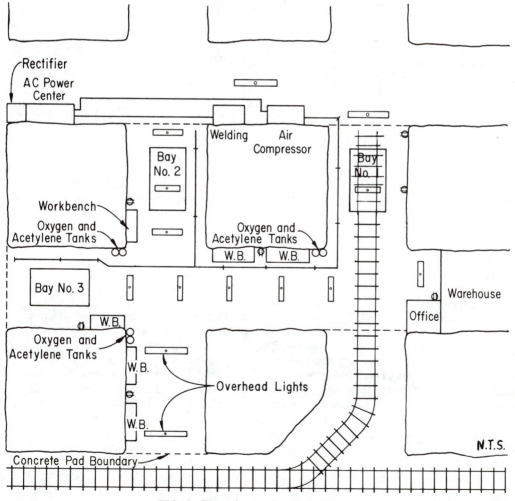

FIG. 9. Plan of an underground shop.

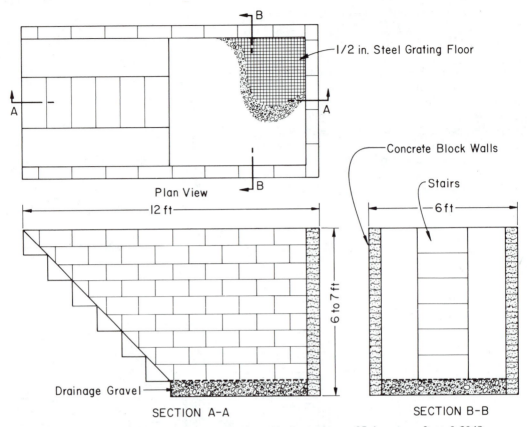

FIG. 10. Profile of a typical repair bay. Metric equivalents: in. × 25.4 = mm; ft × 0.3048 = m.

until a final shop design can be drawn. However, if one looks at Fig. 9, a crew of four workers usually can accomplish all the preliminary work in about 40 man-days. This estimate includes arranging for and delivering needed excavation equipment (approximately eight man-days), pouring and curing the floor (15 man-days), securing and sealing the ribs and roof (15 man-days), plus painting and drilling the hanger holes (two man-days).

The equipment placed in a shop again depends on the desires of management. Most shops will contain shop benches, vises, grinding tools, welding and torch equipment, table saws, and hand tools. In most cases, these tools will be locked up during idle shifts, so adequate storage areas must be planned in advance. Depending on the extent of the shop facilities, more tools and equipment may or may not be planned during design and construction steps.

Scheduling the work is less difficult than imagined. The initial grading and cleanup is done on an idle mine day (because of ventilation problems) and the rest of the tasks are scheduled for convenience by a CPM, Gantt, or bar chart. Project scheduling will be discussed in some detail in Chaps. 8 and 9.

Assembly

Assembly of the shop facilities requires four steps: laying the foundation; securing and sealing the roof and ribs; installing lighting, welding, air and water requirements; and installing additional fabricating facilities. As mentioned earlier in the planning task, the assembly of the shop's shell (floor, walls, and roof) takes an estimated 32 man-days of work for a four man crew. The first task is laying the foundation.

The foundation of the shop is poured for several reasons. The first is the sealing effect it provides to soft mine bottoms. Secondly, the poured floor is usually level and can easily be cleaned over and over without wearing down. Finally, a poured floor will reduce much of the tripping and stumbling hazards associated with uneven mine floors.

The floor is first framed out with 25.4 × 152.4 mm (1 × 6 in.) lumber and well supported along its length with stakes. To provide support and drainage before the concrete is poured, 50.8 mm (2 in.) of 25.4 × 9.5 mm (1 × 3/8 in.) stone is spread throughout the framed area. If rail transportation of large equipment is planned and rail is laid throughout the shop, then the framing should be done so that future rail maintenance work on the track can be accomplished. This means leaving access to the trackbed and rail plates.

A concrete mixer and raw materials are next brought to the site. The standard ratio of 1:3:6 [1 cu ft (0.03 m^3) of cement, 3 cu ft (0.08 m^3) of sand, 6 cu ft (0.17 m^3) of stone] given by O'Rourke (1940) for general mass construction with a water-cement ratio of six is recommended because of cost and lack of age control. Roof bolts, wire mesh, or other reinforcing materials are sometimes placed in the frames to help support the concrete. The concrete is mixed and transported in wheelbarrows to the required spots in the frames and dumped. Concrete is not dumped in the bays for a floor. Steel grating is used instead and the gravel remains uncovered for drainage.

Concrete has initial setting times of between two and four hours, which gives the mixers plenty of time to work the concrete in and finish it. Hoes, rakes, and cement finishing tools are used to do the job. The mixers must pour and level the floor at the top edge of the 1 × 6 frame.

Once the floor is poured and dried, the roof and ribs are coated. Here, wire mesh has been bolted to the ribs and roof in an offset pattern (Fig. 11). This bolting should have been done with a stoper or jackleg drill before the floor was poured. The driller will drive a hole at least 0.9 m (3 ft) deep into the roof or ribs. Extra holes are drilled in both the roof and ribs on a regular basis [approximately every 0.9 m (3 ft)] for future use. Roof bolts [762 mm (30 in.)] are inserted in these extra holes, but not tightened.

Next, a sealant is sprayed on the bolted wire mesh to coat the shop area, 9.5 mm (3/8 in.) thick, and seal it from air and water weathering. The coating can be troweled down to provide a smoother surface and equalize the thickness of the coverage throughout the shop. Latex-based paints can then be used to provide a contrasting background and smoother surface for cleaning.

The next step is installing the lighting, air, and water lines to make the shop serviceable. Utility hangers for light fixtures and cables are placed in the extra roof holes and tightened (Fig. 12). Fluorescent instead of incandescent types are recommended for area lighting. Task lighting of certain operations (grinding, bit sharpening, etc.) may be desirable. Rib bolts with header boards carry the 12.7 mm (1/2 in.) piping or hosing for the other air lines. These lines must be marked accordingly.

Equipment such as benches, shelves, part bins, and the like are bolted with metal straps to the available utility roof bolts found in the extra rib holes. Welding, oxygen, and acetylene tank holders can be fashioned and bolted to the ribs. Steel floor grating is placed in the bays above the gravel on a concrete frame which also tends to support the sealed walls of the bay.

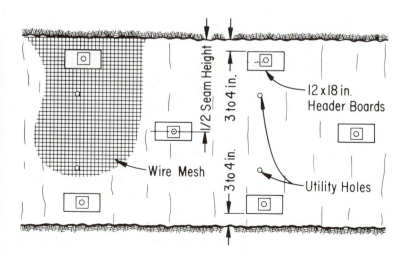

FIG. 11. Wire mesh rib bolting pattern for shop ribs. Metric equivalent: in. × 25.4 = mm.

FIG. 12. Sister Hook type utility hanger (courtesy of Ohio Brass Co., *Catalogue No. 70*, Mansfield, OH).

The final group of tasks falls under the general heading of constructing the fabricating facilities of the shop. This basically means putting the finishing touches on the shop skeleton by providing such things as an air compressor to drive the compressed air system, a movable chain jack to transport heavy objects, washing sinks with drainage, and supplying the vices, grinders, saws, and other tools needed for the shop.

Maintenance

Once the shop is running, a maintenance schedule should be drawn up just like any other construction project. Here the tasks the construction engineer should include are:

1) *Scheduling for effective operation,* which includes delivering, organizing, and picking up of mining equipment for repair. This also includes a regular cleanup day where the shop is hosed down, cleaned, and inspected for safety. Also, the shop is organized for effective operation. Joint efforts among the production, maintenance, and support staffs are desirable for maintaining shop standards.

2) *Resealing the skin* of the shop after a few years is a necessity. Over time, weathering, ground stresses, and material age will weaken the skin. An inspection program on a nine month basis should prevent much potential trouble.

3) *Additional shop area,* based on changing mine needs.

In most mines over time, an underground shop reverts to nothing more than a morgue for old or damaged parts. This is unfortunate since well-maintained shop facilities can aid in efficient coal production in the same manner as the mine's ventilation, haulage, drainage, and transportation systems can and do.

SPECIAL PROJECTS

This section deals with special projects found in underground coal mining. They are called special projects because for the most part, they are not constructed as a matter of course during the mine development; yet many older mines have planned and installed these projects as an integral component of the mine production system. As examples of the third type of project, these tasks usually are performed in partnership with a contractor, but with the mine as the leading participant. The construction engineer's role now becomes one of a project engineer who is primarily involved with overseeing a major project design and installation. Although the distinction in title for the engineer may be small, the difference in responsibilities is large. The construction engineer's share in making the basic decisions instead of following design and construction steps increases sharply. The construction engineer in some instances will become a project manager, responsible for the successful completion of the task. It is these additional responsibilities which make this discussion of special projects unique.

The projects to be covered here, as mentioned earlier, are a rotary dump, an underground crusher station, and an underground foreman/field engineering office. Although design details are considered too site specific to discuss, several basic factors common to these mining installations should be covered. These factors are necessary for the construction engineer in his role as a project manager, as well as project engineer.

Installing a Rotary Dump

A rotary dump design is shown in Fig. 13. This drawing illustrates the elements which compose the basic rotary dump. Older mines with rail haulage frequently use an underground dump to transfer coal from the rail system to a surge bin for transport out of the mine. As with surge bin construction, outside contract help usually is required to initially install the dump at the desired location.

Traditionally, rotary dump facilities are reserved for mainline haulage in deep operations

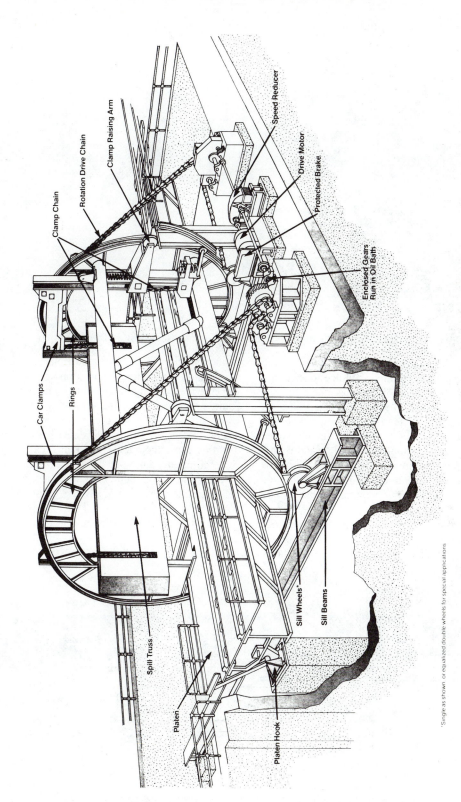

FIG. 13. Rotary dump design (courtesy of Heyl & Patterson, Inc.).

*Single as shown or equalized double wheels for special applications

FIG. 14. A rotary dump being used at a drift type mine (courtesy of Difco, Inc.).

located in the shaft bottom entries. However, they can also be associated with drift operations (Fig. 14). Underground dumps must be designed and planned in advance of the mine because an elaborate haulage system usually is required for effective operation, as shown in Fig. 15. Adequate approach, dump, and return facilities must be present with the rotary dump itself to ensure smooth operations and a flow of production. A dump can either be air or hydraulically powered.

Even though the manufacturer provides the dumping system as a package, a mine must review and provide several design considerations for installation. The first is sufficient foundations for the dumping unit and approach runway. Fig. 16 shows a dumping unit requiring point foundations typical of these turnkey facilities. Conventional pedestal concrete footings are poured to support the steel beams if soft mine bottoms are encountered. In many cases, the mine company can work from design drawings of the dumper for foundation specifics, similar to the example drawing of a rotary dump in Fig. 17.

The excavation requirements are the second consideration for the mine. Adequate pits have to be made to install the dump on the same

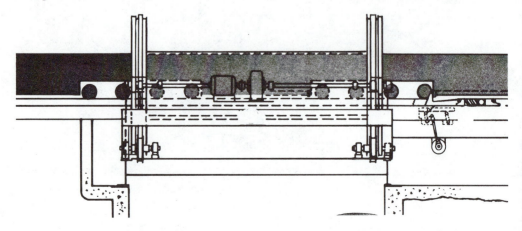

FIG. 16. Point foundations required in a rotary dump design (Anon., 1969).

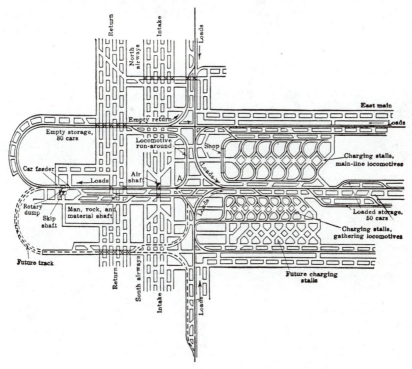

FIG. 15. Shaft-bottom layout, WV coal mines (Peele, 1943).

elevation line as the mine track system, as well as providing a place for the dump motor and controls. The floors of the approach pits are usually poured reinforced concrete to help stabilize the structure.

The third consideration is the runway to the dumper itself. It must be designed and built to provide a straight path for the cars into the dump, as shown in Fig. 18. If it is a single-car dump, the runway distance should be at least as long as two cars. Multiple car dumps require even longer runways.

Finally, the mine must make provisions to seal and reinforce the dump site for mine life usage. This means sealing the roof, ribs, and floor from weathering, water, and ground pressures (Fig. 19). Environmental and safety enhancements such as lights, control booth, water sprays for dust and guarding are also included in these considerations.

The construction engineer can organize his responsibilities into four general categories. These can briefly be defined as the following:

1) *Analyzing the mine requirements* for the

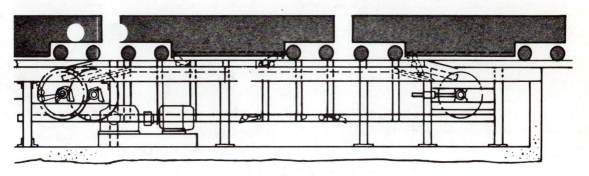

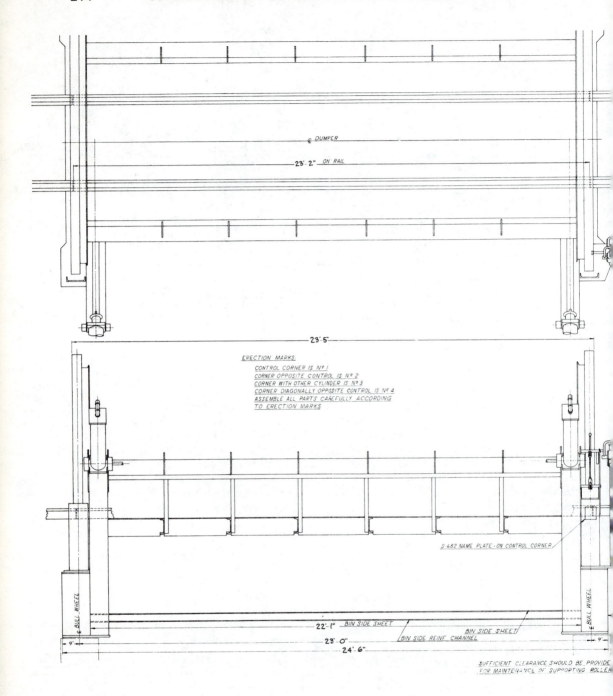

FIG. 17. Design of a dump foundation (courtesy of Difco, Inc.). Metric equivalents: in. × 25.4 = mm; ft × 0.3048 = m; gal × 3.785 412 = L.

SURGE BINS, SHOPS, AND SPECIAL PROJECTS

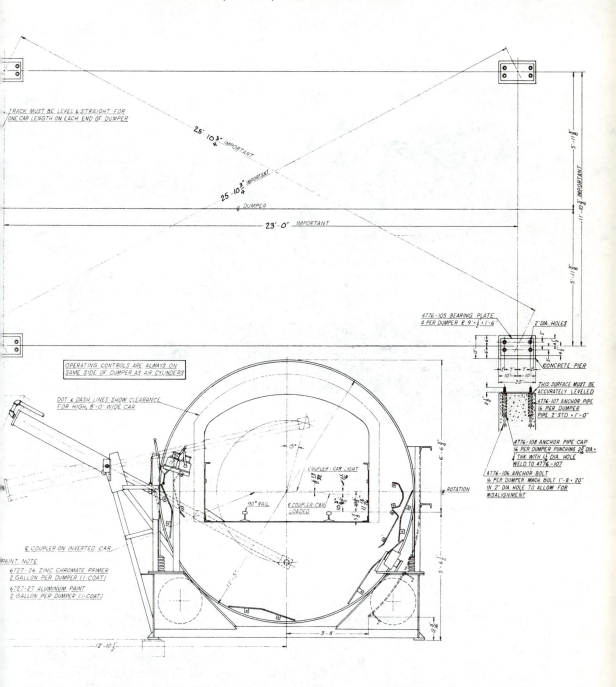

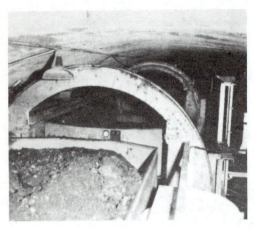

FIG. 18. Dump runway must be straight into and out of the chute (courtesy of Difco, Inc.).

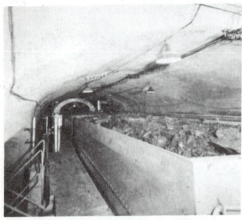

FIG. 19. Sealing the dump site is very important for long life (courtesy of Difco, Inc.).

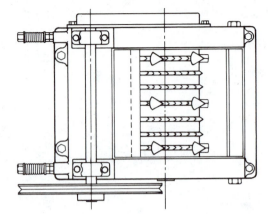

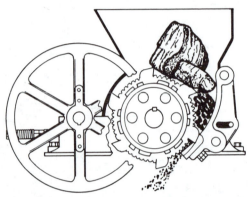

FIG. 20. Sectional view of single-roll crusher (courtesy of Jeffrey Manufacturing Co.).

dumping system chosen to be placed underground. Factors such as location, capacity, cost, auxiliary systems, alternative designs, and excavation requirements are all considered.

2) *Preparation* by the engineer to establish site orientation, construction project schedules, delivery of the required materials, and coordinate necessary construction equipment.

3) *Installation* tasks performed by the contractor need to be monitored and inspected. Interaction and coordination of mine and contractor workers will need direction. Deliveries of supplies and materials for construction to the contractor will have to be brought underground and placed at the construction site.

4) *Operation and maintenance* activities center around the effective performance of the dump during the operating life. Routine, preventive, and "brush-fire" maintenance schedules of needed tasks should be drawn up and rigorously followed.

There are many more tasks involved with the installation of a rotary dump, but in most cases, the contractor will take the leading role. It is sufficient to say the construction engineer will be a key individual in the successful installation of the dump by providing the technical and construction coordination between the mine and contractor.

Installing the Underground Crusher Station

An underground crusher station is another special project found in older mines. Here, the mine system usually requires additional breaking force in the haulage system because of higher than normal amounts of hard rock in the ROM product. A crusher station provides this additional breaking force without interrupting the flow of production from the operating units. In most cases, the station is placed on the discharge end of a surge bin (see Fig. 3) and is termed a feeder-breaker. With drift mines, where the ROM material goes directly out of the mine, the crusher is usually located off-site at the preparation plant or load-out facility.

Like the dump, this structure is, in all likelihood, contracted out to experienced manufacturing firms who will design and fabricate a feeder-breaker for the mine. Therefore, the construction engineer will assume a role as project engineer or manager. His main function is one of motivation, communication, and quality control. Mine labor will be used to excavate and install the crusher (if it is a retrofit) unless there is a contract for turnkey services.

In coal mines, the feeder-breaker acts in the same manner as the grizzly level in a hard rock mine. It regulates and sizes the mine feed to the plant. A feeder-breaker reduces the wear of the slope belt or hoisting equipment as well as providing a coarse size reduction capability.

The roll crusher is the most common type currently used in underground situations. The best known type is the single roll crusher, which breaks the ROM material against a fixed plate (Fig. 20). The roll crusher will reduce the ROM size to 254 mm (10 in.) or less. The rolls are spring-loaded and can be adjusted to conform to many different operational requirements. These crushers are rugged in design, reliable, simple in construction, and easy to repair.

A feeder-breaker can be conveniently installed in the haulage system between the surge bin and slope belt. As the feed is drawn from the bin, it enters the crusher. It falls from the feeder-breaker onto the belt for transportation out of the mine. In this case, the feeder-breaker is mounted on a stand made of I beams and supporting struts. The legs of the stand straddle the receiving belt line to allow the crushed product to dump onto the belt line (Fig. 21). A discharge

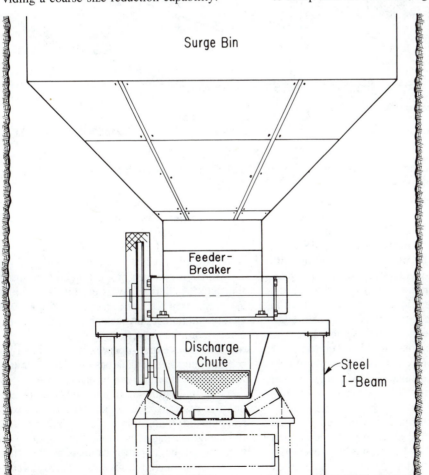

FIG. 21. Crusher stand between bin and belt line.

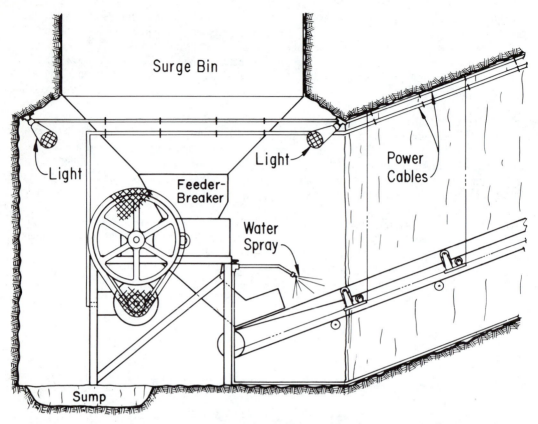

FIG. 22. Lights, power, and water connections around the crusher area (simplified for clarity).

chute may be used to help with the transfer of the ROM material from the crusher to the belt.

The design and installation concerns of the construction engineer in this type of project focus on the interfacing required among the bin, crusher, and belt. Chutes, exceeding the angle of repose, must be of solid construction with adequate linings to withstand abrasion and wear. Power and water connections to activate the crusher and reduce float dust must be placed in safe and strategic locations (Fig. 22). In addition, lights, inspection hatches, and access to high maintenance areas are also required.

In concluding this discussion, it should be pointed out that the underground feeder-breaker is a well used piece of equipment in today's mines. Because mining is becoming more difficult all the time, the underground feeder-breaker should be considered in all mines where a high reject is expected, where faults are repeatedly encountered, or the top, floor, or middle rock is extremely hard.

Building an Underground Foreman/Field Engineering Office

The final topic concerns the merits of building an underground foreman/field engineering office. Many mines are beginning to install just such structures for a number of different reasons. A central office underground can provide coordination between production and construction activities during any given shift. It provides a storage area for items required by the 1969 Health and Safety Act such as first aid supplies, rescue equipment, reflective signs, smoke indicators, and extra supervisory equipment. It provides an intermediate base for belt monitoring equipment which can be easily observed by workers underground. In addition, classes and training sessions can be held in the area with teaching aids being stored in the office. Finally, field engineering files for active construction projects can be stored in cabinets at the office to prevent damage but give easy access to construction foremen. It makes good sense to have

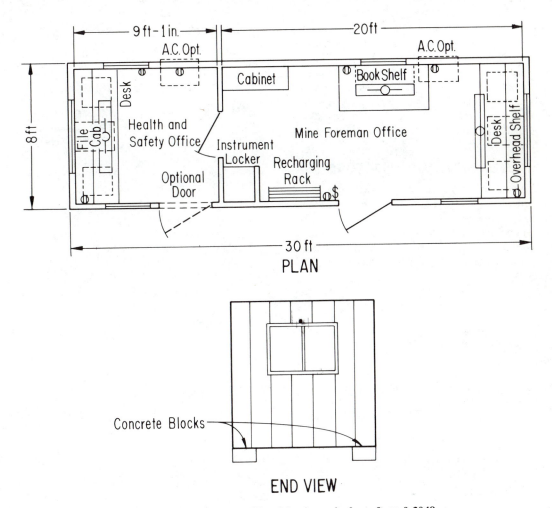

FIG. 23. Sketch of a typical office. Metric equivalent: ft × 0.3048 = m.

a centrally located office for a base of command in a large mine (greater than 1.0 million tons annually) with far-flung production, construction, and maintenance activities.

There are two approaches to constructing an office underground: prefabrication and conventional mine construction. Both achieve the same purpose, although the prefab office is usually portable, prewired, and more costly than the conventional approach.

A prefab approach to a mine office requires a minimum height and width. In most cases, at least 2.4 m (8 ft) in height is required, with a minimum width of 4.5 m (15 ft) necessary. A typical office layout is shown in Fig. 23. The space is broken into two main areas as a division between mine production and project engineering responsibilities.

One advantage of a prefab office is the ease of installation. If possible, a mine office can be transported along with other mine supplies to the appropriate site. Power can be easily run to the unit which is prewired for convenience. No foundation is usually necessary other than leveling the site and providing drainage.

A conventionally constructed office can easily be installed by mine workmen in between two pillars. The floor is poured concrete, the roof and ribs shotcreted for protection. A front and back wall are constructed with concrete block. A door and window are placed in each block wall to provide access and visibility. Lights are strung and a telephone placed in the office to communicate with the mine and surface. Other items are placed by management at their discretion. In some instances, paneled walls and a suspended ceiling are installed to give a better appearance to the office and provide more comfortable working conditions.

The advantages of this approach over the prefab type are cost and durability of the conventional office. Over the mine life, the conventional office approach will tend to remain usable and withstand weathering effects with regular maintenance.

EQUIPMENT MANUFACTURERS

Although extensive details were not given throughout this chapter, the many topics covered require equipment and contractors with experience. Table 1, taken from the Buyer's Guide by McGraw-Hill Mining Services, summarizes some of the firms involved with this type of work.

REFERENCES AND BIBLIOGRAPHY

Anon., 1969, *Nolan Bulletin No. 104,* The Nolan Co., Bowerston, OH.

Beall, J.V., 1980, "The Role of an Engineering and Construction Contractor," *Mining Engineering,* Vol. 32, No. 10, Oct., pp. 1476–1478.

Cummins, A.B., and Given, I.A., 1973, *SME Mining Engineering Handbook,* AIME, New York.

Difco, Inc., 1978, *Mine Cars,* Findlay, OH.

Difco, Inc., 1981, Private communication.

O'Connell, E.J., 1980, "Modern Engineering and Construction Methods in the Execution of Mineral Projects," *Mining Engineering,* Vol. 32, No. 10, Oct., pp. 1482–1487.

O'Rourke, C.E., ed., 1940, *General Engineering Handbook,* 2nd ed., McGraw-Hill, Inc., New York.

Peele, R.J., 1943, *Mining Engineer's Handbook,* 3rd ed., John Wiley & Sons, New York.

TABLE 1. Equipment Manufacturers

Bin & hopper outlets (nonplugging)

Kalenborn
Lancaster Steel Products
Long-Airdox (Div. of Marmon Group)
Long-Airdox Construction
Maschinenfabrik Buckau R. Wolf Aktiengesellschaft
McNally Pittsburgh Mfg. Corp.
Mid-East Conveyor Co.
Rexnord Inc.
Solids Flow Control Corp.
Vibra Screw

Consulting services

Acurex Corp., Energy & Environmental Div.
Americal Coal Training Inst.
Babcock-Moxley Ltd.
Bechtel Corp.
Berger Associates Corp.
Boggess, B.L. Consulting Group
Boyd, John T. Co.
British Coal International
DuBois Chemicals, Div. of Ehemed Corp.
duPont de Nemours, E.I. & Co. Inc.
Ecology and Environment, Inc.
Energy Sciences & Consultants, Inc.
Environmental Research & Technology, Inc.
Envirosphere Co.
Envirotechnic, Inc.
Evergreen Weigh Inc.
Fluor Mining & Metals, Inc.
Ford, Bacon & Davis Utah, Inc.
GEOMIN
Gilbers Associates Inc.
Kaiser Engineers, Inc.
Krupp Indrstrie-und Stahibau
Mars Mineral Corp.
Matthew Hall Ortech Ltd.

Nicholson Engineered Systems
Research Cottrell, Inc.
Statdard Laboratories, Inc.
Thyssen Groups of Companies
Weir, Paul, Co. Inc.

Contractors

Allen & Garcia Co.
Armco Inc.
Badger Construction Co.
Boggess, B.L. Consulting
British Coal International
Cementation Co. of America
Cementation Mining Ltd.
Cpstain Mining Ltd.
Daniels Company
Dravo Corp.
Fairfield Engineering Co.
Fluor Mining & Metals, Inc.
Foraky Ltd.
GEOMIN
Hawker Siddeley Electric Export Ltd.
Kaiser Engineers, Inc.
Lively Mfg. & Equipment Co.
Long-Airdox Construction Co.
Matthew Hall Ortech Ltd.
Morrison-Knudsen Co. Inc.
Roberts & Schaefer Co.
Robinson & Robinson Div., NUS Corp.
Ruttmann Companies
Thyssen Groups of Companies
Zeni Drilling Co.

Dumps

A-T-O Inc.
Card Corp.
Connellsville Corp.

Difco, Inc.
Dorr Oliver Long, Ltd.
FMC Corp.
Heyl & Patterson, Inc.
Jenkins of Retford Ltd.
Kanawha Mfg. Co.

M.A.N. Maschinenfabrik Augsburg-Nurnberg
McNally Pittsburgh Mfg. Corp.
Mining Equipment Mfg. Corp.
Nolan Co.
Sanford-Day/Marmon Transmotive

Section 2
PRINCIPLES OF CONSTRUCTION ENGINEERING MANAGEMENT

It is important as engineers and potential managers of underground construction projects to realize knowing *how to* is not enough. Effective management is composed of numerous elements of leadership, worker motivation, and delegating authority, as well as having knowledge of the required tasks. It also involves using effective cost estimating procedures and taking a broad view of the tools available for project performance during installation, operation, and maintenance tasks.

This section of the book delves into the various aspects of project management. The objective of Chapters 8, 9, and 10 is to introduce and familiarize the engineer with the basic management principles involved. Chapter 8 looks at the three basic aspects of project management, including worker motivation, leadership skills, and organizational management. Chapter 9 covers the computer design and performance tools available for construction engineering work. Finally, Chapter 10 discusses project cost analysis techniques for managers. Although individual textbook treatment of these topics is the usual presentation, I have chosen to include the basics because of their importance to sound installation of underground projects.

For the next three chapters, I refer often to the term *project*. In these instances, I am loosely defining both everyday support construction tasks and any special type of mining task which may arise. The reader should not become confused by the use of this singular word for a host of potential support work.

8

Construction Project Management

Many times during the construction of a project, whether large or small, the engineer will become involved with management of the work and workforce. In some cases, mine management will turn over the direction and responsibility of doing the work to the engineer. Because of this, it is important to know some of the basic principles of human resource applications—among other management concerns.

This chapter, therefore, looks at the fundamental principles of human resource management. Three broad discussions are presented to the reader. These are: (1) organization, communications, and responsibilities of management; (2) analysis of worker motivation; and (3) supervisory leadership.

Each topic encompasses the current thinking for coal mining and mining management. Much of the information in this chapter comes from *Practical Coal Mine Management,* by this author and published by John Wiley & Sons, Inc., which discusses in depth many of the ideas presented here. Interested persons are directed to this text for more information on management skills.

ORGANIZATION, COMMUNICATIONS, AND RESPONSIBILITIES OF MANAGEMENT

This discussion reviews the current structure and responsibilities of underground coal mine management. It is divided into three major topics: (1) construction management organization, (2) vertical and horizontal communication and (3) project responsibilities. Each focuses on what principles or aspects of project management the engineer should be aware of before, during, and after construction of a project.

Construction Management Organization

Construction management can be thought of in mathematical terms as a subset of the mine management. It occupies a niche in the organizational chart under the mine production management structure (Fig. 1). The shift mine foreman or mine construction engineer reports directly to the mine manager, with secondary communication links to the mine engineer and technical staff.

For many mines, a project manager position is temporary, used only on demand, because large projects such as the ones discussed in Chap. 7 are generally few and far between in any given mine. Although a great deal of construction takes place on a daily basis, there is no designated project manager.

The traditional approach (for working without the project manager) was for all construction activities to fall under the direction of the production management. This approach has been quite successful in the past for two reasons. The first is that past projects (prior to 1969) in mine construction were generally localized, with little emphasis on effective performance or safety standards. The general approach could be summed up with the attitude of doing "enough to get by." The second reason was a combination of factors: technology, geologic conditions, and construction materials which all tended to lower the importance of effective construction on the overall mine economic picture. If, for example, an overcast leaked, then the mine system did with a little less air or purchased a larger fan.

Today, of course, these reasons have no place in modern coal mining. This is why construction activities should not be placed under the management of the production personnel at the mine. Regardless of how the production managers view

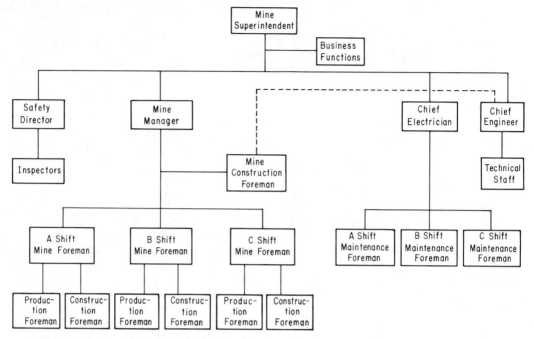

FIG. 1. The mine construction foreman does have a niche in the mine management structure.

construction activities, their attention and concern will be dominated by the production system and production system problems. Therefore, the organizational structure at the mine must be reorganized to elevate the construction end of coal mining to its important position along side of production and maintenance activities.

To this end, there are three approaches to organizing project management at the mine level. The best of the three and the one reviewed here is called the *matrix approach*.

The Matrix Approach: The organizational structure which best lends itself to a coal mine management system is called a *matrix*. As shown in Fig. 2, the matrix structure allows the project manager to cut across all pertinent mining functions (i.e., production, maintenance, technical support, and so on) and utilize the talent within these functions. The project manager reports directly to the mine superintendent or manager and is responsible for the success of the project. The functional manager also reports to the mine superintendent, but is responsible for the quality of work produced in his particular area. For example, the chief electrician controls all the maintenance work at the mine. When a project manager requires some type of maintenance work done, the project manager will assign the personnel available to him and the chief electrician sees that the work is done correctly and on time.

There are many advantages to the matrix organization. It balances the time, cost, and performance requirements of the various mine supervisory functions and maximizes the problem-solving abilities of the project manager. However, it does require that more attention be paid to administrative duties, more monitoring of the project status be done, and it can cause conflicts between the functional manager and the project manager. Considerable cooperation may be necessary between these individuals during any given project.

There are certain characteristics of a matrix organization which are required for successful implementation at the mine level. They can be summarized as follows:

1) One supervisor or engineer must spend his full time as the project manager, for both daily activities and special projects.

2) Good communication channels, both horizontal and vertical, must be present and used effectively to maintain project commitments between managers.

3) There must be a method for resolving managerial conflicts which is both quick and effective.

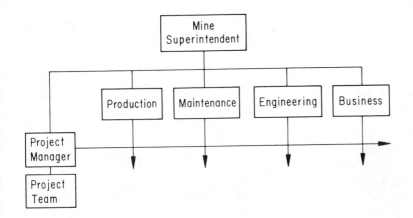

FIG. 2. Matrix organization allows project responsibility in a lateral and vertical direction.

4) An open and honest atmosphere should exist with all managers willing to negotiate mine resources.

5) The project manager must be allowed to operate as a separate entity on all mine matters except for administrative functions.

The basic ideas presented above can make the matrix structure very successful in any underground mine. The commitment by mining management to effective and efficient support systems can be enhanced greatly by developing this separate arm of supervision.

Some mines have begun to implement the matrix structure in a modified form. Here, a construction superintendent reporting to the mine manager controls, directs, plans, and follows through on all support projects at the mine. Although his role is still placed behind the mine production functions, it is clearly more effective than project direction from the production supervisors.

Communications for Project Management

One key to successful project management is effectively communicating ideas, desires, goals, and so on between the project manager, project team, functional managers, and others. This means recognizing the channels of communication which exist within the company management and also the barriers to this communication.

The channels of communication are divided into two basic types, vertical and horizontal. Vertical communications are further subdivided into downward and upward communications. Downward communication from managers to subordinates includes face-to-face, or supervisor-subordinate information exchanges throughout the chain of command. This includes management bulletins, policy and procedure manuals, letters to employees, written performance appraisals, employee handbooks, and so on. A manager usually decides what information will or will not be passed on to his subordinates, and this is followed by the decision on how to transmit the information. Different channels will reflect the importance of the information. For example, a decision to change material or equipment vendors is put on the general company bulletin. This information, while relevant to all future projects, is not important enough to warrant another communication channel such as a confidential memo. Upward information channels include written memos, suggestion boxes, grievances, planned employee meetings, and informal conversations. Additional channels include performance reports and attitude surveys.

Horizontal communication, while being very important, has been largely ignored in the past as an effective communication channel. Such links as interdepartmental memos, staff conferences, employee interaction, and labor-management relations are part of the horizontal communication network. Both formal and informal contacts can be utilized by a project manager.

There are three types of barriers to communication: technical, semantic, and human barriers. Technical barriers include mechanical failures, physical obstructions, technological malfunctions, tangible obstacles, and other common space and distance parameters of the communication process. In the past, technological barriers made up the majority of problems. Today, however, its importance has decreased dramatically with the advent of modern communication systems.

Semantic barriers are problems concerning the question of *what* is being communicated. Often, both written and spoken words mean different things to different persons in different situations. Semantic communication barriers are basically evident within written and oral language expressions, but meaning problems also may be inherent within the use of other forms of signs, symbols, and gestures. This is especially true in coal mining, which boasts its own subset of written and oral language.

Human barriers are considered to be the main forms of communication problems today because they can never be eliminated and they are extremely difficult to reduce in frequency, even slightly. They occur because of biases, perceptions, emotions, competencies, and sensual abilities of the persons involved. These personal differences will dramatically affect the form of communication and the amount of information passed through the system. Technical and semantic barriers can be reduced and sometimes eliminated, but human barriers cannot be removed easily because of the inherent nature of management by humans.

Responsibilities of Management

There are several central ideas of current management structures which will be discussed here. They relate to the responsibilities assumed by the project manager and project team during the course of the project construction. These aspects include:

1) The project manager and his responsibility to the company and to the workers.
2) Managing as a project leader.
3) Managing and project control responsibilities.
4) Safety and training responsibilities.

A construction engineer can be called upon at any time to provide one or more of these aspects in filling a needed position in the organizational structure. Being able to develop and use such skills are an important requisite of today's engineer.

The Project Manager Responsibilities: The project manager is a newcomer to the management team in coal mining. He is the leader of a group of individuals (called the project team) united with a common goal. Project management is only successful when the project manager and his team have a clear understanding of the project objectives and the factors needed to achieve these objectives. The factors for completing any project objective can be summarized as the following:

1) *Project direction,* which encompasses the responsibilities of the project manager through the completion of the project.
2) *Project planning,* which covers the project design, scheduling, labor, and construction steps. The project team takes a major role in project planning on a day-to-day basis.
3) *Project evaluation* as a part of project followup, evaluating the performance of the project in terms of preconstruction objectives.
4) *Project reporting* of all work progress with major construction steps such as preparation, assembly, etc. Cost reporting and progress-to-date reports make up the bulk of these reporting tasks. Milestone, Gantt, or bar charts are very useful for reporting purposes.

These functions, when applied correctly, provide the best communication and coordination system within the mine management.

The question now is: what are the basic company responsibilities of a project manager? Remember, project management has evolved more through necessity than desire. *By definition, project management is the planning, scheduling, directing, and controlling of company resources for a short-lived project.* A project manager, therefore, has responsibilities to the company for:

1) Producing the desired project with the available resources within the constraints of time, cost, and technology.
2) Meeting contractual/desired performance objectives.
3) Making all required decisions, whether alone or with the advice of others.
4) Acting as both an external (quality control) and internal (decision manager) communications focal point.
5) Resolving any conflicts during project duration.

These responsibilities define broad powers which are granted by higher management to the project manager to get the job done. However, the project manager often is in conflict with traditional levels of company management, since granting much of the authority to a project manager requires taking authority from other managers. A project manager, therefore, may spend much of his productive time in organizing and planning company resources in direct competition with other managers.

A project manager also has responsibilities to the employees under him. His primary concern in this case must be the health and safety of the worker in the working place. To this end, his responsibilities can be summarized as:

1) Giving the best training possible toward completing the project.
2) Providing the tools and supplies necessary to complete the project.
3) Providing guidelines for acceptable and unacceptable work.
4) Respecting the worker as an individual, and not just a measure of output.
5) Clearly showing confidence in the employee's ability to do the required work.

Even so, these responsibilities cover just a few fundamentals of worker motivation a project manager should understand on a project.

Managing as a Project Leader: As mentioned earlier, very few mines have full-time project managers. The most common approach to filling this position in times of need is to assign the extra responsibilities to the mine manager, mine foreman, or chief engineer. This is a relatively easy choice for most mines. However, as a project leader, choosing qualified persons for the project team becomes harder. If outside staffing from other company offices, universities, and consulting services is unavailable, then the choices may be limited. Before choosing a staff, two major questions on staffing must be considered:

1) What people and skills are required?
2) What organizational structure will be best for completing the project?

These questions relate to the nature of the project and the steps required to complete it. For example, a sample project could be to evaluate and upgrade the overcast performance in a given mine, initially assigned to the chief engineer. As project manager, he must coordinate and exercise control over the project functions of: (1) planning, (2) organizing, (3) staffing, (4) controlling, and (5) directing. His project team, therefore, must be chosen to integrate and facilitate all the above functions with the project goals.

Project Functions.

Planning is simply decision-making based on future events. It determines what must be done, by whom, and by when, to complete the assigned task. To accomplish this, nine major steps, or questions, must be answered during the planning phase:

1) *Objective:* What is the project goal and when must it be completed?
2) *Program:* How will the project be completed to meet the objective?
3) *Schedule:* What is the plan of action for individual or group functions and when should they be completed?
4) *Budget:* How much should be spent to achieve the objective?
5) *Forecast:* What will happen by a certain time?
6) *Organization:* What are the positions required to achieve the project objective?
7) *Policy:* What are the general guidelines for decision-making and individual actions?
8) *Procedure:* What is the method for carrying out the policy?
9) *Standard:* What is the acceptable performance level?

Planning is a continuous process, dynamic in its function throughout the project. Planning can be composed of steps which invoke logical thinking and organization. The successful project manager, who is particularly adept at the planning stage, can minimize future trouble spots by good planning.

Organizing and *Staffing* (i.e., the project team) have been discussed earlier. The matrix system appears to be the best approach for mining management. Staffing for projects can be a combination of internal and external skills from various technical, business, and mining functions. For example, surface construction at an underground mine may require environmental evaluation skills which are beyond the scope of the technical staff at the mine. The planning function of the project leader will be able to foresee these needs and aid in developing the best organization and staffing requirements for the project.

Controlling is a three-step function of measuring the progress toward the project objective. During the project, controlling functions also evaluate what remains to be done and what steps are required to reach the objective. A project manager must have a sound understanding of the standards and cost control policies of the project so the measuring, evaluating, and correcting steps of project controlling can be compared to planning estimates for a performance review.

Directing is the carrying out of the plans for achieving the project objective. Directing includes:
1) Staffing the project positions.
2) Training each individual and group.
3) Supervising the day-to-day activities.
4) Delegating authority to fully utilize the abilities of others.
5) Motivating others to do their work.
6) Counseling others with project problems.
7) Coordinating the project activities.

It is important to note that project directing is the key component to active project construction. The success of the previous functions requires a strong commitment to directing the project as a project leader and decision-maker.

Managing and Project Control Responsibilities: Mine management is continually seeking new and better control techniques for dealing with the various project tasks. To date, there are four such methods which have had success in the mining industry. These are:
1) Program Evaluation and Review Technique (PERT)
2) Critical Path Method (CPM)
3) Line of Balance (LOB)
4) Gantt chart

A fifth technique, called zero-based budgeting, is relatively new but is gaining acceptance in the industry as a control on project accounting.

PERT techniques were developed by the US Navy during the late 1950s to help manage their large military programs. These techniques basically force managers and project teams to organize the entire project as a series of work networks, composed of either events or activities (Fig. 3). These networks help identify the problem areas and event interdependencies which can snag a project. A major advantage of PERT is the fact that the very detailed planning required for PERT evaluation may otherwise be overlooked during project planning by the project team. Other advantages of PERT include assigning deadline probabilities, evaluating program changes, and providing a vehicle for organizing massive amounts of data. For small projects, PERT is not recommended because of its massive data requirements and expensive upkeep. It is excellent for major construction or rebuilding projects in mining such as prep plants, rebuilding major equipment, or constructing a project such as a surge bin and crusher station.

Critical path method (CPM) analysis is a technique refined from the PERT network approach. Basically, it focuses on the one event and activity path which is critical to the timely completion of the project. By critical, I mean the activities requiring the most time to complete. Therefore, by tracing these activities through a project, the critical path can be identified. For CPM, the emphasis is on activities, not events, (as in a PERT analysis). CPM analysis also considers both the time and cost of each activity, especially for activities on the critical path. CPM works better than PERT in mining situations where:
1) There are well-defined, short-lived projects, such as building overcasts, sumps, and track switches.
2) There is one dominant organization involved, namely the mine management, which is responsible for all project functions.
3) There are relatively small uncertainties with design, construction, and scheduling of the project. This is typical of support construction projects underground.

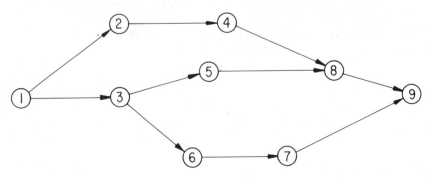

FIG. 3. PERT diagram.

4) The project is confined to one geographic location in the mine.

Mining projects using CPM include both support construction and production, as well as maintenance activities. With support activities, CPM can be best used to analyze such items as ventilation network construction between mains and submains, drainage systems, and belt line/track extension programs.

Line of Balance (LOB) techniques are a third outgrowth of the PERT system. LOB techniques are used to monitor small projects (like drilling hanger holes for messenger wire). Using Line of Balance analysis correctly can lead to:

1) Lower project costs and project time frames.
2) Better coordination and planning between the project team.
3) Less idle time between project activities.
4) Better scheduling of any contractor or consultant activities.
5) Better troubleshooting procedures.
6) Better decision-making procedures.

Many common work elements in typical mining projects can fit the criteria for using Line of Balance techniques, if properly approached.

Gantt charts have been the standard project control technique for a number of years. They fall in the same category as milestone, logic, and bubble charts, so they need no explanation here. The main problem with these techniques is that they cannot show event and activity interdependency. These interdependencies are critical to understanding the project progress at any given time. However, they are simple to understand, require no formal training in their development, and cost practically nothing to develop.

All the techniques discussed here are available to mine management for effective control. Of course, only one or two should be explored for use at any one time, since a great deal of redundancy and wasted effort would occur if all the analysis techniques were established at a single mine. Generally, CPM is recommended for all ordinary mine construction activities for use by the project engineer, project manager, and project team. However, mine supervisors overseeing the construction work as a portion of their production responsibilities should try LOB techniques to aid their efforts. LOB techniques are a shirt-pocket type of approach which can easily be transferred to the underground frontline supervisor.

Before ending this discussion, one may ask the general question: "Why use these techniques at all and is it that important?" The answers are simply because these techniques are proven and, yes, it is very important. The bottom line is, of course, cost control, but time control is also an added benefit to a formal program of project analysis. Mine management must exercise all the tools available to them for increasing the efficiency of their operations. If questions or problems arise over establishing one of these analysis techniques at the mine, then management should seek the outside help of consultant, university, or another company which has the skills to set up a sound and beneficial program.

Safety and Training Responsibilities: For a mining manager, there are probably no responsibilities more important to the worker than safety and training. The coal industry as a whole has taken a great deal of criticism in the past for neglecting both responsibilities. This condition is changing rapidly as more and more companies implement necessary programs.

This discussion looks at the responsibilities of the mine management organization toward safety and training. Many texts have been written, many consulting companies provide services, and many mining schools teach safety and safe management to interested persons. Highly recommended is the Mine Safety and Health Administration's Safety Academy in Beckley, West Virginia. They provide an extensive clearinghouse on safety literature as well as low-cost instructional courses designed specifically for the mining industry. Subject matter is available in a brochure upon request from the Mine Safety and Health Administration, P.O. Box 1166, Beckley, WV 25801.

Safety. Safety is a fundamental right of the worker under supervision. It makes good sense, therefore, to do everything within reason to make the working place safe from accidents. "An ounce of prevention," as the saying goes, "is worth a pound of cure." This has never been more true than in the hostile environments of underground mining.

As a project manager, it is essential to understand some of the basic causes of accidents and the actions needed to prevent them. This discussion, therefore, is divided into two topics, defining mining accidents and hazards; and mining accident prevention fundamentals. Again, these discussions will serve only to introduce the topics to the construction engineer or project manager, since much better discussions abound in other literature.

Mining Accidents and Hazards. Mining accidents are caused mainly by workers performing unsafe acts. The second largest cause of accidents and injuries is unsafe conditions. Both causes account for about 98% of the total mining accidents in underground coal mines.

Unsafe acts is a broad category describing much of the human failure which occurs with accidents, by both supervisors and workers. Some of the categories of unsafe acts include:

1) *Operating without authority* such as jumping on a moving tram, operating a shuttle car or roof bolter without proper training, or using hand tools for other than designed purposes.

2) *Failing to secure* items on moving vehicles, locking out electrical circuit breakers, or following roof and rib control plans.

3) *Eliminating safety devices* by bypassing or bridging out disconnects, breakers, fuses, and so on, or failing to keep them in good shape.

4) *Working on unsafe equipment* by not properly blocking equipment, locking out electrical equipment, or properly using welding or torch equipment.

Many other unsafe acts can be listed and discussed, but the idea is the same. A supervisor must be aware of all possibilities in assessing the worker's (and other supervisor's) behavior and attitudes toward eliminating unsafe acts.

Unsafe conditions are conditions found in the working place which cause or contribute to the cause of an accident. Some unsafe conditions found in coal mining include:

1) *Poor or missing guarding* on moving equipment, stairs, platforms, and scaffolding. The 1969 Health and Safety Act requires guarding on all moving parts. Sometimes this guarding wears out and becomes ineffective.

2) *Unsafe physical conditions* such as loose roof or ribs from weathering, vibration, or improper roof and rib control.

3) *Stumbling hazards* in the working place from trash, loose tools, gob, and loose coal.

4) *Unsafe ventilation* of the mine can increase the hazards from gas, dusts, and poor vision.

5) *Defective equipment and materials* from being worn, cracked, broken, rusty, bent, or torn. Hand tools are particularly susceptible to this.

These conditions, while closely watched on production sections, usually are neglected on construction projects. A project manager, supervisor, or construction engineer should be aware of these conditions and inspect quite frequently for slips in their status.

Mining Accident Prevention Fundamentals. There are systems in-place in coal mining which aid management in preventing accidents. Probably the best known system has been adapted from the military and aerospace industry. It is called *system safety engineering*.

System safety engineering has grown from three general factors, i.e., changing technology, mining economics, and managerial demands.

Mine technology is changing by becoming increasingly complex and specialized. No longer does one worker learn to operate one piece of equipment (e.g., a shuttle car) and automatically be assumed to run them all. Roof bolting, for example, is different from operating a track motor.

Economics dictates more system safety when accidents produce a greater mining cost in terms of manpower and production losses, penalties for negligence, and liability settlements. Finally, upper management (and labor) demands for safety in the working place to minimize loss and maintain standards is becoming a mandate for most mine managements.

So what is a definition of system safety? According to J. E. Rankin of the MSHA Academy, system safety is "the optimum degree of safety that can be achieved within the constraints of system effectiveness." Therefore, the engineering of system safety requires the "application of scientific and management principles for the timely recognition, evaluation, and control of hazards within a system." The three fundamental principles that system safety engineering is based on are:

1) Accidents are the result of interacting causes within a defined system. In this case, the overall defined system is the confines of the mining environment.

2) Individual accidents can be analyzed to logically identify all causes and their interaction leading to the accident. Mine safety departments are mandated to vigorously follow this principle in the course of their duties.

3) Solutions, based on the analysis, can be identified and developed to control the causes of any accident. Mine management should be

cognizant of the fact that the burden of responsibility for implementing accident solutions falls on their shoulders, and not on the safety departments.

The system safety engineering approach requires five definite steps to effectively achieve the objectives of the program. These five steps define the life cycle of any given program and can be generally defined as:

1) *Concept* where system safety engineers, (1) establish standard safety criteria, (2) evaluate design alternatives, and (3) establish program requirements and schedules for implementation.

2) *Design* where engineers review and evaluate the prepared design and conduct hazard analysis resulting in effective program designs which will reduce identified system risks.

3) *Development* steps include monitoring subsequent development activities, conducting hazard analysis design in the previous step, and continuing to reduce hazard risks.

4) *Operation* involves the inspections, maintenance and performance review, updating hazard analysis programs, investigating accidents, and correcting new hazards.

5) *Disposal* defines the program maintenance procedures which will monitor and maintain the standards of the program.

The key to these steps boils down to the knowledge, training, and dedication of the system safety engineer and mine management. Cooperation between parties is fundamental for the program to work effectively.

Training. Training is a second big responsibility for a project manager or supervisor. Training serves several purposes, all of which are beneficial to management. Training increases the worker's ability to perform the required task. It also helps instill a correct safety attitude in the worker, and training establishes a constructive rapport between the supervisor and employee. Finally, effective training helps decrease the cost of production (by reducing the learning curve starting point) and increases the net worth of the employee to the company.

One may wonder, if training is so important, why isn't there more training going on in the mines? The fact is, quite a lot of training is taking place in coal mining, but little of it is effective. Effective training requires systemizing the approach, training the supervisors as potential instructors, and following through to ensure retention.

Systemizing the approach basically says a method of teaching should be established and followed by everyone. The approach should be composed of teaching tools which have been time-tested and found effective, such as a job analysis chart (Fig. 4) or training timetable. A job analysis chart is a table which organizes and summarizes both the important steps (what the employee must do) and key points (what the employee must know) for the particular task. The training timetable is a second tool organizing and recording the employee's training and program progress to date (Fig. 5). By taking every applicable mining job and developing a standard approach using the correct tools, job training will get a big boost.

Training the supervisors is a step which cannot be underestimated. Many times, good training systems are destroyed by poor teaching methods. Not everyone is born to be a teacher, so proper training is a must. Haphazard or careless presentations will neutralize the instructional process and be quite ineffective. Training the supervisor means instructing him in the proper methods of teaching and the proper ways of using the instruction tools. Each supervisor should know the *four-step* method of job instruction. This method is really composed of five steps (by including a preparation step). These are:

1) *Preparation* of the supervisor to teach a job task. This includes establishing the objectives of the instruction, arranging the information to be taught, and preparing the place of instruction.

2) *Orientation* of the learner. Here, the supervisor must explain the purpose of the training, and judge the learner's past experience in the job to be taught.

3) *Presentation* of the job tasks by the instructor to the learner. Presentation is very straightforward, following the job analysis chart.

4) *Demonstration* of the job task by the learner to the instructor. All confusing points and questions should be cleared up during this step.

5) *Follow-through* by the instructor on the training of the learner. The instructor helps with any problems and clears up any questions the learner might have.

Each supervisor should become comfortable with this method before actual job instruction occurs.

Following-through is a management commitment to maintaining effective training methods.

This includes extensive data collection, charting learning curves for each individual for each task, and a continuous program of reinstruction and retraining.

As a separate issue, supervisory training is also a training task which should not be ignored at the mine level. It involves the development of programs for mine supervisors and management personnel to learn the abstract concepts of worker behavior and theories of management in a systematic manner. Suffice it to say there are many different methods for teaching these concepts, including understudies, short courses, job rotation, sensitivity training, coaching-counseling, and business games which can be utilized for the instructional process. Mining companies should seek professional advice when they are unsure as to which approach would be best.

In concluding this discussion on management responsibilities, several points need to be stressed. These can be summarized as follows:

1) A clear approach to project organization should contain the necessary elements of structure, communications, and control to be successful.

2) The matrix organizational structure appears to be the best project management approach for mining.

3) Project analysis techniques such as CPM and LOB are recommended as an important component to project control.

4) Safety and training responsibilities are too important to be applied haphazardly. Effective and proven techniques such as system safety engineering and the four-step method of teaching must be implemented and maintained at the mine level.

5) Management training for supervisors is a key ingredient for the success of all project goals and programs.

6) Professional advice should be obtained for all program components with which mine management is unfamiliar.

ANALYSIS OF WORKER MOTIVATION

Worker motivation is a problem every manager must face. Construction project management faces additional problems because of the caliber and training of the workers usually avail-

Task:_____

Date:_____ Instructor:_____

_____

Key Points (what the employee must know)   Required Steps (what the employee must do)

    1._____   • _____
                                                                    • _____
                                                                    • _____
                                                                    • _____
    2._____   • _____
                                                                    • _____
                                                                    • _____
                                                                    • _____
    3._____   • _____
                                                                    • _____
                                                                    • _____

FIG. 4. Job analysis chart.

Name:_____ Employment date:_____
                                             Start:_____
Job Classification:_____ End:_____

Date	Instructor	Task Number	Comments
____	_____	_____	_____
____	_____	_____	_____
____	_____	_____	_____
____	_____	_____	_____
____	_____	_____	_____

Job Classification:_____ Date Start:_____
                                                    Finish:_____

Date	Instructor	Task Number	Comments
____	_____	_____	_____
____	_____	_____	_____
____	_____	_____	_____
____	_____	_____	_____
____	_____	_____	_____

Job Classification:_____ Date Start:_____
                                                    Finish:_____

Date	Instructor	Task Number	Comments
____	_____	_____	_____
____	_____	_____	_____
____	_____	_____	_____
____	_____	_____	_____
____	_____	_____	_____

FIG. 5. Employee training timetable.

able and assigned for mine construction work. Often, a manager will be forced to use a mix of unskilled trainees, unmotivated, but experienced workers, and skilled veterans. To best utilize these talents assigned to him, the project manager must understand some of the fundamentals of worker motivation.

There is no single definition of worker motivation that everyone agrees upon. The general idea is that worker behavior toward any project or task is motivated. However, no consensus as to what that behavior or motivation consists of can be agreed upon by the experts.

The fundamental theory on the motivation of individuals was advanced by A. H. Maslow in his work, *Motivation and Personality*. Maslow makes the case that each human individual has a hierarchy of needs which must be fulfilled. These needs, in order, are: (1) physical, (2) security, (3) social, (4) self-esteem, and (5) self-actualization. Each need must be fulfilled before moving to the next level, as shown in Fig. 6. First, the physical needs of hunger, thirst, sex, and sleep are fulfilled. The obtaining of food, water, shelter, and clothing as a security need is fulfilled next. The third level consists of social needs, and fulfilling one's desires for love, companionship, cooperation, and belonging. The fourth level of need is self-esteem, where an individual requires acceptance from peers, family, and friends. The last level, self-actualization, describes the level at which a worker strives to do the best job possible. The desire for further training, education, good working habits, and a safe working attitude are extensions of self-actualization. It is the last, or highest, level of motivation for the worker that mine management should strive to fulfill and maintain.

The points to remember in Maslow's theory, as well as other motivational theories, are:

1) All needs exist in a worker to some degree.

2) A need which is completely fulfilled will not act as a motivator. Conversely, an unfilled need may not produce the expected (or desired) behavior.

3) Programs designed by management to fulfill worker needs must be individualized and dynamic so unfulfilled needs can be quickly substituted for fulfilled ones.

A worker can be satisfied on the job by management if the needs of the worker are identified and met. Some *worker satisfiers* on the job are:

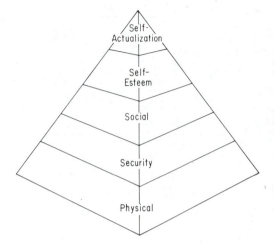

FIG. 6. Maslow's hierarchy of needs.

Money, in all forms (overtime, bonuses, gifts, etc.).

Job position and power.

Job security.

Competition with other workers.

Working conditions.

These satisfiers, however, assume management has made a commitment to using self-actualization motivation. Two other motivational approaches are generally available to mine management and they are (1) paternalistic and (2) reward and punishment. Neither theory is a stranger to current mining operations.

Paternalistic motivation unconsciously creates a "one big happy family" feeling among workers and the supervisor by operating on the buddy system. It requires an adult- (the foreman) to-child (the workers) communication channel, which is not recommended. A supervisor spends a great deal of time babysitting in the paternalistic approach. Current management plays down this type of behavior approach as a practical theory.

Reward and punishment is a straightforward approach to motivational theory. It basically says a supervisor will reward the workers for good or acceptable work, and punish those workers for bad work. This approach, quite frankly, can be self-defeating in the long term. It may create a dictatorship style which can be very detrimental. In many cases, both the paternalistic and reward and punishment approaches are the result of the supervisor's lack of understanding of worker behavior. Unfortunately, these two approaches are the most common types of supervisory techniques found in mine management.

A final point to this discussion concerns worker morale. Morale is an individual's mental and emotional state. A worker's morale can be affected by the physical and system conditions imposed on him by such things as height of the working place, presence of water or gas, roof and rib conditions, treatment from supervisors, company policies, and fellow workers.

Since management can do little to change the physical conditions of the mine, they must concentrate on the system conditions. This means training the supervisor, shaping the company policies, and organizing the work force to help maintain and improve worker morale. A project manager should be constantly aware of the morale of his workers. Poor morale can jeopardize job schedules, quality control, lead to higher costs, and create an unsafe working place.

One technique for a project manager to use to maintain worker morale is active listening. Active listening is the ability of a person to 1) understand what someone is saying to him, and 2) paraphrase the meaning of the other's speech back to that individual. In doing this, the conflicts in communication (the key to worker morale) between the manager and worker can be reduced greatly by eliminating most misunderstandings.

Active listening, surprisingly enough, receives signals from three types of communication channels—words, tone of voice, and body language. Experts suspect that words convey only 7% of a person's meaning, the tone of voice conveys 38% of his meaning, and body gestures provide the final 55% of meaning to a person's idea. This is why it is so important for a successful project manager to study and use active listening techniques to ensure his meaning is clear and not misunderstood, and that he receives the correct message from his workers and subordinates.

Active listening can be learned by anyone. It takes practice and patience. Professional advice should be obtained if a program on this and other management skills is established at the mine.

SUPERVISORY LEADERSHIP

The final discussion under consideration is about supervisory leadership. A supervisor is a leader by virtue of his title; however, *leadership skills* must be acquired or studied and learned by a person. Leadership is simply the ability to effectively direct the activities of others. In this case, the others refers to the mine workers on the project. This discussion of supervisory leadership is divided into two broad topics. The first discussion concerns leadership and coordination, and the second looks at the directing function of management on a given project.

Leadership and Coordination

Leadership is made up in part of a series of functions done to enhance the flow of men, materials, and equipment to a specific company objective. Coordination of these functions is necessary to integrate all diverse mine activities to this single goal. Therefore, a project manager should understand and be familiar with four general leadership functions necessary for dealing with project workers to achieve his project goals:

1) *Planning* the work activities of the project.
2) *Organizing* the necessary men, materials, and equipment.
3) *Directing* the work activities of the project.
4) *Integrating* the efforts of others with the project.

All these functions (except directing) have been covered and emphasized in this chapter as essential tools and basic skills needed for an effective project manager. A construction engineer turned project manager can no longer afford to look at only the "nuts and bolts" portion of the work. His designs, timetables, costs, labor, and other project details must be considered as parts of the whole project. In addition, his work should reflect some study of human resource aspects particular to the underground coal industry and coal mining.

The Directing Function of Management

The most active function for a project manager to undertake is the directing function. Directing, as defined earlier, is the carrying out of the plans and procedures for achieving the project objective. It is composed of seven steps (Fig. 7) which are separately defined, yet interact to a great extent during the project. As you recall, the steps were staffing, training, supervising, delegating authority, motivating, counseling, and coordinating. Directing the project team, supervisors, and in some cases, the workers themselves, is not an easy task. It requires a great deal of training, experience, and common sense.

Staffing is a step which encompasses the choices (if any) of the project team and supervisory staff.

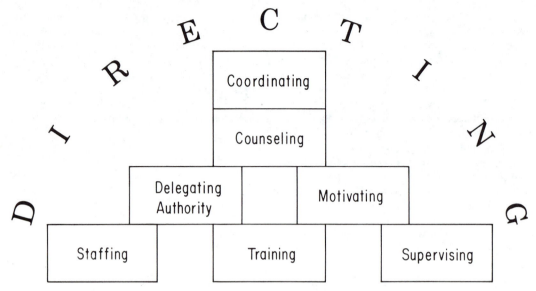

FIG. 7. The seven steps of directing.

A project manager must use his judgment and insight in an effort to place the right person in each position. Getting a buddy a job because he is a friend may not be a good idea. A project manager must think of the project objective in choosing the staffing assignments.

Training steps refer to both supervisory and labor people working on the project. Training must extend from not only a productive viewpoint, but more importantly, a safety viewpoint. Using a systematic approach in conjunction with proven training tools is well advised during the entire project.

Supervising steps refer to the project manager's day-to-day instructions to others, his visits (or residence) at the job site, and giving guidance and discipline to his supervisors so they can fulfill their duties and responsibilities. Daily meetings, memoranda, and progress reports all are part of this supervising effort.

Delegating authority goes along with the supervising step of day-to-day managing of a project. Here, delegating authority to others is done to best achieve the productive capabilities of everyone involved with the project. Authority to make decisions on miscellaneous construction details, individual worker assignments, material delivery positions, and so on should be left to the front-line supervisors.

Motivating others on the project is a step which starts on the first day of the project and ends with the last day. It requires an understanding of worker motivational concepts, and the best approaches for satisfying worker needs. Worker morale should be maintained at a high level in order to create a positive working environment and achieve the project goals.

Counseling is a persuasion skill needed by the manager for maintaining the morale of the team. It may consist of holding private discussions, group interviews, or team meetings to discuss individual problems or ways to improve the work. It also entails discussion of personal problems, or ways to realize personal goals in both the supervisory personnel and workers.

Finally, *coordinating*, as mentioned earlier, is the step of smoothing the project activities into a continuous flow of work. Coordinating functions are dependent upon the project team and the communication channel chosen for the coordinating effort.

Project managers must be firm and decisive in their actions, and move forward rapidly whenever directing functions are necessary. It is better, when in the directing function, to use the KISS (keep it simple, stupid!) rule for giving directives. One of the biggest reasons why the project team or workers will or will not follow a directive is dependent upon the amount of respect they have for the project manager. It is imperative, therefore, the project manager never issue an order which cannot be enforced. However, orders should be issued in such a manner that they are complied with immediately. Oral directives and insignificant orders should be made in terms of requests or suggestions.

This concludes the discussions on construction project management. I have tried to touch upon the more important aspects of this difficult, but challenging, science. Much of the fundamental elements of management will be used more in the future than at the present time in the coal industry. Effective training, safety, and a complete understanding of worker behavior has yet to receive wide acceptance by mine management.

REFERENCES AND BIBLIOGRAPHY

Anon., 1968, *Resource Management*, MDI Publications, Management Development Institute, Wayne, PA.
Anon., 1973, *Systems Analysis and Design Using Network Techniques*, Prentice-Hall, Englewood Cliffs, NJ.
Anon., 1981, "Available Software," *Mining Engineering*, Vol. 33, No. 11, Nov., pp. 1591–1593.
Allen, L. A., 1969, *The Professional Manager's Guide*, Louis A. Allen Assoc. USA.
Anthony, R. N., 1965, *Planning and Control Systems: A Framework for Analysis*, Div. of Research, Graduate School of Business Administration, Harvard Univ., Boston, MA.
Anthony, R. N., and Mouton, J. L., 1975, *Management Accounting*, Richard D. Irwin, Homewood, IL.
Argyris, C., 1960, *Understanding Organizational Behavior*, Richard D. Irwin, Homewood, IL.
ARINC Research Corp., 1969, *Guidebook for Systems Analysis/Cost Effectiveness*, US Dept. of Commerce, National Bureau of Standards, Washington, DC.
Barish, N. N., 1962, *Economic Analysis: For Engineering and Managerial Decision Making*, McGraw-Hill, New York.
Baumgartner, J. S., 1963, *Project Management*, Richard D. Irwin, Homewood, IL.
Baumol, W. J., 1965, *Economic Theory and Operational Analysis*, 2nd ed., Prentice-Hall, Englewood Cliffs, NJ.
Blake, R. R., and Mouton, J. L., 1964, *The Managerial Grid*, Gulf Publishing Co., Houston, TX.
Bideaux, R. A., 1981, "Minicomputer Applications for Today's Mining Problems," *Mining Engineering*, Vol. 33, No. 11, Nov., pp. 1584–1588.
Britton, S. G., 1981, *Practical Coal Mine Management*, John Wiley & Sons, New York.
Britton, S. G., 1981, "Cutting Costs of Mine Support Systems," *Coal Mining & Processing*, Vol. 18, No. 12, Dec., pp. 56–58.
Bucklen, E. P., et al., 1969, "Computer Applications in Underground Mining Systems," R&D Report No. 37, U S Office of Coal Research, Washington, DC.
Butler, A. G., 1973, "Project Management: A Study in Organizational Conflict," *Academy of Management Journal*, Vol. 16, Mar., pp. 84–101.
Cleland, D. I., and King, W. R., 1972, *Management: A Systems Approach*, McGraw-Hill Inc., New York.
Couger, J. D., and Knapp, R. W., 1974, *System Analysis Technique*, John Wiley & Sons, New York.

Datz, M. A., and Wilby, L. R., 1977, "What is Good Project Management," *Project Management Quarterly*, Vol. 8, No. 1, Mar.
Davis, S. M., and Lawrence, P. R., 1977, *Matrix*, Addison-Wesley, Reading, MA.
Drucher, P. F., 1975, "Managing for Business Effectiveness," *Business Classics, Harvard Business Review*, Boston, MA.
Emery, J. C., 1969, *Organizational Planning and Control Systems*, MacMillan, New York.
Friend, F. L., 1976, "Be a More Effective Program Manager," *Journal of Systems Management*, Vol. 27, Feb., pp. 6–9.
Gaddis, P. O., 1959, "The Project Manager," *Harvard Business Review*, Harvard Univ., May–June, pp. 89–97.
Gran, P. F., 1976, *Systems Analysis and Design for Management*, DunDonnelley, New York.
Grant, E. L., and Ireson, W. G., 1970, *Principles of Engineering Economy*, 5th ed., Roland Press Co., New York.
Herzberg, F., 1975, "One More Time: How Do You Motivate Employees?" *Business Classics, Harvard Business Review*, Harvard Univ., Boston, MA.
Holland, T., 1969, "What Makes a Project Manager?" *Engineering*, Vol. 207, Feb., p. 262.
Horowitz, J., 1967, *Critical Path Scheduling—Management Control Through CPM and PERT*, Roland Press, New York.
Houre, H. R., 1973, *Project Management Using Network Analysis*, McGraw-Hill, New York.
International Congress for Project Planning by Network Analysis, 1969, *Project Planning by Network Analysis*, North-Holland Pub. Co., Amsterdam, Holland.
Kast, F. E., and Rosenzweig, J. E., 1974, *Organization and Management: A Systems Approach*, 2nd ed., McGraw-Hill, New York.
Kelley, W. F., 1969, *Management Through Systems and Procedures: A Systems Concept*, New York.
Kerzner, H., 1977, "Systems Management and the Engineer," *Journal of Systems Management*, Oct., pp. 90–107.
Kerzner, H., 1979, *Project Management: A Systems Approach to Planning, Scheduling, and Controlling*, Van Nostrand Reinhold, New York.
Maslow, A. H., 1954, *Motivation and Personality*, Harper, New York.
Martin, C. C., 1976, *Project Management: How to Make It Work*, Amacom, New York.
Martino, R. L., 1968, *Project Management*, MDI Publications, Management Development Institute, Wayne, PA.
McGregor, D., 1967, *The Professional Manager*, McGraw-Hill, New York.
McGregor, D., 1960, *The Human Side of Enterprise*, McGraw-Hill, New York.
Moder, J. J., and Phillips, C. R., 1970, *Project Management with CPM and PERT*, 2nd ed., Van Nostrand Reinhold, New York.
Optner, S. L., 1965, *Systems Analysis for Business and Industrial Problem Solving*, Prentice-Hall, Englewood Cliffs, NJ.
Pearson, R. G., 1971, "Human Factors Engineering," *Industrial Engineering Handbook*, McGraw-Hill, New York.
Pegels, C. C., 1976, *Systems Analysis for Production Operations*, Gordon and Science Pub., New York.
Prelaz, L. J., et al., 1964, "Optimization of Underground

Mining," Research and Development Report No. 6, Vols. 1, 2, 3, NTIS No. PB 166291-1-2-3, Springfield, VA.

Quirin, G. D., 1967, *The Capital Expenditure Decision*, Richard D. Irwin, Homewood, IL.

Ramani, R. V., and Manula, C. B., 1975, "A Master Environmental Control and Mine System Design Simulator for Underground Coal Mining, Executive Summary," Vol. 1, Report No. PB 255 421/AS, US Bureau of Mines, NTIS, Springfield, VA.

Richards, M. D., and Greenlow, P. S., 1966, *Management Decision Making*, Richard D. Irwin, Homewood, IL.

Rodgers, C. R., and Roethlisberger, F. J., 1975, "Barriers and Gateways to Communications," *Business Classics, Harvard Business Review*, Harvard University, Boston, MA.

Rosenbaum, B., 1979, "Supervisory Skills Training Program," Developed for Monterey Coal Co. by MOHR Development Inc., New York.

Scott, F. E., 1981, "MSHA Academy: America's Mine Safety College," *Coal Mining and Processing*, Vol. 18, No. 10, Oct., pp. 46–51.

Shannon, R. E., 1972, "Matrix Management Structures," *Industrial Engineering*, Vol. 4, Mar., pp. 26–29.

Sikula, A. F., 1977, *Personnel Management*, John Wiley & Sons, New York.

Sleyman, K. J., 1971, "Cost Estimating," *Industrial Engineering Handbook*, McGraw-Hill, New York.

Toellner, J., 1977, "Project Estimating," *Journal of Systems Management*, May, pp. 6–9.

Vaughn, D. H., 1967, "Understanding Project Management," *Manage*, Vol. 9, No. 9, pp. 52–58.

Weinberg, G. M., 1975, *An Introduction to General Systems Thinking*, John Wiley & Sons, New York.

Weiss, A., 1979, *Computer Methods for the 80's*, AIME, New York.

Wilson, F., 1963, *Manufacturing, Planning and Estimating Handbook*, McGraw-Hill, New York.

Whitehouse, G. E., 1973, "Project Management Techniques," *Industrial Engineering*, Vol. 5, Mar., pp. 24–29.

9

Design Tools for Construction Engineering

Up to now, the *hands-on* approach to construction has been stressed. This chapter reviews the basic tools available for construction design and monitoring of a construction project. It is divided into three discussions, all under the general title of computer aids, since the trend for using these tools is via the computer.

In the context of our discussions, the focus for computer aids will be on the use of evaluation and analysis techniques for coal mine construction applications, including mainframe programs, CPM techniques, as well as mini and microcomputer applications.

Although the presentation of material in this chapter may be detailed at times, by no means does it cover the entire field of computer applications for design or project scheduling and monitoring. It is only an attempt to illustrate the possibilities that are available to construction and mine engineers. Interested persons can pursue computer applications for coal mining by reviewing *Computer Methods for the 80's*, by Alfred Weiss, editor, published by SME-AIME.

COMPUTER AIDS

The biggest help for the construction engineer in the way of design comes from the advent of computer technology to the coal mining industry. Computer technology consists of two portions, software and hardware. These portions are further divided into two categories based on a number in internal design characteristics which can be simply defined in terms of size as either large or small. Large refers to mainframe computers with large programs and sizeable storage capabilities. Small refers to the current influx of mini and microcomputers, and applicable software which has exploded onto the mining scene at the mine or engineering office.

This discussion of computer aids is divided into three separate topics. These are (1) traditional mining mainframe software; (2) CPM uses for mine construction scheduling; and (3) minicomputer software. The software of main interest here is that which can be applied to planning, scheduling, and evaluating underground support projects and their construction.

Traditional Mining Software

These software packages, called models, were the first approaches to looking at coal mining by computer simulation. The large digital computers developed in the 1960s and early 1970s were used to run these models which still have uses today. Support system models were developed to aid in examining mine transportation, ventilation, pumping, and haulage networks. For the most part, the development of these models followed the general approach consisting of applying mathematical representations to mining operations to optimize production support.

Transportation Models: Rail transportation models for construction planning and scheduling of men, material, and supply distributions were not developed as a separate model from production oriented models. However, limited planning uses could be manipulated from the original objectives of evaluating the production potential of the rail system by varying specific inputs. Two early transportation models were the RAILSIM program from Virginia Polytechnic Institute and the rail haulage subsystem of the Master Design Simulator (MDS) from Pennsylvania State University. The RAILSIM program operates on an event-oriented simulation approach, whereas the MDS is a simulator operating on time-slicing logic. Neither program could isolate or focus on the construction scheduling function to use in short-range mine support planning. Their use, however, is to analyze the design of the system to better estimate required material weights, switch sizes, and material amount. They are not recommended for construction planning per se.

Ventilation Models: Because ventilation is so critical to efficient mining practices, much attention has been paid to developing a model which would solve the required network balancing problems. The best known ventilation model is the mine ventilation subsystem (MVS) of the Penn State MDS program (Ramani, et al., 1975). For construction purposes, the MVS program is excellent because it permits up to 32 variables on installation factors to be inputs to the program. By changing only these variables in any single network, the effects of different material sealing, structure, and auxiliary components of a stopping, overcast, door, or regulator can be analyzed. The MVS subsystem is designed on the Wang-Hartman approach of resistance buildup. The basic characteristics of the MVS subsystem can be summarized as the following:

1) Uses procedures which facilitate the multi-analysis processing of variations of a system and/or different systems without reloading the program.

2) Has five types of airway loops to reduce ventilation system to schematic analysis.

3) Calculates the Natural Ventilating Pressure (NVP); after fan stoppage, this could be worked over to give flow values.

4) Permits verification of the accuracy of simulation network.

5) Details six resistances to be studied from any source.

6) Uses three points on second degree polynominal equation for Head-Volume, HP = Volume curve.

7) Types of problems solved:
   a. Characterizes fans in parallel, and series installations.
   b. Changes fans to new density, speed, block, series or parallel installation.
   c. Calculates methane emissions into the mine atmosphere.

8) N number of fans available for analysis, whether in series or parallel.

9) Stopping description allows 32 stopping construction variables.

Of course, the overall mine design capabilities of the MVS program overshadow the support planning functions. However, by using some simplifying assumptions and known data, this model can function to evaluate the performance of almost any ventilation structure.

A second program is the VPI VENTSIM model. This model uses the Hardy Cross modular approach to network balancing analysis (which tends to eliminate loop overlap). The construction variables built into the program offer six different types, (1) dry stacked, (2) mortar laid, (3) mortar laid and painted, (4) dry stacked and plastered, (5) mortar laid and plastered, and (6) brick (mortar laid). Although this is but a fraction of the MVS system, the VENTSIM does provide a range of default values to begin any analysis work. Other characteristics of VENTSIM are:

1) Uses procedures which facilitate the multi-analysis processing of variations of a system and/or different systems without having to reload the program for each.

2) Has five types of airway loops to reduce ventilation system to schematic computer analysis.

3) Recalculates flow of air after fan stoppage, balances network.

4) Permits verification of network simulation accuracy.

5) Handles airway resistance in three different modes (sources).

6) Types of problems solved:
   a. Leakage volume induced by a given flow requirement.
   b. Operating point for a given fan curve and ventilation network model.
   c. Benefits gained from additional airways, overcasts, and other facilities.
   d. Single variable regulator sizes to force airflow in splits.
   e. Benefits of cleaning up airways.
   f. Benefits of new airshaft.
   g. The study of alternative solutions to specific problems in active mines.

7) Maximum of five fans possible, either variable or fixed speed type.

8) Six stopping variables are given as default values.

Since these programs have been developed, many coal mining companies (e.g., Bethlehem Steel, Pittston, etc.) have extended the capabilities of these models for various mine planning purposes, including construction scheduling and quality control.

Pumping Models: Currently, the only pumping models for mainframe computer application specific to underground mining are again the models from VPI and Penn State. These models analyze the drainage networks by quantifying flow rates and pumping characteristics. For example, the PUMPSIM program from VPI uses the Darcy equation for network analysis. Any changes in the pumping requirements of the mine can be examined. By using this model characteristic, some construction differences in installation and materials can be modeled. A second advantage is in the quick analysis of various flows to project full drainage and surge requirements for a mine on the drawing board.

Haulage Models: Like ventilation, belt haulage models have been widely distributed and worked on as a production analysis tool. BELTSIM, from VPI (Bucklen, et al., 1969), was the first to be developed for public use. BELTSIM is a user-oriented program focusing on the production characteristics of belt haulage using event-simulation techniques. A second production oriented simulation was developed by the National Coal Board (NCB) of England and was labeled SIMBELT. Others followed these models, including SIGUT (Redling, 1975) from Germany and SIMBUNK, again from the NCB. To date, there is no widely used application of these programs in planning and scheduling of belt extensions, transfer points, or belt drive placements.

Because the previous models in transportation, ventilation, pumping, and haulage do not look specifically at the construction engineering of these support systems, mine engineers must turn to another general model for construction planning called CPM, or the Critical Path Method of analysis.

CPM Uses for Mine Construction Scheduling

CPM, as mentioned in Chap. 8, was developed in the late 1950s from the PERT techniques designed by the US Navy. CPM can be done by hand, as in the past, or by computer, which is the current trend. CPM is primarily a tool of the construction industry. It lends itself to smaller, well-defined projects like building overcasts, laying track, or extending a belt line. This unique ability will help CPM gain greater acceptance in mining as time goes on. It allows the construction engineer to achieve improved time and cost control over construction projects which are repeated over and over, and provide accurate data for estimating purposes.

CPM is essentially a project management tool covering the installation phases of mine support systems. It aids the construction engineer in his decision-making process by providing the best alternatives for accomplishing a task. To provide these alternatives, CPM is broken into two analysis steps. The first step is project planning, and consists of:

1) Identifying the work elements needed to complete the job.

2) Establishing the logical sequence of these work elements.

3) Illustrating this sequence graphically in network diagram form.

Once this step is complete, the scheduling step of identifying necessary labor, materials, and supplies also can be estimated. The project planning step is fairly straightforward, and requires only simple calculations. However, doing a number of simple calculations becomes tedious, and hence, the computer application.

To fully understand the CPM techniques, it is essential that some of the necessary calculations, terminology, and logic used in developing the networks be presented to the reader. The following discussions explore CPM uses in mining, and for the purposes of this text, are based on manual procedures, since much of the mining applications in support construction can be easily developed by hand. It is not the purpose of this text to completely discuss CPM and its applications, but only to present the fundamentals and develop appropriate networks for the support projects discussed in Chaps. 2 through 7. More detailed discussions of CPM can be found in O'Brien's *CPM in Construction Management*, 2nd ed., 1971, published by McGraw-Hill, Inc. Further references are included in the Bibliography.

CPM Fundamentals: As mentioned above, the first step in using CPM is project planning. This involves breaking down the total task or construction project into *activities*. An *activity* in CPM is defined as any single, identifiable task or work step of the project. For example, in building an overcast, one activity would be pouring the foundation. These activities depend upon design considerations as they are interpreted by the construction engineer, and should at least consider the following factors:

1) Separate areas of responsibility (such as subcontracted work) which are distinctly separate from that being done by the mine management directly.

2) Different categories of work as distinguished by production or maintenance considerations.

3) Different categories of work because of equipment requirements and subsequent operator labor skills.

4) Different categories of work due to different materials such as concrete, timber, or steel.

5) Distinct and identifiable tasks of construction work such as walls, floors, beams, and welding.

6) Work location within the project necessitating different times or different crews to perform.

7) Construction engineer's breakdown for estimating or cost accounting purposes.

8) Mine schedules or limiting times that may interrupt project construction due to production or other regular activities.

These activity tasks can be as detailed as desired. For example, a concrete foundation for an overcast may be a single activity—pouring a foundation. Or, it also may be broken down into the separate steps necessary to construct it, such as picking the floor level, placing the forms and rebar, pouring the concrete, curing, and cleanup.

As these activities are identified, the logical sequence of interaction must be determined. This logical sequence of activities is called *job logic*, and consists of the time (duration) and sequential order of each task needed to complete the project. Job logic in turn defines the restraints of the activities, i.e., one task cannot start before another one is finished. The example activity of pouring the foundation obviously cannot be done until the roof is graded, the concrete materials are at the site and mixed, and the foundation forms are built. Therefore, the start of this activity is *restrained* by the time required to complete the activities of ordering and delivering the concrete, site orientation, roof grading, roof cleanup, and building the foundation. This fundamental relationship is the basis of the CPM network, and is translated into network diagrams by the use of arrows (→) and circles (○).

Arrows represent the activities of a single task. They usually point left or right, have no fixed length or path, and must always have an identified starting and stopping point. These starting and stopping points are called *events*. Events are commonly drawn as circles, but could be any shape. Events are numbered to distinguish them from other events and to identify the ends of an activity. A simple representation of a CPM network is shown in Fig. 1. Here, the circles are the events and the arrows represent the activity. The head and tail of the activity is commonly referred to as i–j designations, where i is the start (arrow tail) of the activity and j is the end (arrow head) of the activity.

By stringing all the identified job activities together with arrows and circles, a sequence of job logic can be represented as a network diagram for the entire job. For the purposes of CPM, job logic requires that each activity in the network have a definite event to mark its starting point. This event can be either the beginning of the project itself, or the completion of a preceding activity. At no time in CPM should one activity overlap the start of a succeeding activity (Fig. 2). If this happens, it is a signal to the construction engineer that further subdivision of the activities is necessary to clear up the network. No given activity can start until all those activities leading to it have been completed.

In the course of reviewing CPM networks, a construction engineer will, in all probability, come across such items as event numbering, dummy arrow, time-scaling, and float time. These characteristics are very valuable tools in communicating pertinent construction data to the network reader.

Event Numbering. Event numbering is a simple, yet extremely important characteristic of CPM networks. Event numbering is governed by two simple rules:

1) Every activity must have its own i–j designation, identified by the events bounding each activity.

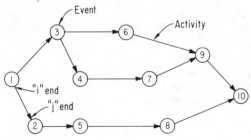

FIG. 1. Basic components of the network diagram.

DESIGN TOOLS FOR CONSTRUCTION ENGINEERING

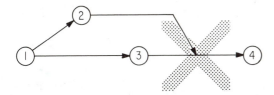

FIG. 2. Activity overlap should not occur in the CPM diagram.

2) Assigning event numbers is always done by putting the larger event number at the head of the arrow.

Two methods are commonly used to number events, horizontal or vertical.

The horizontal method assigns event numbers along a line of activities until they meet at a point of two or more other activities. This is repeated sequentially until all the lines into the junction point are labeled. Then, another horizontal line is followed and the procedure is repeated. Fig. 3 illustrates the horizontal method.

Vertical event numbering consists of moving from top to bottom in a left to right direction. Numbers are assigned sequentially in a vertical fashion so as not to violate the i–j rule of larger numbers being placed at the head of the arrow, as shown in Fig. 4.

Since there is no significance to event numbers other than as a means of identifying an activity, even a random assignment would work for CPM diagrams. However, this practice is not recommended.

Dummy Arrows. A dummy arrow, shown on a CPM diagram as a dashed line, is a way of indicating the completion of one activity is restraining the start of two or more other activities. It represents an interdependency which otherwise would not be shown. For example, Fig. 5 shows a typical CPM network with a dummy arrow between event (2–3). This means the start of activity (3–4) cannot begin until activities (1–2) and (1–3) are completed. If the dummy arrow was not present, one may interpret the network diagram to read no interdependence between the completion of the (1–2) and (1–3) activities, and therefore, activity (3–4) could begin immediately after completion of the (1–3) activity.

A second use of dummy arrows revolves around the practice of assigning related activities individual i–j designations to avoid confusion. If two or more activities share common start and finishing points, they would conceivably have the same number designations (Fig. 6). Dummy arrows to separate events can individualize the activities and still show a dependency.

Time-Scaling. Time-scaling is the practice of assigning the estimate of time needed to complete a single activity. Time-scaling can be expressed in hours, manshifts, mandays, or just workdays. Time-scaling allows the *critical path* (i.e., the bottlenecks of the project) to be traced through the diagram. Usually, the critical path is the single activity path requiring the most time to complete. Let's return to the network shown in Fig. 5. This simple overcast CPM network can be time-scaled and presented as a network diagram for critical path analysis, as shown in Fig. 7. In this case, manshifts have been used as the basic time-scale. It is relatively easy to see the critical path to event (4) is the (0-1-3-4) path of $10\frac{1}{4}$ manshifts. The (0-1-2-4) path is estimated at four manshifts and the (0-1-4) path is estimated at two manshifts to completion. Although this example is straightforward, the complexities of network diagrams and CPM analysis are in the interdependency of the activities and the relative starting and stopping points in the scheduling process. This complex idea is handled by the concept of float time.

Float Time. Float time can be best described as scheduling leeway among the necessary ac-

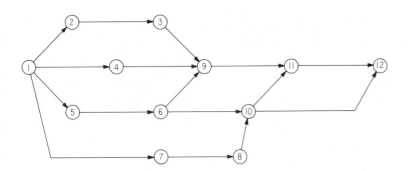

FIG. 3. Typical horizontal numbering system used in the CPM diagram.

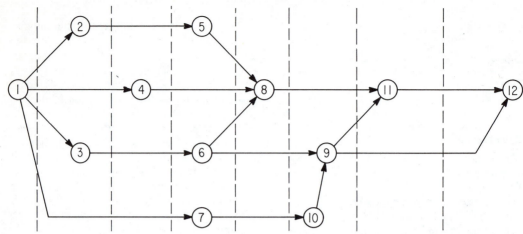

FIG. 4. Typical vertical numbering system for a CPM diagram. The vertical system is more common than the horizontal system.

tivities of a project. It is simple to compute float time. Arithmetically, it is the difference between the early and late dates for a given activity. When float time is available, the starting times of an activity can be delayed, or its duration extended, or both.

To establish the float time of any given project, it is necessary to establish three important timing facts from a time-scaled CPM diagram. These are:

1) Estimating the earliest time an activity can start and finish.
2) Estimating the latest time an activity can start and finish.
3) Estimating the available leeway for scheduling an activity by the difference between 1) and 2).

The critical path defines the route where no float time exists. This means the early and late dates of an activity are the same. If a project falls behind schedule, then in all likelihood, the earliest start date of the activity will fall after the latest time that it can be done to remain on schedule. With this situation, no float time exists, and the difference between these two dates can measure how far behind a project is.

To illustrate float time, Fig. 8 is a typical network diagram using vertical event numbering. Establishing a time-scaled chart from the network is the first priority. This is shown in Fig. 9, with the horizontal axis corresponding to the number of working days for the project. Each path is graphed out to scale according to the time-scaled estimates. The critical path becomes the one(s) with no float time, (with float time shown in the chart as a wavy line). In this example, the activities have been lettered to aid in identifying individual activities. The critical path (hatched line) is 29 days long and is composed of the activities defined by the letters B-E-I-L-P-Q.

Float time can be expressed as either free or total float time. Free float time is defined as the amount of time any individual activity can be delayed without affecting the early start of the following activity. Total float time is defined as the time that any activity can be delayed without

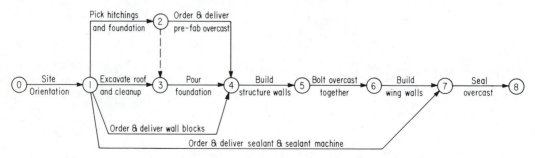

FIG. 5. Dummy arrow between 2–3 shows interdependency of activities.

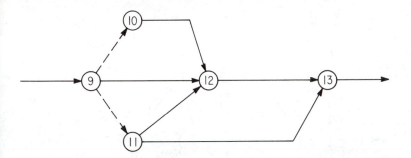

FIG. 6. Dummy arrows separate *i–j* specifications for parallel activities to individualize CPM activities.

affecting the overall project completion schedule. The main difference is that free float time belongs to the individual activity and does not interact with the project. Total float time, however, is shared with all the other activities. This means when the total float time is used up, all remaining activities in the project become a critical path. The point to be made here is that total float time does not belong to any individual activity. Complete activity paths share the same total float time and when any one activity uses up this pool of free time, it is gone.

CPM Networks for Mine Support Construction: This portion of the discussion on CPM networks focuses on the application of CPM to mine support construction planning. The approach is to define the activities for selected projects from Chaps. 2 through 7 and present the CPM network diagram for it. The critical path will be estimated in man-hours in conjunction with various comments upon the activity breakdown for each project.

Transportation. The construction project selected to illustrate a typical transportation project is the installation of a switch. From Chap. 2, the activities for a track switch can be designated as:

(A) Site orientation
(B) Ordering and delivering predesigned switch
(C) Cutting inside rail and unclipping ties
(D) Bolting the frog to the inside rail
(E) Bolting the prebent turnout rail on
(F) Bolting the curved and straight closure rails to the frog
(G) Placing the switch ties and clipping them on
(H) Bolting the switch points on
(I) Hooking up the switch stand
(J) Bolting the outside turnout rail to the frog
(K) Clipping spacer ties on the main and turnout track
(L) Bending outside rail to tie plates
(M) Lubricating and adjusting the switch throw
(N) Blocking and leveling

These activities can be illustrated by the network diagram shown in Fig. 10. The critical path is fairly straightforward. It consists of the activities defined by the letters (B-D-F-G-H-K-M) and is 2 3/4 manshifts long. This means two men in eight hours may not complete the activities on the critical path. This tends to support the conclusion that three to four workers are needed to place the typical switch in a track system in one shift. Further analysis yields the fact that delivery of the material is independent of actual switch construction. While this is true from a management point of view, it still makes up a fundamental link in the critical path of the

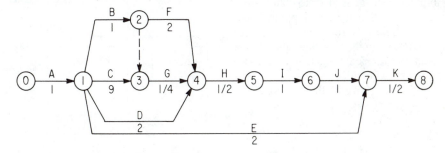

FIG. 7. CPM diagram of overcast construction activities are time-scaled: all figures are scaled in manshifts.

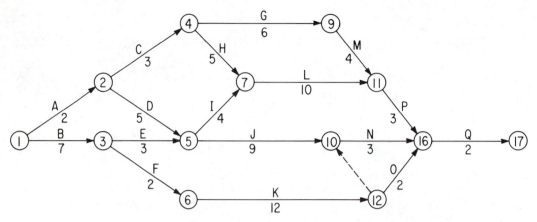

FIG. 8. A CPM diagram showing float time in several activity paths.

project. Experience and judgment should guide the engineer in manipulating these CPM estimates for his own use.

Ventilation. Fig. 7 in this chapter illustrates the diagram of an overcast from Chap. 3 on ventilation construction, so this section will focus on building a block stopping. The activities involved with this project are as follows:

(A) Site orientation
(B) Cleaning the floor and scaling the roof and ribs
(C) Picking the ribs for stopping hitches
(D) Pouring the foundation (if necessary)
(E) Ordering and delivering the sealant and machine
(F) Stacking and wedging the blocks
(G) Sealing the stopping
(H) Cleanup of the site

As a network diagram, stopping construction looks like Fig. 11. The critical path is defined as (B-C-F-G-H) and is composed of 13 manhours. Activity (D) is an optional activity in this case, and therefore is not included in the time-scaling until the decision is made to pour a foundation. If a foundation for the stopping is desired, then it would require another designated task on the CPM diagram requiring two men for half an hour to complete. Thus, it would still not be on the critical path.

Pumping. Chap. 4 covered the pumping construction requirements necessary for large mines. Therefore, it is appropriate to diagram a main sump and pumping station installation. The activities included in this CPM analysis are:

(A) Site orientation of the sump and pump station

(B) Transport of excavation equipment
(C) Excavate sump
(D) Clean up sump
(E) Bolt wire mesh to sump walls
(F) Order and deliver sealant and machine
(G) Seal sump floor and walls
(H) Frame pump foundation
(I) Pour pump foundation
(J) Order and deliver pump, motors, piping, and accessories

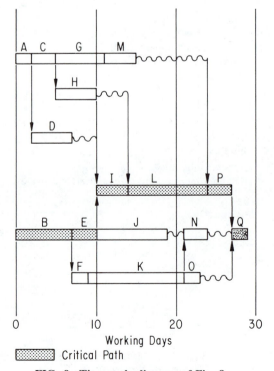

FIG. 9. Time-scale diagram of Fig. 8.

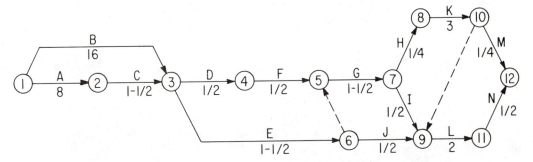

FIG. 10. CPM diagram of a track switch (time-scaled in man-hours).

(K) Bolt block and tackle props and beams to roof
(L) Set and bolt pump and motor
(M) Connect intake piping
(N) Connect discharge piping
(O) Test sump and pump system
(P) Build fireproof housing for controls
(Q) Place guarding and signs
(R) Cleanup of site

Of course, these activities could be shuffled and reidentified as other activities. However, the list above will serve as a general model. The corresponding network diagram is shown in Fig. 12. It should be mentioned that any number of smaller activity networks can emerge from a diagram such as this. For example, the activity path (L-M-N-O) consists of connecting the piping between the sump, the pump, and outside discharge. A network designer could easily break these activities into the component activities of installing pipes, foot valves, gate valves, couplings, and so on. The point is that the activity breakdown of any project should meet the scheduling needs for management by identifying the critical path of the project. The construction engineer will have to use his judgment on the activity breakdown of a given project.

Returning to the pump network diagram, the critical path is defined by the letters (B-C-F-G-M-N-O-R) which sums to an estimated 77 man-hours. This critical path assumes the sump excavation and pump installation are done in one project timeframe.

Haulage. The final CPM diagrams concern projects involving belt line haulage. The examples here are both the installation of the belt drive and the belt extension.

For the belt drive installation, the activities can be listed and identified as:

(A) Site orientation
(B) Ordering and delivering belt drive materials and supplies
(C) Leveling the site
(D) Rough placement of the drive and takeup
(E) Fine tuning drive and takeup placement
(F) Anchoring belt drive and placement
(G) Placement of deluge and detection systems
(H) Hooking up the controls and motor
(I) Tensioning the takeup pressure
(J) Placing the machinery guards
(K) Cleanup of site

The diagram of these activities is shown in Fig. 13. The critical path for this CPM is composed of the activities (B-D-E-F-J-K) with an estimated project time of 25 man-hours.

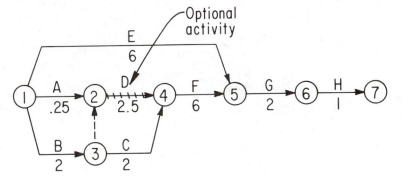

FIG. 11. CPM diagram of a block stopping (time-scaled in man-hours).

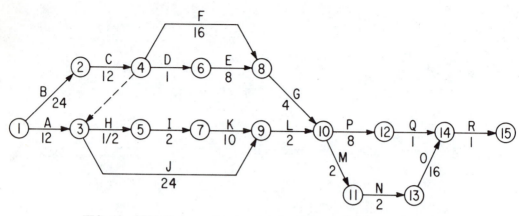

FIG. 12. CPM diagram of a pump station (time-scaled in man-hours).

Turning to the belt line extension, the required activities are listed as:

(A) Site orientation
(B) Ordering and delivering belt for splicing
(C) Drill holes for chain hangers and sailbases
(D) Order and deliver extension supplies
(E) String belt structure
(F) Splice belt into belt line
(G) Breakdown and move tailpiece
(H) Bolt sailbases to roof
(I) String ropes and tension
(J) Place structure and tighten in position
(K) Place tailpiece in new position
(L) Anchor tailpiece
(M) Hook tailpiece to belt line
(N) Train belt
(O) Hook up auxiliary systems
(P) Troubleshoot the extension

Fig. 14 illustrates the network diagram. The critical path for this project includes the activities (A-D-E-H-I-J-O-P). This critical path is estimated at $60^1/_2$ man-hours. With a six-man extension crew, the critical activities should be completed within a prescribed schedule.

Using CPM Techniques via Computer: As illustrated in earlier discussions, most mining projects can use CPM techniques manually. However, some projects like the ones in Chap. 7 may require computer scheduling to maximize the analysis possibilities and minimize the time involved.

Another example is the use of CPM for controlling large efforts involving many support projects in one area, say a main entry—submain entry intersection. Here, in a matter of a few months of construction time, a dozen overcasts, stoppings, plus complete belt, track, and piping systems must be put in place to support new production activities. Over the life of a large mine, this construction setup could be repeated 75 to 100 times.

To illustrate this complexity around a major mine intersection, Fig. 15 is a plan view of a typical main-submain connection before working the submain entries. Rooms are *necked* out during normal main entry development to allow the intake airstream to be rerouted during construction activities. To set up the proper support system for submain development, the intersection will require major construction. An outline

FIG. 13. CPM diagram for a belt drive project (time-scaled in man-hours).

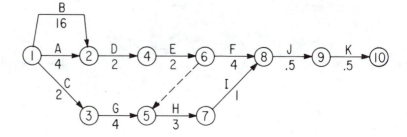

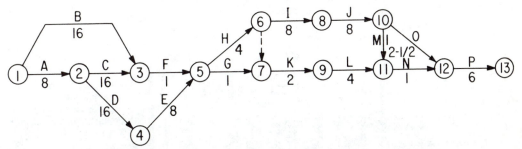

FIG. 14. CPM diagram of a belt line project (time-scaled in man-hours).

of the systems is shown in Fig. 16. It requires 14 overcasts, a 1067 mm (42-in.) belt drive, one No. 2½, 38.5 kg (85 lb) left-hand switch, and all connecting support. To accomplish the scheduling and CPM analysis by hand would be a major undertaking.

If one did a CPM network for developing this system, it would resemble Fig. 17, which shows only one approach to scheduling the entire necessary intersection construction. The real CPM use for analyzing and planning purposes (which requires changing, rearranging, adding, and subtracting time and labor estimates from the basic approach) is so difficult by hand that it

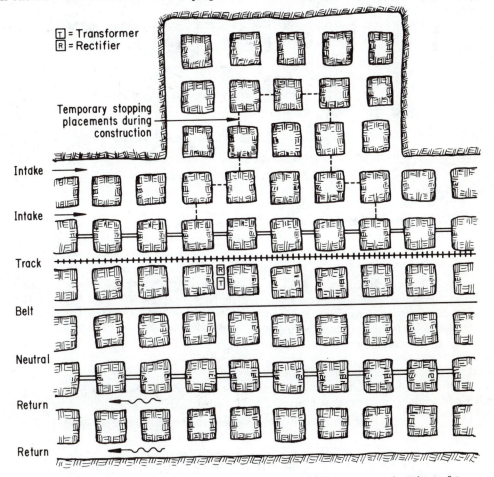

FIG. 15. Typical main-submain entry intersection. Dashed lines indicate locations of temporary stopping requirements during construction.

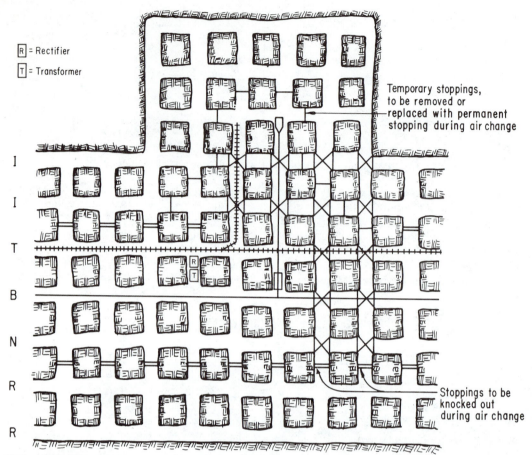

FIG. 16. Outline of production support systems needed for main-submain entry construction. Note: Legend for column at left is I = intake, T = track, B = belt, N = neutral, and R = return.

becomes self-defeating. Therefore, complex projects done by hand are definitely at a disadvantage to computer analysis, as well as being expensive and time-consuming.

To handle these cases, the computer will generate the appropriate scheduling network based on the inputs of project activities and event-timing estimates. Two types of output normally are available to the construction engineer: a network diagram and time-scale chart as a printout form. If the mine or mining company has contracted out the bulk of the construction work (as in the case of a rotary dump or surge bin), then a CPM printout is usually a standard contractual requirement for part of project control.

Typical computer-generated CPM printout may include such things as bar charts by early starts; sorting by identified activity numbers; sorting by float time configurations; estimated cost by identified activity numbers; and scheduling of cost savings by estimated early start and finishing dates. Each is designed to provide a single approach to viewing the project schedule by the user. Different computer runs will allow sensitivity analysis for any combination of start and finishing dates. Computer-generated CPM scheduling can greatly enhance mine management's ability to schedule and control their mine construction systems.

Minicomputer Software for Mine Construction

The final discussion of the computer aids section concerns the development of mini and microcomputers at the mine level. Today, many companies offer some type of CPM software for day-to-day activities without requiring large digital computer support.

The apparent burst of minicomputers onto the coal mining scene is due to a number of factors. The most significant is the fact that, until recently, technology was not at a stage to provide these small powerful computers. One of the first mining applications for computer use occurred for ore reserve analysis and geotechnical data bases. As computer technology developed through the 1970s, the cost of computer central memory (a large cost item) declined significantly at a rate close to 45% per year. In contrast to this trend, the computer industry was able to increase both the available memory capacity and memory speed tremendously. Coupled with these advances was the increasing reliability and quality of telecommunication systems. The costs for these improvements changed very little and even decreased in most cases.

The big breakthrough came in the mid-1970s when two improvements occurred. The first was a hardware development consisting of remote-batching terminals (dubbed *SMART*) which could print, plot, and read highly compressed data, as well as control and buffer the two-way traffic between the main computer and remote setups. The second and most important breakthrough was the development and distribution of the microprocessor, the so-called *computer-on-a-chip*. Its main advantages lay in the size, memory storage, and processing speeds which were available with the chip. In conjunction with the chip in computer services, the chip was impressed into telecommunication systems with great success. It was used to implement error detection and correction codes, which greatly increased the reliability of data transfer over the communications network.

At the mine site, a minicomputer will generally have at least 32 kilobytes of main memory. Random access memory (RAM) is used for data storage, program files, and temporary copying. RAM disks are added to the minicomputer with disk drives, and greatly increase the range of use for the minicomputer.

Turning to software for these minicomputers, the key word is versatility. Software programs can provide a host of analytical services. For construction purposes, many types of programs will optimize project schedules, compute event-timing, and analyze design options. Although much of the present software is for other mining aspects (production, maintenance, geotechnical engineering, and so on), there are several software programs available to aid the construction engineer.

One aspect of minicomputer software should be mentioned. It concerns the compatibility of any program to be run on a given machine. Because of the differences in hardware between manufacturers, many software programs can be used as an off-the-shelf package only with their respective systems. It is important the construction engineer compare his present and future scheduling and design needs with the company hardware now in place.

To conclude this discussion, Table 1 summarizes the commercially available software for use in mining support projects as of Dec. 1, 1981. This table is by no means exhaustive, but is intended only to show the scope and depth of computer industry offerings. Many factors are involved in selecting, buying, and leasing software, so the engineer should do a great deal of research before making any decisions.

REFERENCES AND BIBLIOGRAPHY

Anon., 1968, *Resource Management*, MDI Publications, Management Development Institute, Wayne, PA.

Anon., 1973, *Systems Analysis and Design Using Network Techniques*, Prentice-Hall, Englewood Cliffs, NJ.

Anon., 1981, "Available Software," *Mining Engineering*, Vol. 33, No. 11, Nov., pp. 1591–1593.

Anthony, R. N., 1965, *Planning and Control Systems: A Framework for Analysis*, Div. of Research, Graduate School of Business Administration, Harvard Univ., Boston, MA.

Anthony, R. N., and Mouton, J. L., 1975, *Management Accounting*, Richard D. Irwin, Homewood, IL.

Argyris, C., 1960, *Understanding Organizational Behavior*, Richard D. Irwin, Homewood, IL.

ARINC Research Corp., 1969, *Guidebook for Systems Analysis/Cost Effectiveness*, US Dept. of Commerce, National Bureau of Standards, Washington, DC.

Barish, N. N., 1962, *Economic Analysis: For Engineering and Managerial Decision Making*, McGraw-Hill, New York.

Baumgartner, J. S., 1963, *Project Management*, Richard D. Irwin, Homewood, IL.

Baumol, W. J., 1965, *Economic Theory and Operational Analysis*, 2nd. ed., Prentice-Hall, Englewood Cliffs, NJ.

Blake, R. R., and Mouton, J. L., 1964, *The Managerial Grid*, Gulf Publishing Co., Houston, TX.

Bideaux, R. A., 1981, "Minicomputer Applications for Today's Mining Problems," *Mining Engineering*, Vol. 33, No. 11, Nov., pp. 1584–1588.

Britton, S. G., 1981, *Practical Coal Mine Management*, John Wiley & Sons, New York.

Britton, S. G., 1981, "Cutting Costs of Mine Support Systems," *Coal Mining & Processing*, Vol. 18, No. 12, Dec., pp. 56–58.

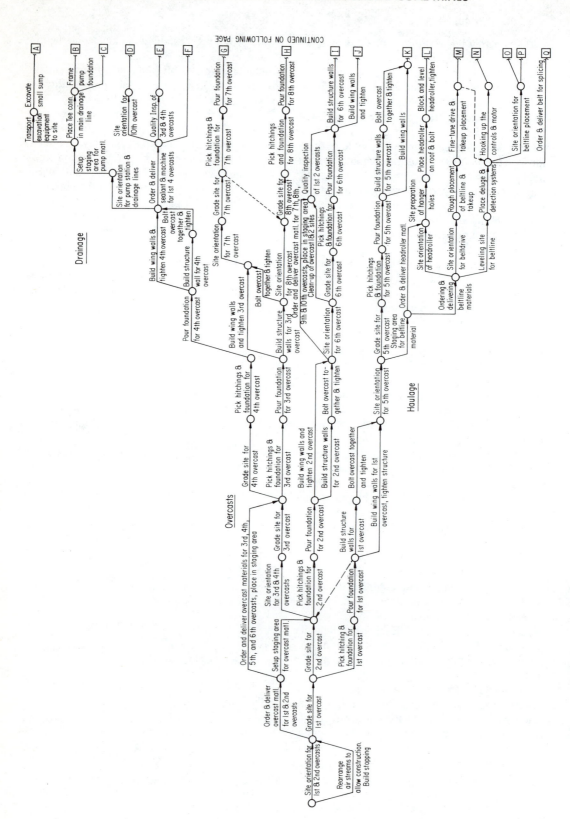

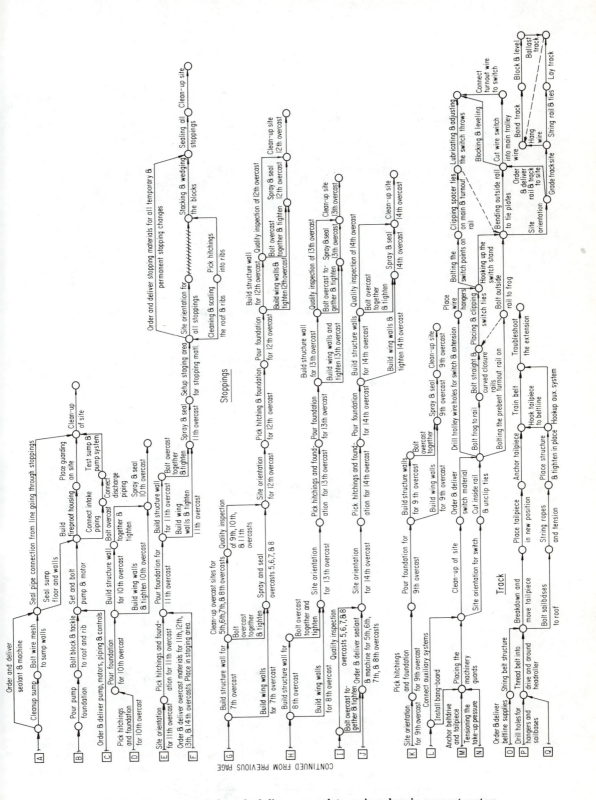

FIG. 17. One approach for scheduling a complete main-submain support system.

Bucklen, E. P., et. al., 1969, "Computer Applications in Underground Mining Systems," R&D Report No. 37, US Office of Coal Research, Washington, DC.

Butler, A. G., 1973, "Project Management: A Study in Organizational Conflict," Academy of Management Journal, Vol. 16, Mar., pp. 84–101.

Cleland, D. I., and King, W. R., 1972, *Managment: A Systems Approach*, McGraw-Hill Inc., New York.

Couger, J. D., and Knapp, R. W., 1974, *System Analysis Technique*, John Wiley & Sons, New York.

Datz, M. A., and Wilby, L. R., 1977, "What is Good Project Management," *Project Management Quarterly*, Vol. 8, No. 1, Mar.

Davis, S. M., and Lawrence, P. R., 1977, *Matrix*, Addison-Wesley, Reading, MA.

Drucher, P. F., 1975, "Managing for Business Effectiveness," *Business Classics, Harvard Business Review*, Boston, MA.

Emery, J. C., 1969, *Organizational Planning and Control Systems*, MacMillan, New York.

Friend, F. L., 1976, "Be a More Effective Program Manager," *Journal of Systems Management*, Vol. 27, Feb., pp. 6–9.

Gaddis, P. O., 1959, "The Project Manager," *Harvard Business Review*, Harvard Univ., May–June, pp. 89–97.

Gran, P. F., 1976, *Systems Analysis and Design for Management*, DunDonnelley, New York.

Grant, E. L., and Ireson, W. G., 1970, *Principles of Engineering Economy*, 5th ed., Roland Press Co., New York.

Herzberg, F., 1975, "One More Time: How Do You Motivate Employees?" *Business Classics, Harvard Business Review*, Harvard Univ., Boston, MA.

Holland, T., 1969, "What Makes a Project Manager?" *Engineering*, Vol. 207, Feb., p. 262.

Horowitz, J., 1967, *Critical Path Scheduling—Management Control Through CPM and PERT*, Roland Press, New York.

Houre, H. R., 1973, *Project Management Using Network Analysis*, McGraw-Hill, New York.

TABLE 1. Typical Software Packages* Available to Coal Mine Management

Type of Mine Support Construction	Program Name	Manufacturer or Supplier	Compatible Hardware	Description
General Mining	COMPASS	Bonner & Moore Houston, TX	Data General ECLIPSE(AOS, AOS/VS)	Divides construction and maintenance projects into labor, materials, and equipment components for planning and scheduling.
Ventilation	MIVENDES	Control Data Corp. Minneapolis, MN	*	Design and evaluates mine networks using a mode system. This allows singular analysis of each branch within the network.
Drainage	PENT	Engineering and Computer Services El Toro, CA	Data General ECLIPSE (AOS)	Mine water system analysis of supply pumps, valves and reservoirs. Allows time simulation for better network evaluation.
Haulage & Drainage	FCP	Foxboro Co. Foxboro, MA	FOX 3 Computer	Process control for flowing materials.
Drainage	*	Geomin Computer Services Corp. West Vancouver, British Columbia	*	Finite element seepage analysis.
Ventilation	*	Geomin Computer Services Corp. West Vancouver, British Columbia	*	Ventilation simulation
Drainage	Aquifer	Thorncroft Manor Ltd. Leatherhead, England	*	Simulates ground water levels.
General Mining	GIFTS	University of Arizona Tucson, AZ	Data General ECLIPSE (AOS)	Finite analysis, static analysis, vibrational analysis, and substructing transient response analysis of mining structures

*Adapted from *Mining Engineering*, Nov., 1981, Vol. 33, No. 11, pp. 1591–1595.

International Congress for Project Planning by Network Analysis, 1969, *Project Planning by Network Analysis*, North-Holland Pub. Co., Amsterdam, Holland.

Kast, F. E., and Rosenzweig, J. E., 1974, *Organization and Management; A Systems Approach*, 2nd ed., McGraw-Hill, New York.

Kelley, W. F., 1969, *Management Through Systems and Procedures: A Systems Concept*, New York.

Kerzner, H., 1977, "Systems Management and the Engineer," *Journal of Systems Management*, Oct., pp. 90–107.

Kerzner, H., 1979, *Project Management: A Systems Approach to Planning, Scheduling, and Controlling*, Van Nostrand Reinhold, New York.

Maslow, A. H., 1954, *Motivation and Personality*, Harper, New York.

Martin, C. C., 1976, *Project Management: How to Make It Work*, Amacom, New York.

Martino, R. L., 1968, *Project Management*, MDI Publications, Management Development Institute, Wayne, PA.

McGregor, D., 1967, *The Professional Manager*, McGraw-Hill, New York.

McGregor, D., 1960, *The Human Side of Enterprise*, McGraw-Hill, New York.

Moder, J. J., and Phillips, C. R., 1970, *Project Management with CPM and PERT*, 2nd ed., Van Nostrand Reinhold, New York.

Optner, S. L., 1965, *Systems Analysis for Business and Industrial Problem Solving*, Prentice-Hall, Englewood Cliffs, NJ.

Pearson, R. G., 1971, "Human Factors Engineering," *Industrial Engineering Handbook*, McGraw-Hill, New York.

Pegels, C. C., 1976, *Systems Analysis for Production Operations*, Gordon and Science Pub., New York.

Prelaz, L. J., et al., 1964, "Optimization of Underground Mining," Research and Development Report No. 6, Vols. 1, 2, 3, NTIS No. PB 166291-1-2-3, Springfield, VA.

Quirin, G. D., 1967, *The Capital Expenditure Decision*, Richard D. Irwin, Homewood, IL.

Ramani, R. V., and Manula, C. B., 1975, "A Master Environmental Control and Mine System Design Simulator for Underground Coal Mining, Executive Summary," Vol. 1, Report No. PB 255 421/AS, US Bureau of Mines, NTIS, Springfield, VA.

Ramani, R. V., Owili-Eger, A. S., and Manula, C. B., 1975, "Ventilation Subsystem," Vol. 10, NTIS Publication No. 225 430/AS, US Dept. of Commerce.

Redling, G., 1975, "SIGUT-Ein Modell fur die Simulation des Grubentriebes uter Tage," *Proceedings*, 13th APCOM Symposium, University of Clausthal, West Germany.

Richards, M. D., and Greenlow, P. S., 1966, *Management Decision Making*, Richard D. Irwin, Homewood, IL.

Rodgers, C. R., and Roethlisberger, F. J., 1975, "Barriers and Gateways to Communication," *Business Classics, Harvard Business Review*, Harvard University, Boston, MA.

Rosenbaum, B., 1979, "Supervisory Skills Training Program," Developed for Monterey Coal Co. by MOHR Development Inc., New York.

Scott, F. E., 1981, "MSHA Academy: America's Mine Safety College," *Coal Mining and Processing*, Vol. 18, No. 10, Oct., pp. 46–51.

Shannon, R. E., 1972, "Matrix Management Structures," *Industrial Engineering*, Vol. 4, Mar., pp. 26–29.

Sikula, A. F., 1977, *Personnel Management*, John Wiley & Sons, New York.

Sleyman, K. J., 1971, "Cost Estimating," *Industrial Engineering Handbook*, McGraw-Hill, New York.

Toellner, J., 1977, "Project Estimating," *Journal of Systems Management*, May, pp. 6–9.

Vaughn, D. H., 1967, "Understanding Project Management," *Manage*, Vol. 9, No. 9, pp. 52–58.

Weinberg, G. M., 1975, *An Introduction to General Systems Thinking*, John Wiley & Sons, New York.

Weiss, A., 1979, *Computer Methods for the 80's*, AIME, New York.

Wilson, F., 1963, *Manufacturing, Planning and Estimating Handbook*, McGraw-Hill, New York.

Whitehouse, G. E., 1973, "Project Management Techniques," *Industrial Engineering*, Vol. 5, Mar., pp. 24–29.

10

Cost Analysis of Underground Support Construction

In the real world of coal mining, the bottom line for almost every support project decision is economic (i.e., what is it going to cost?). Up to this point, I have discussed the methods to improve the performance of typical support projects by better installation, materials, and techniques, thus lowering the impact of the bottom line cost. In this chapter, I shall explore what costs are relevant in mining construction, and how these costs are developed.

The chapter centers around two major discussions: (1) developing cost accounting techniques, and (2) cost-estimating procedures for support construction. Each focuses on the fundamentals of practical cost considerations necessary for decisions on construction projects underground. Since cost accounting, economics, and investment planning are all subjects which warrant complete discussions on their own, this chapter will focus on distilling the important concepts required by a planning engineer into an easily understandable presentation.

DEVELOPING COST ACCOUNTING TECHNIQUES

A fundamental element of the word *cost* is its ability to measure the use of resources by a mining company. Cost measures such things as material quantities, labor hours, and intangible services. Even so, the word *cost* by itself can mean many different things in regard to one project. It is common to see cost modified by a description such as *full cost, differential cost,* or *direct cost.*

It is important to understand two additional elements of the idea of cost. First, a cost is measured in terms of dollars and cents (or other monentary terms). This provides a common denominator from which mining projects can be evaluated and compared with other mining projects. It also provides a singular base from which to forecast project progress and compare results. Secondly, the cost of something relates to a specific purpose, whether completing a mining project, leasing equipment, purchasing land, or mining coal. This purpose can always be identified and catalogued through its cost parameters, thus giving mining management another tool for project control.

To best use this cost tool for project control, some discussion on the basic concepts of what a cost is will be helpful. The fundamental consideration for mining centers on the term *full cost.*

Full cost is a term encompassing all the resources used for an objective. For example, if a mine pays $2500 for a single prefabricated overcast from a manufacturer, then the full cost of the overcast is $2500; that is, the mine used $2500 of their resources (in this case, money) to acquire the overcast.

However, suppose that it is necessary to ask: what is the full cost of producing a ton of coal? This is a much more complex question, one difficult to answer immediately. A coal mine produces thousands of tons of coal each month. Some tonnage may be mined on development (with conventional equipment), some in panels, or some by retreat methods (maybe with a continuous miner or longwall). Clearly, different amounts of resources are used for these different methods, yielding different costs. For this type of situation, the full cost can be broken into its components of *direct costs* and *indirect costs*. The full cost of a ton of coal, therefore, is the sum of its direct costs, plus a fair share of the indirect mining costs of the operation. Let's explore these costs a little further.

Direct costs are simply cost items which can be traced specifically to a project or objective. In coal mining, when the objective is mining coal, direct costs of production include the wages of workers, supplies used in the production of coal (bits, oil, splices, and so on), and equipment used to mine the coal.

Indirect costs are costs which are necessary for the completion of the project or objective without being traced directly to it. For example, no coal mine can operate without support systems like ventilation and haulage. However, the costs of these systems cannot be traced to any single ton of coal, hence, these costs are indirect. Not all costs can be pinned on a single production unit or ton of coal, therefore, a share of these costs are proportioned to each ton of coal by the mine accountants to accurately represent the indirect costs necessary to produce the coal.

Going one step further, it should be pointed out that mine support costs, although considered an indirect cost component to the production of coal, are made up of direct and indirect cost items (Fig. 1). For example, an overcast is composed of certain direct costs (overcast materials, wages of the workmen building the overcast, and the salary of the front-line supervisor) and indirect costs (transportation costs of materials to the site, costs of site orientation by mine surveyors, and the salary of the mine construction engineer). Therefore, reviewing and evaluating costs, especially operating mine costs or project costs, also requires looking at the relationship of these costs to the desired project goal.

The Nature of Project Costs

All project costs can be broken down into direct and indirect costs much like the previous example with the ton of coal and overcast. In addition, these cost categories can be further divided into subcategories describing project materials, labor, or services. The main subcategories are:

Direct material and supply costs.
Direct labor costs.
Services.
Indirect project costs.
Project inventory cost.
General and administrative (G & A) costs.

Direct material costs define the subcategory which includes those materials going directly into the project. For example, the mortar used in the wing walls of an overcast is a direct material cost. In the same light, the concrete block, sealant, bolts, wire mesh, chain, and any other material needed to build a support project and which becomes a permanent part of the project can be considered a direct material cost. Direct material costs are not the same as direct supply

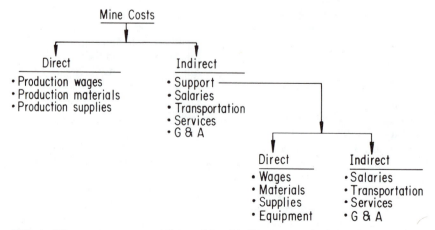

FIG. 1. Mine support costs, while considered indirect costs of mining coal, are made up of direct and indirect costs.

costs. Direct supply costs refer to the materials used to build a project, but which do not become part of the finished project. Such supplies can be best seen in lubricating oil, rock bits, rock dust, air hose, and so on used during project construction.

Direct labor costs relate to the wages of workmen on the particular project. This includes all work in the preparation, assembly, and O & M stages of a project. The salary of the front-line supervisor is also included as a direct labor cost because the supervisor is directly involved with the construction of this particular project.

Service costs define the costs relating to a project which have no physical substance. Some examples include the power for the construction equipment, heat, if necessary, for comfort, and insurance protection against damage. In most cases, service costs are considered indirect, although some portion may be traced to the project.

Indirect project costs refer to all project costs other than services and direct project costs. Such items as the salaries of mine supervisory staff, maintenance costs of the project, as well as transportation of the materials to the construction site are considered indirect project costs. Many times the service costs are included as a total, instead of being broken out, so portions of the mine costs for power, ventilation, depreciation, taxes, and insurance will be attributed to the project.

Project inventory costs are simply the sum of the direct and indirect costs of the project at the time of accounting. Inventory costs are most frequently quoted during the time of project completion, but before project use begins. Often one will see a statement like "81% of the projected expenditures against 75% of the project schedule." This refers to the current inventory cost of the project. The statement says that 81% of the total projected direct and indirect costs have been used to reach the 75% mark in project completion. Inventory costs usually do not, however, include costs unrelated to the project, such as marketing or G & A costs, but are posted to the project at some predetermined interval.

General and administrative costs are those costs borne by a mining company which are incurred but not attributable to another cost category. Such items as mine facilities, engineering, R & D, environmental concerns, and miscellaneous expenses are included in the G & A category.

A portion of these costs will ultimately be attributed to the project and counted as an indirect cost.

Allocating Direct and Indirect Project Costs

This discussion focuses on the general approach to measuring direct and indirect costs of a project. Both direct and indirect costs are measured best when a cost accounting system is in place for project control. Costs have the inherent characteristic of flowing through to the completion of the project. To measure this flow, mining companies commonly use the *process cost system*.

The process cost system basically collects cost data for a specified period of time (say a month) and then compares the applicable production (or work progress) to the total cost. No one cost is identified with any one activity by a workman, unless an activity is composed of this single cost item. Process cost accounting requires a standard recordkeeping system, but requires no elaborate techniques.

Measuring Direct Cost: The first question which comes to mind when measuring direct project costs is: what is classified as a direct project cost? According to the Cost Accounting Standards Board (1973), an item of cost is direct if the specified project was intended to benefit from that cost; or if the specified project caused incurrence of the cost. However, if a single cost is shared by two or more projects but still meets the criteria, then it is considered an indirect cost for each of the projects.

Direct Labor Costs. There are two requirements necessary for measuring direct labor cost. The first is measuring the quantity of labor time (man-hours) expended on the project. The second is verifying the price per time unit for the labor.

Measuring the quantity is fairly easy since daily time sheets are used to record a worker's involvement with the project. Every day worked on the project is kept and used in measuring the direct labor costs of the project.

The price per unit of time in coal mining is most often measured in dollars per work day. Underground labor works a straight eight-hour day and receives a direct wage in the neighborhood of $75. Some companies also add the cost of labor to the basic wage. These costs of labor include such items as social security taxes, pension contributions, unemployment insurance, and other fringe benefits the company must pay because the employee works. These are real costs to the company and their inclusion in the direct labor cost calculations will increase the accuracy of the estimate. Currently, the percentage of cost of support labor benefits totals about 80% of the daily wage rate or a total of $135 per day.

Direct Material Costs. Again, measuring the material costs includes two requirements, quantity and unit price of the material. These costs are measured in much the same way as labor costs. The quantity is specified on requisition or bid sheets for the project. The unit prices are supplied from manufacturers and material suppliers with considerations for bulk sales and other incentives. As with labor, related costs for materials may be added to the unit price to reflect such things as freight costs, moving costs, purchasing costs, interest, and space charges associated with handling the materials.

Allocation of Indirect Costs: One of the more difficult tasks to perform in cost estimating is the allocation of indirect costs to a project. It is usually accomplished by using a roundabout and less accurate process.

The question now arises as to what the accounting difference is between direct and indirect costs. Earlier I defined the allocation difference with these costs. The accounting difference relates to the actual identification and measurement of direct and indirect costs. There are three tests which distinguish the accounting differences of these costs. A cost is considered indirect if it falls under any one or more of the following tests: (1) if it is impossible to directly trace the cost to the project (as with the salaries of the technical mine staff); (2) if it is not feasible to directly trace the cost to the project, within manageable limits (such as the nails, string, or utility tape used during a project); and (3) if management chooses to treat a cost item as an indirect cost, probably out of industrial convention or prior practice. In continuous operations like coal mining, there is an argument which suggests the entire cost of a construction or support project should be considered as an indirect cost of producing coal. However, this type of treatment is left to the discretion of the individual mining company.

Allocating a fair share of the indirect costs of the entire mine to any one single support project can be difficult without a systematic approach. This systematic approach utilizes the overhead rate method to produce a *fair share* indirect cost.

There are three steps required during the calculation of the overhead rate of a given project. These steps produce an overhead rate which is usually expressed as a fixed percentage on a project type basis (i.e., either a transportation, ventilation, haulage, or drainage project). The three steps can be summarized as the following:

1) The first step requires estimating all the project indirect costs which are part of the support construction cost at the mine in the form of indirect labor and supplies. Service costs are also included along with other mine indirect costs for support construction.

2) The second step is to arrange these costs by project type. The initial arrangement is by transportation, haulage, ventilation, and drainage. Later, individual projects can be identified such as track switches, overcasts, sumps, and belt line extensions.

3) The total project type costs are divided by the estimated numbers of man-days required to finish it. This gives an overhead rate in dollars per man-day.

Another accounting approach designed to simplify the calculation effort is to assign a fixed overhead rate for all support construction. This rate should be based on good historical data.

The key step in the overhead rate method is, of course, the first one where decisions are made on what and how much of the mine indirect costs are to be allocated to mine support activities. The original criteria given for direct costs (benefits from or is related to a project) should serve as a guide in identifying total indirect costs. From this total, the direct support project costs can be identified and subtracted from the mine indirect total. This will yield the total indirect support costs to be allocated over the mine projects. However, mine management can form any policy toward this task, just as long as they maintain a consistency of purpose.

Indirect costs, therefore, such as overhead rates, function as an equitable assignment of the total indirect mining costs attributable to support functions. In practice, this rate changes from one project to another, raising the question: why? Some plausible explanations which may be applicable to most mining situations are:

1) Because more labor or more skilled labor was needed on a given project. Therefore, a higher indirect cost is allocated for this.

2) Because one project used more equipment time than another. For example, cutting top for an overcast requires the continuous miner more than grading for a track entry. Therefore, more indirect cost is attributed to the overcast activity.

3) Because one project had a higher direct labor or material cost, the indirect costs are proportioned in relation to these costs to reflect the higher direct cost.

These reasons suggest the basis for allocating indirect costs can be made with labor hours, machine-time, or total project direct costs. For mining, either using the number of labor hours or the total project direct costs for the basis appears better than using the number of machine hours.

To conclude, cost items are termed indirect because they cannot be matched with a project directly, because it is not cost-effective to do so, or because mine management chooses not to do so. Mine as well as service indirect costs are allocated to individual projects by means of an overhead rate. This rate, expressed in dollars per unit of time or as a direct cost, is calculated prior to beginning a project; is based on historical data usually; and is updated at least once a year. The number of man-hours or total direct costs of a project are then multiplied by the overhead rate to give the total amount of indirect cost allocation.

Capital Investment Decisions for Support Projects

When a mining company commits itself to a particular mine design and support construction approach, it makes an investment. In this case, the company commits the funds for support construction in the expectation of future earnings for the mine. In other words, the support systems will be put in place before any coal from the production unit will be seen. An investment is thus the initial capital outlay for a future stream of cash inflows (earnings).

For the most part, investment decisions are made during the planning stages of mine design, on the drawing board. This is also where the most care and effort of the mining engineer should come into play. It is his job to devise and plan the most cost-effective systems and mining approaches to maximize (1) the safety of the work environment, (2) the extraction ratio of the reserve, and (3) the profit potential of the mine. Therefore, the investment of materials, equipment, and labor by a mining company toward any mine design should reflect these considerations in the approach submitted by the mining engineer.

When a mining company is considering whether or not to install a new support project in an existing mine, the essential question to be answered is whether the future cash inflows will be large enough to cover the investment decisions concerning:

Replacement of existing structures or facilities with more efficient ones. A decision to switch from rail haulage to belt haulage may be made in the expectation of cost savings from lower operating costs and higher reliability.

Expansion of the support systems for mine use. Here, decisions for expansion during mine development are made as a matter of course. However, support construction during retreat mining must be closely looked at. Also, decisions such as opening up a remote mine portal in expectation of future mine development in that area is an expansion decision.

Equipment choice of support system materials. With three or four types of overcasts, a myriad of pump types, and rope or channel belt lines, mine management must consider the cost advantages as well as technical advantages of choosing one over the rest.

Buy or lease of support construction equipment. Here, decisions are made on whether or not the investment required to purchase the equipment will earn an adequate return because of the savings of construction costs and avoiding lease payments. An example would be whether to buy or lease a cement mixer for a certain underground project. The cement mixer will reduce the labor costs associated with mixing and pouring concrete. If the mixer will be used over and over again in the mine, the savings from the labor costs may exceed the lease payments and therefore be a better buy. If not, then the reduction in labor costs for the project will help offset the leasing cost.

In all these decision types, two cash flows are compared. The first is in the form of a lump sum, called the *investment*. The investment is conveniently thought of as occurring, for comparison purposes, at a point called time-zero, or in the present. The investment is considered as the present value amount for comparing any future cash flow from the investment.

The second cash flow is the stream of *future*

TABLE 1. Example of Cash Flow for a Cement Mixer

	Year	Amount	Present Value of $1 @ 10%*	Total Present Value
Investment	0	− $300	1.000	$ − 300.00
Future earnings	1	+ 200	0.909	+ 181.80
	2	+ 200	0.826	+ 165.20
			Net present value	+ 47.00

*Discount Rate.

earnings generated by the investment. These earnings can be generated in the form of cost savings or revenue inflows.

To compare these two cash flows, it is necessary to convert or bring back the stream of future earnings to the present value. This is done by multiplying the future earnings for each year by the present value of $1 for that year at the appropriate rate of return. This method is called discounting the future earnings of an investment by the rate of return. The difference between the investment and the discounted future earnings is called the *net present value*.

For example, a cement mixer is purchased for a $300 investment in an effort to reduce labor costs of hand-mixing on a surge bin project, as well as other mine construction projects. What would be the net present value of the investment if it saves $200 per year over its useful life of two years? The construction engineer using a 10% discount rate can set up the following calculations illustrated in Table 1. The mine will make $47 over two years by purchasing a cement mixer to help mix the concrete for the project (Fig. 2). Even though this example appears small, the savings from this type of analysis for an entire mine can be considerable. In this case, the construction engineer will accept the investment for the mixer because of the positive return. This points out a basic decision rule concerning investments: a proposed investment is considered acceptable if the present value of the future earnings equals or exceeds the amount of the investment.

Investment Decision Variables: All through the beginning discussion of investment decisions, four elements of the decision process have been assumed. These were (1) the rate of return (discount factor), (2) the future earnings of the investment by year, (3) the useful life of an investment for which future earnings can be expected, and (4) the total amount of the investment. Each element plays a fundamental role in determining whether an investment will be acceptable. Therefore, it is very important to accurately determine or estimate these elements when looking at an investment.

The *rate of return*, or discount factor, of an investment looks at the future earnings potential brought back to the present value. The discount factor is an estimate of inflation and its effect on the future value of the earnings. It follows that the higher the discount factor, the lower the present value of the future earnings will be, and therefore, fewer investment proposals will be acceptable. The usual discount rate is selected mainly by comparison with other similar companies in the area. The typical rate for mining

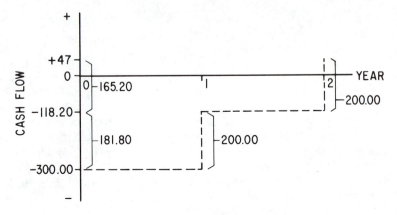

FIG. 2. The purchase of the cement mixer generates $47 in revenues over two years.

firms is somewhere between 10 and 15%. Higher rates are used in certain high profit industries and by most firms in periods of uncertainty over future inflation rates.

The *future earnings* of a proposed investment is essentially the additional cash flows obtained by the company as a consequence of making the investment. The additional cash flow comes from generating more revenues (i.e., the project produces more coal for the company), taking an equipment depreciation charge, or by lowering operating costs (to reduce the expenses of the company). For most support construction in existing mines, the additional revenues from future production will exceed the investment of the support construction. For example, setting up a production panel off the main development entries with all necessary production support systems is an investment which is made to reap the rewards from the production panel output. Or in much the same light the cost of mine development for the main entries is an investment in the future output of the mine. However, efficient construction practices also can be considered as an investment of future earnings from lower operating costs.

To the construction engineer, the potential earnings from lower operating costs is one of the most significant factors for using effective construction practices. Sometimes, structure maintenance activities are undertaken to lower operating costs. This includes replacing poor overcasts with high efficiency types, resealing stoppings, replacing worn track, or cleaning a sump (increasing system efficiency). Again, the potential earnings from this work can be considerable if the construction engineer uses the right planning approach.

The *useful life*, or *economic life*, of an investment measures the number of years that future earnings can be expected. It may or may not coincide with the equipment's or structure's physical life underground (Fig. 3). The period selected for the economic life of an investment is called the *investment horizon*. With unusually long investments, future earnings are set at a specific maximum number of years, such as 10, 15, or 20. Many times, the economic life will be set equal to the life of the mine.

The economic life also ends when the company ceases to make profitable use of the support structure. If a production panel is setup, mined, and abandoned within three years, the economic life of the support structure in relation to that panel also ends in three years. It does not mean, however, the amount of investment should be compared with the earnings of just three years. The IRS gives most support structures a 5 to 10 year depreciable life, regardless of how much they are used or reused.

In most cases, knowledgeable guesses can be made on the useful life of the support system. Technological improvements in designs and equipment will inevitably make current systems less attractive to mining operations for future mine applications. The trick for the mining engineer or planner is to guess which systems, with what improvements, will be affected, and how soon these improvements will come about. It is an educated guess which must be made, since the investment will cease to earn a return when it is replaced by an improved system or piece of equipment.

The *amount of investment* is the funds that a company risks if it accepts the proposed investment. The costs of materials, transportation, installation, and other related costs (training, overtime, etc.) are all part of this capital outlay. The basic amount of the investment under consideration is the incremental amount which is added to the current committed funds for a mine toward completion of the project.

Net Present Value Method of Analysis: Once a construction engineer has determined the probable ranges for the variables of investment decisions, he can use the net present value method

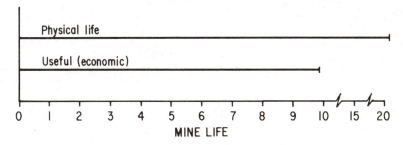

FIG. 3. Physical vs. the economic life of a mainline belt line.

of analysis to determine whether the investment is acceptable or not.

The net present value method has six distinct steps. They can be summarized as:

1) Selecting a rate of return (discount factor). Usually, a mining company chooses a rate between 10 and 15%. This rate can be used over and over for each investment decision if desired.

2) Estimating the amount of investment required. This is the sum of additional outlays of funds (including the costs of borrowing the funds) made at time-zero, less the proceeds from disposing of existing equipment (if applicable).

3) Estimating the economic life of the project. This may or may not correspond to the physical life of the project.

4) Estimating the future yearly earnings of the investment, including a salvage value at the end of the useful project life and all tax incentives and investment credits allowed under the law.

5) Calculating the net present value of the future earnings by multiplying the discount factor by the appropriate yearly earning. The difference between the total future earnings in the present value and the investment is the net present value.

6) If the net present value is zero or positive, the investment is acceptable. If it is negative, the investment is rejected.

As an example of the net present worth method, let us assume the Jolly Giant Coal Company is considering the purchase of a roadheading machine for $136,000. The roadheader is going to be used for main entry development work on cutting and grading. It has an economic life of 10 years, although the mine is hoping for at least 15 years of service. The mining engineer has estimated the yearly future revenues to total $18,500, which appears to be a sizable savings. This includes taking into account the recovered labor wages from moving production equipment, idle shift scheduling, equipment depreciation, operating costs of the machine, and the increased production of coal due to greater equipment availability. Because of a slump in the coal market, the rate of return or discount factor is estimated at 15%.

The calculated cash flow from the investment is shown in Table 2. As one can see, the net present worth method predicts a loss on the investment, given that the four fundamental elements are correct. Therefore, the mine would choose not to invest in the roadheading machine. As seen in Fig. 4, the total future earnings, in present value, would not offset the initial capital investment of $136,000. As mentioned earlier, the net present worth method is a good and simple indicator of acceptable and unacceptable investments.

Payback Method of Analysis: The payback method of analysis is also a simple, straightforward approach to evaluating a proposed investment. The term *payback* refers to the ratio of the investment and future earnings of a project. It is expressed as the number of years over which the investment outlay will be recovered from the future earnings. The payback method is simply the measure of how well the project will pay for itself. If a support project costs $5000 and generates future earnings of $1000 per year, then it has a payback of 5 years.

If we return to the roadheading machine investment, the payback would be expressed as:

TABLE 2. Cash Flow for the Roadheading Machine Investment
(1981 dollars)

	Year	Amount	Present Value at 15%	Cash Flow
Investment	0	$136,000	1.0000	−136,000.00
Future earnings	1	18,500	0.8696	16,087.60
	2	18,500	0.7562	13,989.70
	3	18,500	0.6575	12,163.75
	4	18,500	0.5718	10,578.30
	5	18,500	0.4972	9,198.20
	6	18,500	0.4323	7,997.55
	7	18,500	0.3759	6,954.15
	8	18,500	0.3269	6,047.65
	9	18,500	0.2843	5,259.55
	10	18,500	0.2472	4,573.20
			Total	−43,150.35

$$\frac{\$136,000}{\$18,500 \text{ /yr.}} = 7.35 \text{ years}$$

This basically shows the engineer the simple payback of the roadheader is approximately three-quarters of the economic life, without considering any other economic variables.

As one can see, the payback method is a quick, but crude, technique for analyzing a proposed investment. If the payback period was equal to or greater than the economic life of the roadheader, then the investment would be clearly unacceptable. If, on the other hand, the payback had been relatively less than the economic life (i.e., one half or one-quarter), then the investment may have been attractive.

The drawback to the payback method is the fact it gives no consideration to the time value of money (i.e., inflation) or to the differences in the economic lives of various projects. The tendency to view the payback method as "shorter is better" may eliminate a longer project with a potential for greater earnings.

To try and offset some of the drawbacks of the payback method, and at the same time improve its accuracy, financial experts developed the discounted payback method. This method requires the future earnings be discounted to the investment time-zero and then be divided into the investment funds. This method also provides the year at which the investment will recoup its investment and provide for the rate of return (i.e., inflation). To pinpoint the exact year, the yearly future earnings are discounted and added up until they equal or exceed the amount of investment. The year in which this happens is called the discounted payback year.

An example of the discounted payback method can be seen in the following example. A coal company is contemplating a stopping change in their production panels from cement block stoppings to expanding metal stoppings. The investment of metal stoppings for a three year production panel is $4,500 (cost of metal stoppings minus the investment in the block material) with future earnings estimated at $2,625 per year. The subsequent discounted payback cash flow with a discount factor of 10% can be seen in Table 3. The payback, therefore, is the comparison of the totaled discounted future

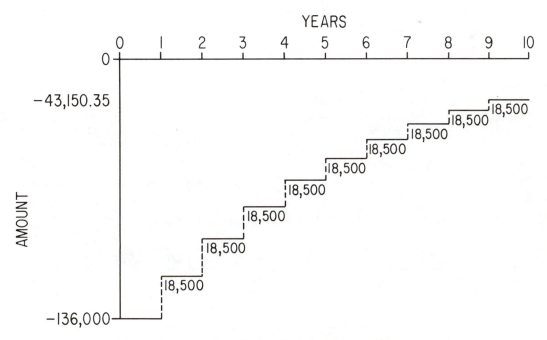

FIG. 4. Total future earnings do not offset the initial capital investment.

earnings with the initial investment. At year two, the total discounted future earnings exceed the investment and so is the payback year. The discounted payback method is preferred by many over the simple payback method.

Until this point, I have referred to investment decisions on a singular basis, i.e., putting in an overcast, buying a machine, replacing some stoppings. It should be noted that these examples are for illustration only, and should not be considered the only types of investment decisions. On the contrary, most investment decisions will revolve around design alternatives on mine-wide construction, such as the decisions between:

1) Transportation alternatives for rail haulage by investigating what weight is necessary, how many switches should be used, whether ballasting and grading should be used, and so on.

2) Ventilation designs for main, submain, and panel entries, revolving around the type, number, and placement of overcasts, regulators, doors, and stoppings.

3) Haulage choices on the designed belt capacity, strength and type of structure, belt drive design and placement, as well as auxiliary controls, detection, and communication devices.

4) Piping, pump, and wastewater layout choices.

Therefore, investment decisions can involve large sums of capital over a 20 to 30 year mine life. The role of the mining engineer during this process is to recommend the best investment decisions by combining the optimum mining approach with the best return on the company investment. Large new underground mines may face operating costs of $3 to $4 million each year, so a 10% reduction in the yearly operating expenses from careful planning and correct decisions among the alternatives can be significant.

The Cost Control Process

Once an investment has been accepted, the funds committed, and bids awarded (if necessary), the construction engineer begins the cost control process as a part of project control. The control process is made up of seven distinct requirements:

1) Detailed planning of the required project tasks.
2) Accurate estimating of time, labor, and material costs.
3) Good communication of project planning and objectives.
4) A straightforward budget.
5) Timely accounting of project progress and cost expenditures.
6) Periodic evaluation and adjustments of time and cost estimates for remaining work.
7) Frequent checks for quality control and regular comparisons of actual vs. scheduled progress of the budget.

These requirements underscore the first objective of cost control as a verification process of comparing the actual project performance to date (in terms of progress and cost) with the projected estimates of performance. This objective also serves to guide future forecasts of costs and performance in making better construction estimates for similar projects.

A second purpose of cost control involves decision-making responsibilities. Managers and engineers working on a project require information on the project status to be able to make decisions on:

1) The project plan, schedule, and budget agreed upon during the planning phase.
2) The resources expended to date and those projected to be expended.
3) The total estimates of required project resources.

These decisions allow informational feedback to reach management, thereby identifying problem areas, major planning changes, and possible contingency plans.

Progress Analysis Planning (PAP): The key

TABLE 3. Cash Flow of Stopping Example

	Year	Amount	Discount Factor	Cash Flow Present Value
Investment	0	4,500	1.0000	−4500.00
Future earnings	1	2,625	0.9091	+2386.39
	2	2,625	0.8265	+2169.56
	3	2,625	0.7513	+1972.16

to the cost control process is the implementation of the progress analysis planning (PAP) method for each project. The PAP method relies on the accurate charting of the project progress measured against the estimated progress during the planning stages. This is done by charting the variance of any schedule, technical performance, or cost item of a project. These deviations are charted and compared together because the cost variance compares only deviations of the present budget, and the scheduling and technical performance compares only the project performance between actual and estimated amounts. To calculate these deviations from project schedules, it is necessary to define three basic variances for budgeted and actual costs for work scheduled and performed. Archibald (1976) defines these variables as:

Budgeted Cost for Work Scheduled (BCWS) is the budgeted amount of cost for work scheduled to be accomplished, plus the level of effort or apportioned effort scheduled to be accomplished in a given time period.

Budgeted Cost for Work Performed (BCWP) is the budgeted amount of cost for completed work, plus budget amounts for level of effort or apportioned effort activity completed within a given time period. This is sometimes referred to as an *earned value*.

Actual Cost for Work Performed (ACWP) is the amount reported as actually expended in completing the work accomplished within a given time period.

Using these definitions, variance equations can be developed to express the relationship between the budgeted and actual costs and schedules. These equations can be expressed as:

Cost Variance = BCWP − ACWP
Schedule/Performance
   Variance = BCWP − BCWS

Both equations can be graphically represented with time (work progress) as the horizontal axis and cost ($) as the vertical axis, as shown in Fig. 5. These equations can be used for any level of project objective whether a subtask, task, or total project.

By plotting both variance equations on one graph, a complete PAP cost/schedule reporting system can be developed. It provides a visual representation from which the construction engineer or project manager can evaluate and measure project costs to completed project work, thus giving a total performance picture.

For example, Fig. 6 is a PAP cost/schedule system showing several common features of project construction. At the time of measurement, there is a cost overrun of the projected budgeted cost and a performance slippage in work progress. However, the cost overrun has not exceeded total contingency funds, so can still be considered manageable. The PAP method of measuring variances, as this example shows, can keep excellent track of project performance on a given reporting basis (daily, weekly, monthly, and so on).

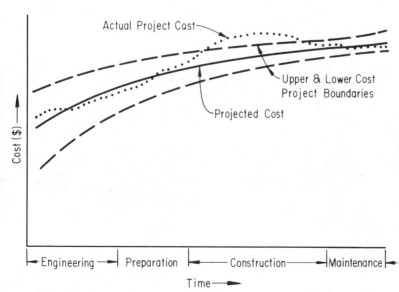

FIG. 5. Variance projections of work progress in terms of cost and time.

COST ESTIMATING FOR SUPPORT CONSTRUCTION

The final discussions of this chapter focus on the cost estimation techniques used in support project planning. Estimating combines both science and art by utilizing skillful analysis and experienced judgment. When done properly, cost estimating will produce an accurate prediction of the eventual project cost. This is accomplished by detailed study of project cost components (labor and material) whose result is a statement of cost that will be used as a basis for making sound planning decisions.

This discussion is divided into six sections. These are identified as (1) general approach to cost estimating, (2) transportation projects, (3) ventilation projects, (4) drainage projects, (5) haulage projects, and (6) mine support projects as a system. The first section covers the general approach to cost estimating for any project. Subsequent topics focus on current cost for selective support projects.

General Approach to Cost Estimating

There are three necessary elements to basic cost estimating. They are timing, historical data, and knowledgeable judgment. Because each project is different, no two elements contribute equally to any project decision. Timing is more important when submitting a competitive bid for construction, and historical data or judgment may carry more weight on feasibility studies.

Timing: Timing is an adjective describing the cost estimator's ability to grasp the concept of *time is money*. Certainly, time itself adds no value to a project, so the project cost changes in direct proportion to the time taken. In other words, the more time it takes to complete a project, the more it costs.

Timing also reflects the estimator's ability to meet a deadline. In competitive bidding situations, proposals, and marketing strategies, being able to deliver accurate cost estimates at a certain date is of great concern. Late estimates do no one any good, so the estimator must have a realistic sense of timing to discipline his efforts for planning purposes.

Historical Data: Each time some task is completed, a supplier submits a quote, or a new cost estimate drawn up to reflect changing conditions. This adds to the potential historical data base of the mine. If this information is collected and organized in a systematic manner, it frequently will eliminate the need to "reinvent the wheel." Considerable amounts of time and ef-

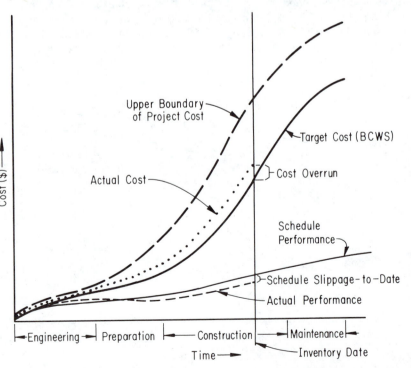

FIG. 6. Project schedule and cost chart showing a total performance picture.

fort can be eliminated if accurate reference material on a project task or component is available in the industrial engineering or mine engineering data files. The greater the amount of historical detail available, the smaller the margin of error between predicted costs and schedules and the actual costs and schedules. In addition, not having this factual base often forces the mine engineering personnel to participate in a guessing game during planning stages.

The most important function of the historical data base is to test and check the validity of new cost estimates. By continually updating and expanding this data base, remarkable results can be achieved in a relatively short time. Computer stored data bases greatly aid in cost estimating procedures by allowing (1) more data to be considered, (2) greater speed in reviewing, and (3) more alternatives to be explored. Computers are not, however, a prerequisite to establishing a good data base. What is a requirement is knowing the basic sources of data available to cost engineers.

Data Sources. The information needed for cost estimating purposes comes from many different functional departments within and from outside mining. Within a company structure, departments contributing cost data at the mine level include mine engineering, maintenance, production, safety, accounting and warehousing (purchasing), and training. From outside mining circles, data sources include manufacturers or equipment suppliers, technical societies, consultants, universities and government agencies.

To summarize the types and sources of data available, Table 4 is an example of some project planning data which can be used in data base development. The approximate time frame for developing a respectable cost estimating data base is normally two to three years. After that initial development phase, continual upgrading and adding to the data base is necessary to keep it current. It is very important to maintain the quality of the data base to ensure good estimates. A lack of confidence, and hence, a disregard of the figures by mine management will result if they are convinced the estimates do not reflect the current situation. Once this happens, the data base is practically worthless, and all time and effort wasted.

Knowledgeable Judgment: There is no substitute in mining for experience or skill in cost estimating. Even with the best of data bases and all the time in the world, the task of accurate cost estimating requires both systematic analysis and study of project specifications during planning. Labor and material escalations, transportation and delivery schedules, safety concerns, legal constraints, training problems, as well as facility and manpower requirements are but a

TABLE 4. Types and Sources of Cost Estimating Data

Type of Data	Sources
General design specifications	Mine engineering or purchasing department
Quantity and schedule of production	Request for estimate or purchasing department
Assembly or layout drawings	Mine engineering or maintenance department or equipment supplier
General construction plans and list of proposed subassemblies of project	Mine engineering department or manufacturers
Detail drawing and bill of material	Mine engineering or purchasing department
Test and inspection procedures and equipment	Maintenance, safety, engineering, or government agencies
Packaging and transportation requirements	Shipping department of equipment supplier
Equipment routings and operation sheets	Maintenance, mine engineering, or manufacturers
Detail machine and equipment requirements	Manufacturers or equipment suppliers
Operation analysis and workplace studies	Mine engineering, training, and safety departments
Standard time data	Special charts, tables, time studies, and technical books and magazines
Material release data	Purchasing department or equipment suppliers
Subcontractor cost and delivery data	Mine engineering or purchasing department
Area and maneuvering requirements	Mine engineering department
Historical records of previous cost estimates	Mine engineering department, universities, consultants or government agencies
Current costs of items presently in production	Purchasing department or accounting and warehouse

Adapted from: Wilson, F., 1963, *Manufacturing Planning and Estimating Handbook*, McGraw-Hill, New York.

few of the factors which must be considered during estimating. Experience in production, engineering, and maintenance aspects of mine management will increase the accuracy of the figures, so drawing from these functions as necessary is especially important.

Preparing Cost Estimates: There are a few basic steps to follow in preparing a cost estimate for a project. In Chap. 4, Fig. 14 presented a calculation sheet for estimating the cost of a drainage network. This calculation sheet assumed there was an accurate data base to provide material and labor costs, as well as construction time estimates. Mining engineers, turned cost estimators, are expected to have a general knowledge of the project tasks in mining as well as some familiarity with construction principles. However, they need to be able to float between the mine functions (production, maintenance, etc.) for details and to provide a sounding board for tentative estimates.

When the engineer or estimator begins work on an unfamiliar project of which he has limited knowledge or meager data, he must use some basic estimating techniques to guide him in obtaining a good estimate. The steps are summarized as follows:

1) By using detailed drawings and prints, the engineer must first make a manual take-off of materials and calculate both fabrication and purchased components. This is done for both internal (company) and external (consultant) designs if such estimates are not already done.

2) Once the materials are identified, the costs of each are developed. Purchases material price can be taken from vendor quotes and price lists, including any quantity price breaks and handling charges. Fabricated materials are estimated from estimates of raw materials, labor time, and labor costs per unit of time.

3) Detailed planning and project scheduling must be provided or estimated by the engineer. CPM techniques facilitate development of manpower requirements for the entire project. Labor can be estimated in any unit of time (usually man-days in mining).

4) Direct labor costs are then calculated on the basis of the project plans and scheduling. These labor costs are the time multiplied by the workman's wages. Burden (indirect cost) factors are added to strengthen the estimate accuracy.

5) The final cost estimates are composed of the direct material and labor costs, plus the indirect costs as a factor of the total direct cost. G & A and contingency costs are added to complete the estimate.

Normally, a summary cost sheet is prepared for presentation to interested parties and for higher management. When done properly, cost estimates will accurately predict the costs and time frame for any project in mining. These estimates will serve as a base in planning future mining projects and add to the historical data file.

The upcoming discussions of cost estimating on mining support projects will consist of addressing key support cost issues, as well as providing tables illustrating some basic analysis steps for selected projects. In some cases, summary tables will be used to avoid confusion. The mining engineer or cost estimator should be aware of the potential trade-offs associated with different materials, designs, and support systems before presenting a final mining design to management. An important ingredient, therefore, is accurate cost estimating of support construction on a mine-wide basis. All costs in the following examples are in 1981 dollars, except where noted.

Transportation

Transportation in this text has centered around the use of track systems for the purpose of transporting men, materials, and mining supplies throughout the mine. I have electrified this track system, standardized the weight at 50.4 kg/m³ (85 lb per cu yd), and designated steel clip-on ties and stone ballast to provide structure and support.

This description barely scratches the surface of the spectrum of materials, designs, and applications of rail transportation in underground mines. Each mine is different, so the approach to mining should also be individualized. For example, high coal mines [greater than 1.8 m (6 ft)] may choose ballast and wooden ties [203.2 × 203.2 mm × 1.8 m (8 in. × 8 in. × 6 ft)] to support their track systems, while low coal mines [less than 914.4 mm (36 in.)] may do without ballast or choose to use battery-powered rubber-tired vehicles. The point to consider is that mines are site specific, so the controllable mining elements of a particular site (i.e., the mine design, management, labor, and to some extent, materials) must be approached to reflect the particular site.

To illustrate the cost estimating procedure for transportation projects, I have chosen the task of laying a track switch. Initially, the mine or construction engineer knows a complete switch can be purchased for about $2500 (Chap. 2). The other materials, therefore, can be supplied from the mine stocks of ties, rails, and tools. The switch is delivered to the mine staging area (warehouse) and can easily be transferred to the underground site as needed.

The detailed planning of the switch tasks is the same as the CPM network discussed in Chap. 9 and shown in Fig. 10 from that chapter. A complete switch can be installed in an estimated two shifts with a track crew of four workers. The breakdown of labor by task is shown in Table 5. From this breakdown the labor cost can be developed by adding the direct labor wages and fringe benefits separate from the salaried personnel. This can be seen in Table 6. Notice the surveyors are not eligible for fringe benefit additions to their wages in this estimate.

From these costs a project subtotal can be obtained by adding the direct material and labor cost. To this subtotal one adds the burden factors of engineering, G & A, and contingency factors. These factors are developed from historical data and can be estimated as a percentage of direct costs. For our purposes, these can be expressed as:

Engineering— 5%
G & A— 2%
Contingency—10%

Each burden cost is added to the subtotal before the next is estimated. Therefore, G & A is 2% of the direct costs plus engineering, and the contingency fee is 10% of the total direct costs, engineering, and G & A costs added together.

Therefore, the switch can be summarized as the following:

Direct Materials	$2,500.00
Direct Labor	599.28
Subtotal	3,099.28

TABLE 5. Labor Breakdown for a Switch

CPM Task	Labor Category	Number of Men	Man-hours Required	Total
Site orientation (A)	Surveyors	2	4	8
Ordering and delivery of switch (B)*	General inside	1	8	8
	Motorman	1	8	8
Cutting inside rail and unclipping ties (C)	Trackman	3	1/2	1 1/2
Bolting frog to inside rail (D)	Trackman	2	1/4	1/2
Bolting prebent rail on (E)	Trackman	3	1/2	1 1/2
Bolting curved and straight closure rails (F)	Trackman	2	1/4	1/2
Placing and clipping switch ties (G)	Trackman	3	1/2	1 1/2
Bolting switch points on (H)	Trackman	1	1/4	1/4
Hooking up switch stand (I)	Trackman	1	1/2	1/2
Bolting outside turnout rail on (J)	Trackman	2	1/4	1/2
Clipping spacer ties to turnout and main rails (K)	Trackman	3	1	3
Bending outside rail (L)	Trackman	2	1	2
Lubricating switch throw (M)	Trackman	1	1/4	1/4
Blocking and leveling switch (N)	Trackman	1	1/2	1/2

*To be delivered during the working shift; estimates will vary.

Engineering (5%)	154.96
Subtotal	3,254.24
G & A (2%)	65.08
Subtotal	3,319.33
Contingency (10%)	331.93
Total	3,651.26

As one can see, the investment of one switch can become significant when one considers an average large mine of 0.9 to 1.8 Mt (one to two million st) annually using rail transportation will install no less than 75 new switches during an average life.

Ventilation

Ventilation construction is approached in a similar manner as transportation. A complication with ventilation projects since the 1969 Safety Law is the sheer number of ventilation devices needed underground. Therefore, construction estimating of, for example, an overcast for submain development is better handled if it is expanded to a project size encompassing all the overcasts needed to divert the air, rather than simply adding the construction time and cost estimates of one together with the required number of times for an estimate. An 11 overcast submain project has scheduling time-frames which are very different from building a single overcast.

The mining engineer will need to recognize this economy of scale in mining construction when he is faced with the entire mine design.

Ventilation construction also presents some interesting cost alternatives to the construction or mining engineer for consideration. Because the whole ventilation system is based on the operation of a surface-mounted mechanical fan, a baseload cost is inherent in the mine design approach. Therefore, the mine engineer must strive to choose materials and designs which will maximize both the cost-efficiency of the individual or group of ventilating devices, and the cost performance of the main mine fan. This entails considering such items as double stoppings, high efficiency sealants, equalizing overcasts, smoothing main airways, angled intersections (less than 90°), and solid (vs. hollow) concrete blocks. The trade-off in higher initial cost comes in the longer life and lower operating costs from the ventilating device and main mine fan. Low leakage, high performance stoppings, overcasts, and so on will pay for themselves many times over. The key to their performance is, of course, sound installation practices (Chap. 3) coupled with management's desire to maintain their performance.

In this section we will look at the construction of a single prefabricated overcast. The materials, none of which need to be fabricated by the mine, are costed out as $2600. Notice this example uses a prefabricated overcast which, during the previous discussions, was mentioned for its high performance and long life abilities.

The ventilation labor, by task, is given in Table 7. This labor breakout follows conventional union worker classifications.

From the detailed CPM network shown in Fig. 7 from Chap. 9, the project estimate of 12 manshifts is derived. The labor costs, developed in Table 8, can be directly compared to this network by task if desired.

Added to the direct costs of the overcast are the burden costs. To reflect the greater involvement of the engineering design and planning in an overcast project, a $7^{1}/_{2}$% factor will be used. Project G & A remains at 2% and contingency drops to 8% to reflect tighter project control and better labor estimates.

The summarized cost estimate for an overcast can be presented as:

TABLE 6. Labor Cost Estimate for Switch Construction

Labor Category	Number of Man-hours	Number of Manshifts	Cost per Manshift*	Fringe Cost(55%)	Total
Surveyors	8	1	$95.00	—	95.00
Motorman	8	1	71.76	34.47	111.23
General inside labor	8	1	71.18	39.15	110.33
Trackman	$20^{1}/_{2}$	2.56	71.18	39.15	282.72
					599.28

*As per wage agreement of 1978, Appendix A-Part I, unadjusted for COLA.

TABLE 7. Labor Breakdown for the Overcast

CPM Task	Labor Category	Number of men	Man-hours Required	Total Man-hours
Site orientation (A)	Surveyors	2	4	8
Pick hitchings and foundations (B)	General inside	2	4	8
Excavate and cleanup roof (C)	Miner operator	1	16	16
	Miner helper	1	16	16
	Roof bolter	2	8	16
	Shuttle car operator	1	8	8
Deliver wall block (D)	Motorman	1	8	8
	General inside	1	8	8
Deliver sealant and machine (E)	Motorman	1	8	8
	General inside	1	8	8
Deliver prefab overcast (F)	Motorman	1	8	8
	General inside	1	8	8
Pour foundation (G)	General inside	1	2	2
Build structure walls (H)	Precision mason	2	2	4
Bolt overcast together (I)	General inside	2	4	8
Build wing walls (J)	Precision mason	2	4	8
Seal overcast (K)	Precision mason	2	2	4
			Total	146

Direct Materials	$2,600.00
Direct Labor	2,081.25
Subtotal	4,681.25
Engineering (7½%)	351.09
Subtotal	5,032.34
G & A (2%)	100.65
Subtotal	5,132.99
Contingency (8%)	410.64
Total	5,543.63

Of all support construction projects required in underground mines, ventilation construction remains the most challenging in terms of sound installation and project control. Most often ventilation construction is a neglected and poorly maintained area, thus allowing costly inefficiencies to permeate the system. As a mine grows older, the ventilation problems grow larger. It is due to these considerations that planning mining engineers should choose designs and materials carefully to ensure the best system will perform safely and effectively. For the same reasons, construction engineers should be extra careful when installing every ventilation device, whether in the mainline, the panel, or submain entries.

Drainage

Drainage projects provide the most latitude in project engineering of any support system. Many

TABLE 8. Labor Cost Estimate for Overcast Construction

Labor Category	Number of Man-hours	Number of Manshifts	Cost per Manshift*	Fringe Costs (55%)	Total
Surveyors	8	1.00	$95.00	—	$95.00
Miner operator	16	2.00	79.72	43.85	247.14
Miner helper	16	2.00	76.48	42.06	237.08
Roof bolter	16	2.00	79.72	43.85	247.14
Shuttle car operator	8	1.00	73.54	40.45	113.99
Precision mason	16	2.00	73.54	40.45	227.98
Motorman	24	3.00	71.76	39.47	333.69
General inside	42	5.25	71.18	39.15	579.23
					2,081.25

*As per wage agreement of 1978, Appendix A-Part I, unadjusted for COLA.

more variations in design and equipment are available to the engineer for cost estimating purposes. Unfortunately, this can be both help and hindrance in the development of accurate estimates.

The most significant advantages of drainage construction stem from its space requirements and flexibility. The main system is composed of small diameter piping, which fits virtually everywhere in the mine, and of intermittent sumps and pumping stations whose location is not fixed to one entry or entry intersection. Although significant differences in piping and pump costs remain between equipment manufacturers, other factors allow a selective comparison to be made.

The project under scrutiny here is construction of the main pumping station and main mine sump. Unlike the transportation and haulage projects, major fabrication of materials and equipment is necessary. The sump rib bolting and strut and beam placement tasks are considered both an installation task and fabrication requirement (Chap. 4).

The direct materials needed for the pump can be obtained from vendor price lists and the mine purchasing departments. The costs are estimated as $3000 for the pump and $1000 for the fabricated materials. Estimates for mine materials such as sealants, beams, struts, and rib bolts are based on a sump excavated to dimensions of 2.1 × 76.2 × 6.1 m (7 × 250 × 20 ft) and the maneuvering requirements of the site.

The direct labor costs can be developed by task for the entire project by reviewing the CPM network diagram for planning shown in Fig. 12 of Chap. 9. The costs associated with this are developed from manshift estimates shown in Table 9 by labor classification and work task. These costs are developed in Table 10 for the entire drainage project. Both material and labor costs are added together to obtain a subtotal direct cost estimate.

From this subtotal, the project burden estimates of engineering, G & A, and contingency are derived. To reflect the additional engineering required for planning and design of the station and pump, an estimate of 10% of direct costs is used. Project G & A is set at 2% of the sum of direct and engineering costs. Finally, contingency costs are estimated at 10% of total costs. The final cost estimate for the pump project can be summarized as the following:

Direct Materials	$4,000.00
Direct Labor	2,041.94
Subtotal	6,041.94
Engineering (10%)	604.19
Subtotal	6,646.13
G & A (2%)	132.92
Subtotal	6,779.06
Contingency (10%)	677.91
Total	7,456.96

Haulage

Haulage represents another rigid engineering design project with fairly definable tasks. The burden factors, therefore, can be lowered to reflect the engineer's confidence in the estimate. Belt haulage has been the main focus of the discussions in this text because it is the preferred method of materials handling. Belt haulage offers great flexibility in material and design choices and can be applied to almost every type of mining situation.

The example project under consideration is the 914.4 mm (36 in.) panel belt line extension typical of the dynamic haulage system. The length of the extension is designed for 91.4 m (300 ft) or two 45.7 m (150 ft) anchor setups. No fabrication of materials other than a tailpiece bunker is necessary.

The materials for the move are costed out to an estimated total of $9,400 for one setup. These costs are obtained from either belt manufacturers, CEMA representatives, or the mine purchasing department. This estimate does not include the tools or working supplies needed during the extension.

The belt line workers needed, by task, are shown in Table 11. Practical scheduling dictates this crew be divided into two groups. One group of four works on the belt line and splicing tasks while the other group of two workers break down and move the tailpiece. The worker classifications are presented by task and show a mixture of necessary skills.

Project planning of the belt line extension can be accomplished with a CPM diagram similar to Fig. 14 shown in Chap. 9. From this diagram and the detailed scheduling by the project team,

TABLE 9. Labor Breakdown for Pump Station

CPM Task	Labor Category	Number of Men	Man-hours Required	Total Man-hours
Site orientation (A)	Surveyors	2	6	12
Transport excavating equipment (B)	Miner operator	1	8	8
	Miner helper	1	8	8
	General inside	1	8	8
Excavate sump (C)	Miner operator	1	4	4
	Miner helper	1	4	4
	Shuttle car operator	1	4	4
Cleanup sump (D)	General inside	1	1	1
Bolt wire mesh to sump walls (E)	General inside	2	4	8
Deliver sealant and machine* (F)	Motorman	1	8	8
	General inside	1	8	8
Seal sump floor and walls (G)	Precision mason	2	2	4
Frame pump foundation (H)	General inside	2	1/4	1/2
Pour pump foundation (I)	Precision mason	1	1	1
	General inside	1	1	1
Deliver pumps, motors, piping, and accessories (J)	Motorman	1	8	8
	General inside	2	8	16
Bolt block and tackle props to roof (K)	General inside	2	4	8
	Roof bolter	2	1	2
Set and bolt pump (L)	General inside	2	1	2
Connect intake piping (M)	General inside	2	1	2
Connect discharge piping (N)	General inside	2	1	2
Test sump and pump system (O)	Electrician	1	8	8
	Pumper	1	8	8
Build fireproof housing for controls (P)	Precision mason	2	4	8
Place guarding and signs (Q)	General inside	2	1/2	1
Cleanup of site (R)	General inside	2	1/2	1
				145.5

*As per wage agreement of 1978, Appendix A-Part I, unadjusted for COLA.

TABLE 10. Labor Cost Estimate for Pump Construction

Labor Category	Number of Man-hours	Number of Manshifts	Cost per Manshift*	Fringe Cost (55%)	Total
Surveyor	12	1.50	$95.00	—	$142.50
Miner operator	12	1.50	79.72	43.85	185.36
Miner helper	12	1.50	76.48	42.06	177.81
General inside	58.5	7.31	71.18	39.15	806.79
Shuttle car operator	4	0.50	73.54	40.45	56.99
Motorman	16	2.00	71.76	39.47	222.46
Precision mason	13	1.63	73.54	40.45	185.24
Roof bolter	2	0.25	79.72	43.85	30.89
Electrician	8	1.00	79.72	43.85	123.57
Pumper	8	1.00	71.18	39.15	110.33
					2,041.94

*As per wage agreement of 1978, Appendix A-Part I, unadjusted for COLA.

TABLE 11. Labor Breakdown for a Belt Line Extension

CPM Task	Labor Category	Number of Men	Man-hours Required	Total Man-hours
Site orientation (A)	Surveyors	2	4	8
Deliver belt for splicing (B)	Motorman	1	8	8
	General inside	1	8	8
Drill holes for chain hangers and sailbases (C)	Roof bolters	2	8	16
Deliver extension structure (D)*	Motorman	1	8	8
	General inside	1	8	8
String belt structure (E)	General inside	2	4	8
Splice belt into belt line (F)	Beltman	1	1/2	1/2
	General inside	1	1/2	1/2
Breakdown and move tailpiece (G)	General inside	2	1/2	1
Bolt sailbases to roof (H)	General inside	4	1	4
String ropes and tension (I)	General inside	4	2	8
Place structure and tighten (J)	General inside	4	2	8
Place tailpiece in new position (K)	General inside	2	1	2
Anchor tailpiece (L)	General inside	2	2	4
Hook tailpiece to belt line (M)	General inside	2	1/2	1
Train belt (N)	General inside	1	1	1
Hookup auxiliary systems (O)	Beltman	1	1/2	1/2
	General inside	2	1	2
Troubleshoot extension (P)	General inside	6	1	6
				102.5

*Delivered during normal shift; estimates will vary.

an accurate estimate of needed time in manshifts is completed. The costs of this labor are estimated in Table 12. The total direct costs for a belt line extension are estimated as $36.09 per 0.3 m (ft) of belt line.

The burden factors of engineering, G & A, and contingency costs are added to the total direct costs to derive a project cost. Engineering is estimated at 5% of direct costs, G & A at 2%, and contingency fees equal to 5%. These lower burden factors reflect the higher direct cost of the project and the simple design and engineering required for the move-up. The project cost can be summarized as:

Direct Material	$ 9,400.00
Direct Labor	1,427.38
Subtotal	10,827.38
Engineering (5%)	541.37
Subtotal	11,368.75
G & A (2%)	227.37
Subtotal	11,596.12
Contingency (5%)	579.81
Total	12,175.93

Again, the cost estimate for a 91.4 m (300 ft) production panel move-up gives some insight into the tremendous cost of materials handling underground. It also underscores the fact that poor planning and sloppy installation practices can be a large unnecessary cost item over the life of a mine. Since many mines operate with over 8 km (five miles) of belt, this is at least 88 setups, and the larger the belt width, the greater the cost per 0.3 m (ft). The potential for cost waste through the loss of some items when transporting extension materials, overrunning required time estimates for the extension, failing to train and maintain the belt line, or other ways should be recognized and reduced by management.

TABLE 12. Labor Cost Estimate for Belt Line Extension

Labor Category	Number of Man-hours	Number of Manshifts	Cost per Manshift*	Fringe Costs (55%)	Total
Surveyors	8	1.00	$95.00	—	95.00
Motorman	16	2.00	71.76	39.47	222.46
Roof bolter	16	2.00	79.72	43.85	247.14
Beltman	1	0.13	71.18	39.15	14.34
General inside	61.5	7.69	71.18	39.15	848.44
					1,427.38

*As per the wage agreement of 1978, Appendix A-Part I, unadjusted for COLA.

Mine Support Projects as a System

The final discussion is concerned with the cost estimates of mine support projects as a system. The use of the word *system* implies that transportation, haulage, drainage, and ventilation projects act as one in supporting the production system of the mine (Fig. 7). This is essential for any active mine or new mine design in order to maximize their performance and minimize their costs. The key to recognizing the effects of the support system on mine production lies in (1) collecting and evaluating historical mine data from all available sources and (2) using safe innovative approaches to mine design.

For example, Fig. 8 shows the typical main-entry-panel-entry intersection with the installed production support system for a modern mine. This design is perfectly acceptable—it conforms to federal regulations, state requirements, and company desires. To the cost and mining engineer concerned with support construction however, this design approach does have problems. It contains 11 overcasts, 16 metal or concrete stoppings, six man doors, 731.5 m (2400 ft) of water piping (plus valves and cutoffs), a 914.4 mm (36 in.) belt drive, 335 m (1100 ft) of track, and 685.8 m (2250 ft) of 1219.2 or 914.4 mm (48 or 36 in.) belt. The estimated support construction cost for this area is around $265,000 (1981 dollars). This cost includes all materials (approximately $187,500), labor (approximately $50,000), and burden costs ($27,000). If one considers a typical 0.9 to 1.8 Mt (one to two million st) per year mine will setup and breakdown approximately 60 or more intersections, the projected cost is almost $14 million (1981 dollars) in capital and operating costs.

By utilizing the innovative design of split ventilation, this system can be redesigned to function similar to the system shown in Fig. 9. Here, only five overcasts are used. This is offset by the increase in required stoppings, but stoppings are much cheaper in both materials and labor. There are 35 metal or block stoppings, 10 man doors, and approximately the same amount of piping, belt, track, and other support equipment. Besides reducing the difficulties with the scheduling of cutting and grading activities, the cost of this design comes in at approximately $185,000 (1981 dollars), or a reduction of 30% over the first approach.

Although this example is subject to a number of other technical and economic considerations, the message is clear: mining conventions must be tempered with site specific data and a fresh approach to cutting costs of support system construction and operation.

SUMMARY

Cost estimating is a mix of science and art. It is very important to construction engineering in mining when cost becomes an essential component to the planning and design process. The construction engineer in most cases will need to acquire these skills of cost estimating to maintain project control of the mine support system.

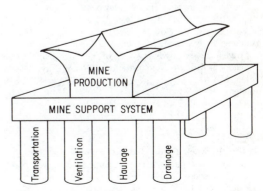

FIG. 7. All support projects act as one to hold up the mine production.

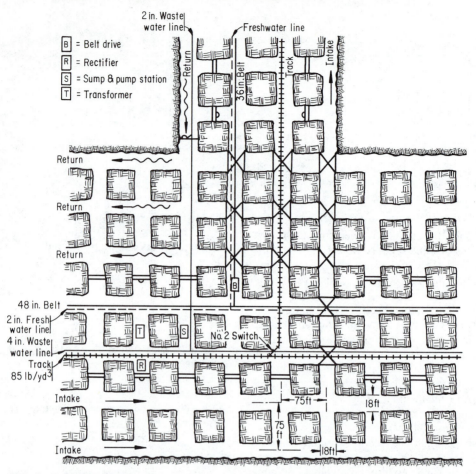

FIG. 8. Typical main panel entry intersection with the installed production support systems. Metric equivalents: in. × 25.4 = mm; ft × 0.3048 = m; lb per cu yd × 0.593 276 = kg/m^3.

REFERENCES AND BIBLIOGRAPHY

Anon., 1968, *Resource Management,* MDI Publications, Management Development Institute, Wayne, PA.

Anon., 1973, *Systems Analysis and Design Using Network Techniques,* Prentice-Hall, Englewood Cliffs, NJ.

Anon., 1981, "Available Software," *Mining Engineering,* Vol. 33, No. 11, Nov., pp. 1591–1593.

Anthony, R. N., 1965, *Planning and Control Systems: A Framework for Analysis,* Div. of Research, Graduate School of Business Administration, Harvard Univ., Boston, MA.

Anthony, R. N., and Mouton, J. L., 1975, *Management Accounting,* Richard D. Irwin, Homewood, IL.

Archibald, R. D., 1976, *Managing High-Technology Programs and Projects,* John Wiley & Sons, Inc., New York, p. 176.

ARINC Research Corp., 1969, *Guidebook for Systems Analysis/Cost Effectiveness,* US Dept. of Commerce, National Bureau of Standards, Washington, DC.

Barish, N. N., 1962, *Economic Analysis: For Engineering and Managerial Decision Making,* McGraw-Hill, New York.

Baumgartner, J. S., 1963, *Project Management,* Richard D. Irwin, Homewood, IL.

Baumol, W. J., 1965, *Economic Theory and Operational Analysis,* 2nd ed., Prentice-Hall, Englewood Cliffs, NJ.

Britton, S. G., 1981, *Practical Coal Mine Management,* John Wiley & Sons, New York.

Britton, S. G., 1981, "Cutting Costs of Mine Support Systems," *Coal Mining & Processing,* Vol. 18, No. 12, Dec., pp. 56–58.

Butler, A. G., 1973, "Project Management: A Study in Organizational Conflict," *Academy of Management Journal,* Vol. 16, Mar., pp. 84–101.

Cleland, D. I., and King, W. R., 1972, *Management: A Systems Approach,* McGraw-Hill Inc., New York.

Couger, J. D., and Knapp, R. W., 1974, *System Analysis Technique,* John Wiley & Sons, New York.

Datz, M. A., and Wilby, L. R., 1977, "What is Good Project Management," *Project Management Quarterly,* Vol. 8, No. 1, Mar.

Davis, S. M., and Lawrence, P. R., 1977, *Matrix,* Addison-Wesley, Reading, MA.

Drucher, P. F., 1975, "Managing for Business Effectiveness," *Business Classics, Harvard Business Review,* Boston, MA.

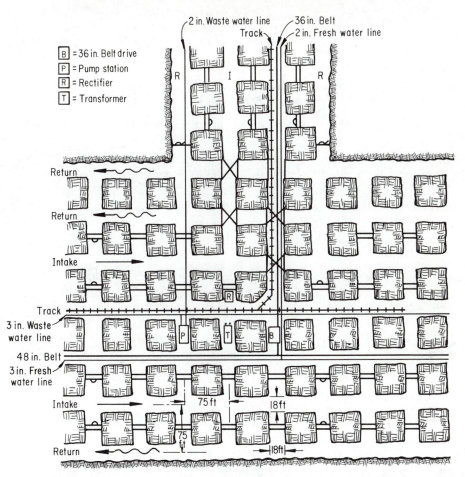

FIG. 9. Redesigned support systems for Fig. 8 produce a reduction in cost of over 30%. Metric equivalents: in. × 25.4 = mm; ft × 0.3048 = m.

Emery, J. C., 1969, *Organizational Planning and Control Systems*, MacMillan, New York.

Friend, F. L., 1976, "Be a More Effective Program Manager," *Journal of Systems Management*, Vol. 27, Feb., pp. 6–9.

Gaddis, P. O., 1959, "The Project Manager," *Harvard Business Review*, Harvard Univ., May-June, pp. 89–97.

Gran, P. F., 1976, *Systems Analysis and Design for Management*, DunDonnelley, New York.

Grant, E. L., and Ireson, W. G., 1970, *Principles of Engineering Economy*, 5th ed., Roland Press Co., New York.

Holland, T., 1969, "What Makes a Project Manager?" *Engineering*, Vol. 207, Feb., p. 262.

Horowitz, J., 1967, *Critical Path Scheduling-Management Control Through CPM and PERT*, Roland Press, New York.

Houre, H. R., 1973, *Project Management Using Network Analysis*, McGraw-Hill, New York.

International Congress for Project Planning by Network Analysis, 1969, *Project Planning by Network Analysis*, North-Holland Pub. Co., Amsterdam, Holland.

Kast, F. E., and Rosenzweig, J. E., 1974, *Organization and Management; A Systems Approach*, 2nd ed., McGraw-Hill, New York.

Kelley, W. F., 1969, *Management Through Systems and Procedures: A Systems Concept*, New York.

Kerzner, H., 1977, "Systems Management and the Engineer," *Journal of Systems Management*, Oct., pp. 90–107.

Kerzner, H., 1979, *Project Management: A Systems Approach to Planning, Scheduling, and Controlling*, Van Nostrand Reinhold, New York.

Maslow, A. H., 1954, *Motivation and Personality*, Harper, New York.

Martin, C. C., 1976, *Project Management: How to Make It Work*, Amacom, New York.

Martino, R. L., 1968, *Project Management*, MDI Publications, Management Development Institute, Wayne, PA.

McGregor, D., 1967, *The Professional Manager*, McGraw-Hill, New York.

McGregor, D., 1960, *The Human Side of Enterprise*, McGraw-Hill, New York.

Moder, J. J., and Phillips, C. R., 1970, *Project Management with CPM and PERT*, 2nd ed., Van Nostrand Reinhold, New York.

Optner, S. L., 1965, *Systems Analysis for Business and Industrial Problem Solving,* Prentice-Hall, Englewood Cliffs, NJ.

Pearson, R. G., 1971, "Human Factors Engineering," *Industrial Engineering Handbook,* McGraw-Hill, New York.

Pegels, C. C., 1976, *Systems Analysis for Production Operations,* Gordon and Science Pub., New York.

Quirin, G. D., 1967, *The Capital Expenditure Decision,* Richard D. Irwin, Homewood, IL.

Ramani, R. V., and Manula, C. B., 1975, "A Master Environmental Control and Mine System Design Simulator for Underground Coal Mining, Executive Summary," Vol. 1, Report No. PB 255 421/AS, US Bureau of Mines, NTIS, Springfield, VA.

Richards, M. D., and Greenlow, P. S., 1966, *Management Decision Making,* Richard D. Irwin, Homewood, IL.

Rodgers, C. R., and Roethlisberger, F. J., 1975, "Barriers and Gateways to Communication," *Business Classics, Harvard Business Review,* Harvard University, Boston, MA.

Rosenbaum, B., 1979, "Supervisory Skills Training Program," Developed for Monterey Coal Co. by MOHR Development Inc., New York.

Scott, F. E., 1981, "MSHA Academy: America's Mine Safety College," *Coal Mining and Processing,* Vol. 18, No. 10, Oct., pp. 46–51.

Shannon, R. E., 1972, "Matrix Management Structures," *Industrial Engineering,* Vol. 4, Mar., pp. 26–29.

Sikula, A. F., 1977, *Personnel Management,* John Wiley & Sons, New York.

Sleyman, K. J., 1971, "Cost Estimating," *Industrial Engineering Handbook,* McGraw-Hill, New York.

Toellner, J., 1977, "Project Estimating," *Journal of Systems Management,* May, pp. 6–9.

Vaughn, D. H., 1967, "Understanding Project Management," *Manage,* Vol. 9, No. 9, pp. 52–58.

Weinberg, G. M., 1975, *An Introduction to General Systems Thinking,* John Wiley & Sons, New York.

Wilson, F., 1963, *Manufacturing, Planning and Estimating Handbook,* McGraw-Hill, New York.

Whitehouse, G. E., 1973, "Project Management Techniques," *Industrial Engineering,* Vol. 5, Mar., pp. 24–29.

Appendix

**Mining Equipment Manufacturers
(1980 *Coal Age* Buyer's Guide)**

A

ACF Industries, Inc., 2300 3rd Ave., P.O. Box 547, Huntington, WV, 25710

A.C.R. Equipment Co. Inc., Parts Div., 19615 Nottingham Rd., Cleveland, OH, 44110

AD-X Corp., P.O. Box 272, Littleton, CO, 80160

AEC Inc., P.O. Box 106, 531 E. Marylyn Ave., State College, PA, 16801

A & K Railroad Materials, Inc., P.O. Box 30076, Salt Lake City, UT, 84130

ALPS Wire Rope Corp., 2350 Lunt Ave., Elk Grove, IL, 60007

A & M Welding and Mfg. Co., 411 Guyandotte Ave., Mullens, WV, 25882

AMF Inc., 777 Westchester Ave., White Plains, NY, 10604

AMP Special Industries, Div. of AMP Products Corp., Valley Forge, PA, 19482

A-S-H Pump, Div. of Envirotech Corp., P.O. Box 635, Paoli, PA, 19301

ATO Construction Equipment Div., P.O. Box 10263, Charleston, SC, 29411

ATO Hydraulic Products, 527 W. Rich St., Columbus, OH, 43215

Abex Corp., P.O. Box 1126, Wall St. Station, New York, NY, 10268

Abex Corp., Denison Div., 1220 Dublin Rd., Columbus OH, 43216

Abex Corp., Friction Products Group, 3001 W. Big Beaver Rd., Troy, MI, 48084

Abex Corp., Jetway Div., P.O. Box 9368, Ogden, UT, 84409

Abex Corp., Railroad Products Group, Valley Rd., Mahwah, NJ, 07430

Acco Industries Inc., American Chain Div., 454 E. Princess St., York, PA, 17403

Acco Industries Inc., Crane & Monorail Systems Div., Box 140, Fairfield, IA, 52556

Acco Industries Inc., Industrial & Automation Systems Div., 41225 Plymouth Rd., Plymouth, MI, 48170

Acco Industries Inc., Mining Sales Div., P.O. Box 15537, Pittsburgh, PA, 15244

Acco Industries, Page Welding Div., P.O. Box 976, Bowling Green, KY, 42101

Acker Drill Co., Inc., P.O. Box 830, Scranton, PA, 18501

Acme Machinery Co., Box 2409, Huntington, WV, 25725

Acodur GmbH & Co., P.O. Box 901168, 2100 Hamburg 90, Germany

Acrow Corp. of America, 396 Washington Ave., Carlstadt, NJ, 07072

Acurex Corp., Autodata Div., 485 Clyde Ave., Mountain View, CA, 94042

Acurex Corp., Energy & Environmental Div., 485 Clyde Ave., Mountain View, CA, 94042

Adalet-PLM div. Scott & Fetzer Co., 4801 W. 150th St., Cleveland, OH, 44135

Adams Equipment Co., Inc., 8421-25 Wabash, St. Louis, MO, 63134

Adhesive Engineering Co., 1411 Industrial Rd., San Carlos, CA, 94070

Advance Car Mover Co., Inc., 112 N. Outagamie St., P.O. Box 1181, Appleton, WI, 54912

Advanced Mining Systems, Robinson Plaza Two, 6600 Steubenville Pike, Pittsburgh, PA, 15205

Aerial Surveys, Inc., 4041 Batton N.W., North Canton, OH, 44720

Aerofall Mills Ltd., 2640 So. Sheridan Way, Mississauga, Ont., Canada, L5J 2M4

Aeroquip Corp., 300 S. East Ave., Jackson, MI, 49203

Aeroquip Tomkins-Johnson Div., 2425 W. Michigan Ave., Jackson, MI, 49202

Aero Service Div., Western Geophysical Co. of Amer., P.O. Box 1939, Houston, TX, 77001

Aerospatiale Helicopter Corp., 1701 W. Marshall Dr., Grand Prairie, TX, 75051

Aggregates Equipment Inc., 9 Horseshoe Rd., Leola, PA, 17540

Air Correction Div., UOP Inc., P.O. Box 5440, Norwalk, CT, 06856

Air-Lert, Inc., P.O. Box 342, Proctorville, OH, 45669

Air Products & Chemicals, Inc., Box 538, Allentown, PA, 18105

Airtek, Inc., 76 Clair St., No. Huntingdon, PA, 15642

Aitken Products, Inc., P.O. Box 151, Geneva, OH, 44041

Alabama State Docks, P.O. Box 1588, Mobile, AL, 36601

Albright Mfg. Co., Inc., 8117 N. Austin Ave., Morton Grove, IL, 60053

Alcoa, 1501 Alcoa Bldg., Pittsburgh, PA, 15219

Alcoa Conductor Products Co., Div. Aluminum Co. of America, 510 One Allegheny Sq., Pittsburgh, PA, 15212

Alcolac, Inc., 3440 Fairfield Rd., Baltimore, MD, 21226

Aldon Company, The, 3410 Sunset Ave., Waukegan, IL, 60085

Alemite & Instrument Div., Stewart-Warner Corp., 1826 Diversey Pkwy., Chicago, IL, 60614

Alison Control Inc., 35 Daniel Rd., West Fairfield, NJ, 07006

Allegheny Ludlum Steel Corp., 2004 Oliver Bldg., Pittsburgh, PA, 15222

Allen-Bradley Co., 1201 S. Second St., Milwaukee, WI, 53204

Allen & Garcia Co., 332 S. Michigan Ave., Chicago, IL, 60604

Allen Group, Inc., 534 Broad Hollow Rd., Melville, NY, 11746

Allentown Pneumatic Gun Co., P.O. Box 185, Allentown, PA, 18105

Allied Chemical Corp. Chemicals Co., P.O. Box 2064R, Morristown, NJ, 07960

Allied Colloids Inc., 1015 Shary Circle, Concord, CA, 94518

Allied Steel & Tractor Products, Inc., 5800 Harper Rd., Solon, OH, 44139

Allied Weight Systems Inc., 1821 University Ave., St. Paul, MN, 55104

Allis-Chalmers, P.O. Box 512, 1125 S. 70th St., Milwaukee, WI, 53201

Allis-Chalmers, Crushing & Screening Equipment, P.O. Box 2219, Appleton, WI, 54913

Allis-Chalmers, Engine Div., P.O. Box 563, Harvey, IL, 60426

Allmand Bros., Inc., W. Highway 23, Holdrege, NE, 68949

Alloy Rods Div., Chemetron Corp., Member Allegheny Ludlum Metals Group, P.O. Box 517, Hanover, PA, 17331

ALMEG, P.O. Box 11430, Kansas City, MO, 64112

Alnor Instrument Co., 7301 N. Caldwell Ave., Niles, IL, 60648

Alten Speed Reducer Div., Alten Foundry & Machine Works, Inc., P.O. Box 550, Lancaster, OH, 43130

Amco Plastics, Whalley Rd., Barnsley, S. Yorks, England, S75 1HT

MINING EQUIPMENT MANUFACTURERS

Amerace Corp., Elastimold Div., Esna Park, Hackettstown, NJ, 07840

American Air Filter Co., Inc., 215 Central Ave., Louisville, KY, 40277

American Alloy Corp., Pyramid Parts Div., 3000 E. 87th St., Cleveland, OH, 44104

American Alloy Steel, Inc., 2070 Steel Dr., Tucker, GA, 30084

American Can Co., Chemical Products, American Lane, Greenwich, CT, 06830

American Coal Training Inst., P.O. Box 7460, Monroe, LA, 71203

American Commercial Barge Line Co., P.O. Box 610, Jeffersonville, IN, 47130

American Cyanamid Co., Industrial Chemicals Div., Berdan Ave., Wayne, NJ, 07470

American Hoist & Derrick Co., 63 South Robert St., St. Paul, MN, 55107

American Industrial Leasing Co., 201 N. Wells St., Chicago, IL, 60606

American Meter Div., Singer Co., The, 13500 Philmont Ave., Philadelphia, PA, 19116

American Minechem Corp., P.O. Box 231, Coraopolis, PA, 15108

American Mine Door Co., Box 6028, Station B, Canton, OH, 44706

American Mine Research, Inc., P.O. Box 1628, Bluefield, WV, 24701

American Mine Supply Co., 404 Frick Bldg., Pittsburgh, PA, 15219

American Optical Corp., 14 Mechanic St., Southbridge, MA, 01550

American Optical Corp., Safety Products, 14 Mechanic St., Southbridge, MA, 01550

American Pulverizer Co., 5540 W. Park Ave., St. Louis, MO, 63110

American Standard, Industrial Products Div., 8111 Tireman Ave., Dearborn, MI, 48126

American Tractor Equip. Corp., P.O. Box 1226, Oakland, CA, 94604

Amerind-Mac Kissic, P.O. Box 111, Parker Ford, PA, 19457

Ametek, 26531 Ynez Rd., Temecula, CA, 92390

Ampco Metal Div., Ampco-Pittsburgh Corp., P.O. Box 2004, Dept.1723, Milwaukee, WI, 53201

Amsco Div., Abex Corp., 389 E. 14th St., Chicago Heights, IL, 60411

Amvest Leasing and Capital Corp., P.O. Box 5347, Charlottesville, VA, 22905

Anaconda Industries, Wire and Cable Div., Greenwich Office Park 3, Greenwich, CT, 06830

Analytical Measurements, Inc., 31 Willow St., Chatham, NJ, 07928

Anchor Conveyors Div., Standard Alliance Indus., Inc., 6906 Kingsley Ave., P.O. Box 650, Dearborn, MI, 48121

Anchor Coupling Co., Inc., 342 N. Fourth St., Libertyville, IL, 60048

Anchor/Darling Valve Co., 1650 S. Amphlett Blvd., Suite 319, San Mateo, CA, 94402

Anderson Electrical Connectors, Square D Co., Box 455, Leeds, AL, 35094

Anderson Mavor (USA) Ltd., 303 Progress St., Cranberry Ind. Park, Zelienople, PA, 16063

Anderson Power Products, Inc., 145 Newton St., Boston, MA, 02135

Anixter Bros., 4711 Golf Rd., Skokie, IL, 60076

Anixter Mine & Industrial Specialists, 5040 E. 41st St., Denver, CO, 80216

Ansul Co., The, 1 Stanton St., Marinette, WI, 54143

Anvil Attachments, Inc., 10100 Brecksville Rd., Brecksville, OH, 44141

Apache Powder Co., P.O. Box 700, Benson, AZ, 85602

Apollo Technologies Inc., One Apollo Dr., Whippany, NJ, 07981

Applied Mining Equipment & Services Co., 909 Camden Rd., Huntington, WV, 25704

Applied Science, Box 158, Valencia, PA, 16059

Aquadyne, Div. of Motomco, Inc., 267 Vreeland Ave., P.O. Box 300, Paterson, NJ, 07513

ARCO Chemical Co., 1500 Market St., Philadelphia, PA, 19101

Armco Autometrics, 4946 N. 63rd St., Boulder, CO, 80301

Armco Inc., 703 Curtis St., Middletown, OH, 45043

Armstrong-Bray & Co., 5366 Northwest Hwy, Chicago, IL, 60630

Armstrong Bros. Tool Co., 5200 W. Armstrong Ave., Chicago, IL 60646

ARO Corp., One Aro Center, Bryan, OH, 43506

Arus-Andritz, 1010 Commercial Blvd. So., Arlington So. Industrial Park, Mansfield, TX, 76063

Asbury Industries, Inc., 4351 William Penn Hwy., Murrysville, PA, 15668

ASEA, Inc., Dept. MA, 4 New King St., White Plains, NY, 10604

Ashland Chemical Co., P.O. Box 2219, Columbus, OH, 43216

Associated Research, Inc., 8221 N. Kimball, Skokie, IL, 60067

Associates Commercial Corp., 55 E. Monroe St., Chicago, IL, 60603

Astrosystems, Inc., 6 Nevada Dr., Lake Success, NY, 11042

Athey Products Corp., P.O. Box 669, Raleigh, NC, 27602

Atkinson Dynamics, 10 West Orange Ave., So. San Francisco, CA, 94080

Atkinson Industries, Inc., 116 E. 1st St., Pittsburg, KS, 66762

Atlantic Track & Turnout Co., 270 Broad St., Bloomfield, NJ, 07003

Atlas Car & Mfg. Co., 1140 Ivanhoe Rd., Cleveland, OH, 44110

Atlas Copco, Inc., 70 Demarest Dr., Wayne, NJ

Atlas Copco MCT AB, Fack, S-102 60 Stockholm, Sweden

Atlas Hoist & Body Inc., 7500 Cote de Liesse Rd., Montreal, Que., Canada, H4T 1E8

Atlas Powder Co., 12700 Park Central Pl., Ste. 1700, Dallas, TX, 75251

Atlas Railroad Construction Co., P.O. Box 8, Eighty Four, PA, 15330

A-T-O Inc., 4420 Sherwin Rd., Willoughby, OH, 44094

Aurora Pump, Unit of General Signal, 800 Airport Rd., N. Aurora, IL, 60542

Austin, J. P. Assoc., 300 Mt. Lebanon Blvd., Pittsburgh, PA, 15234

Austin Powder Co., 3735 Green Rd., Cleveland, OH, 44122

Auto Crane Co., P.O. Box 45548, Tulsa, OK, 74145

Auto Specialties Mfg. Co., P.O. Box 8, St. Joseph, MI, 49085

Auto Weigh Inc., P.O. Box 4017, 1439 N. Emerald Ave., Modesto, CA, 95352

Automatic Sprinkler Corp., P.O. Box 180, Cleveland, OH, 44147

Automation Products, Inc., 3030 Max Roy St., Houston, TX, 77008
Autoverin, Inc., P.O. Box 3395, College Station, Fredericksburg, VA, 22401
AVCO Industrial Engines Operation, 17220 Park Row, Houston, TX, 77084

B

BPB Instruments Inc., 1609-B Allens Lane, Evansville, IN, 47710
BTR Trading, Inc., 1581 Stone Ridge Rd., Stone Mountain, GA, 30083
Babcock & Wilcox, 161 East 42nd St., New York, NY, 10017
Babcock Contractors, Inc., 921 Penn Ave., Pittsburgh, PA, 15222
Babcock Corrosion Control Ltd., 119-120 High St., Eton, Windsor, Bershire, England, SL4 6AN
Babcock Hydraulic Handling Ltd., 10/18 Spurgeon St., London, England, SEI 4YP
Babcock Hydro-Pneumatics Ltd., 10/18 Spurgeon St., London, England, SEI 4YP
Babcock Minerals Engineering Ltd., Retford, Notts, DN22 7 AS, England
Babcock-Moxey Ltd., Winglos House, Bristol Rd., Gloucester, England, GL1 5RX
Babcock Roof Supports Ltd., Aidan House, Tynegate Precinct, Sunderland Rd., Gateshead, Tyne & Wear NE8 3HY, England
Bacharach Instrument Co., 625 Alpha Dr., Pittsburgh, PA, 15238
Badger Construction Co., Div. of Mellon-Stuart Co., 1425 Beaver Ave., Pittsburgh, PA, 15233
Baker Industries, Inc., P.O. Box 668, Hartselle, AL, 35640
Balderson Inc., Box 6, Wamego, KS, 66547
Baldor Electric Co., Fort Smith, AR, 72902
Baldwin, J. A. Mfg. Co., Kearney, NB, 68847
Banner Bearings, P.O. Box 546, Princeton, WV, 24740
Barber-Greene Co., 400 N. Highland Ave., Aurora, IL, 60507
Barber Industries, P.O. Box 5280, Sta. A, Calgary, Alta., Canada, T2H 1X6
Barnes Engineering Co., 30 Commerce Rd., Stamford, CT, 06904
Barnes & Reinecke, Inc., 2375 Estes Ave., Elk Grove Village, IL, 60007
Barrett Battery Inc., 3317 Lagrange St., Toledo, OH, 43608
Barrett, Haentjens Co., Box 488, Hazelton, PA, 18201
Bateman Coal Technology, 8 Emery Ave., Randolf, NJ, 07801
Batteries Inc., Box 275, Sprague, WV, 25926
Bausch & Lomb, 1400 N. Goodman St., Rochester, NY, 14602
Bausch & Lomb, SOPD Div., Optics Ctr., 1400 N. Goodman, Rochester, NY, 14602
Baylor Company, P.O. Box 36326, Houston, TX, 77036
Bearcat Tire Co., 5201 W. 65th St., Chicago, IL, 60638
Bearing Service Co., 500 Dargan St., Pittsburgh, PA, 15224
Bearings, Inc., 3600 Euclid Ave., Cleveland, OH, 44115
Bechtel Corp., 50 Beale St., San Francisco, CA, 94105
Beckman Instruments, Inc., 2500 Harbor Blvd, Fullerton, CA 92634
Becorit (G.B.) Ltd., Hallam Fields Rd., Ilkeston, Derbys, U.K.
Beebe Bros., Inc., 2724 Sixth Ave. S., Seattle, WA, 98006

Bekaert Steel Wire Corp., 3200 W. Market St., Akron, OH, 44313
Belleville Wire Cloth Co., Inc., 135 Little St., Belleville, NJ, 07109
Bell Helicopter Co., P.O. Box 482, Fort Worth, TX, 76101
Bemis Co., Inc., 800 Northstar Center, Box 84A, Minneapolis, MN, 55402
Bepex Corp., 10225 Higgins Rd., Rosemont, IL, 60018
Berger Associates Corp., P.O. Box 2116, Columbus, OH, 43216
Berger Industries, Inc., P.O. Box 31, Edison Industrial Center, Rte. 1, Metuchen, NJ, 08840
Berntsen Cast Products, Inc., P.O. Box 3025, Madison, WI, 53704
Berry Davidson Co., 92 rue Bonte-Pollet, Lille, France, 59020
Bertea Industrial Products, 3115 Airway Ave., Costa Mesa, CA, 92626
Bessemer & Lake Erie R.R., P.O. Box 536, Pittsburgh, PA, 15230
Bete Fog Nozzle, Inc., 324 Wells St., Greenfield, MA, 01301
Bethlehem Steel Corp., Martin Tower, Bethlehem, PA, 18016
Betz-Converse-Murdoch, Inc., 1 Plymouth Mtg Mall, Plymouth Meeting, PA, 19462
Betz Laboratories, 4636 Somerton Rd., Trevose, PA, 19047
BICC Limited, 811 Parkway View Dr., Pittsburgh, PA, 15205
BIF, a unit of General Signal, 1600 Division Rd., West Warwick, RI, 02893
Biach Industries Inc., P.O. Box 280-L, Cranford, NJ, 07016
Biddle Co., James G., Township Line & Jolly Rds., Plymouth Meeting, PA, 19462
Big Sandy Electric & Supply Co., Inc., P.O. Box 2099, South US 23, Pikeville, KY, 41501
Bigelow-Liptak Corp., P.O. Box 837, Southfield, MI, 48037
Bindicator, 1915 Dove St., Port Huron, MI, 48084
Binkley Co., Building Prod. Div., 12115 Lackland Rd., St. Louis, MO, 63141
Bird Machine Co., Inc., Neponset St., South Walpole, MA, 02071
Bison Instruments, 5708 W. 36th St., Minneapolis, MN, 55416
Bitcon Ltd., 1706 Olive St., St. Louis, MO, 63103
Bixby-Zimmer Engrg. Co., 961 Abingdon St., Galesburg, IL, 61401
Blaw-Knox Equipment, Inc., P.O. Box 11450, Pittsburgh, PA, 15238
Bofors America, Inc., 23 Progress St., Edison, NJ, 08817
Boggess, B.L., Consulting Eng. & Mine Development Group, 7649 West 32nd Ave., Wheat Ridge, CO, 80033
Bolton R.B. (Mining Engineers) Ltd., Castleside Trading Estate, Consett,Co. Durham DH 8 8HB, England
Bonded Scale & Machine Co., 6700 Tussing Rd., Dept. CA, Columbus, OH, 43227
Boston Industrial Products Div., American Biltrite Inc., P.O. Box 1071, Boston, MA, 02103
Boston Insulated Wire & Cable Co., Bay St., Boston, MA, 02125
Bostrom Div., UOP Inc., 133 W. Oregon St., P.O. Box 2007, Milwaukee, WI, 53201
Bowdil Co., Box 470, 1018 Boylan Ave., S.E., Canton, OH, 44701
Bowman Distribution, Barnes Group, Inc., 850 E. 72nd St., Cleveland, OH, 44103

Boyd, John T. Co., Oliver Bldg., Pittsburgh, PA, 15222
Boyles Bros. Drilling Co., 1624 Pioneer Rd., Salt Lake City, UT, 84104
Brad Harrison Co., 600 E. Plainfield Rd., LaGrange, IL, 60525
Braden-Goodbary Corp., P.O. Box 2679, Tulsa, OK, 74101
Braden Steel Corp., Metal Bldg. Div., P.O. Box 2619, Tulsa, OK, 74101
Brand-Rex Co., Electronic & Industrial Cable Div., P.O. Box 498, Willimantic, CT, 06226
Branford Vibrator Co., The, Div. of Electro Mechanics, Inc., 150 John Downey Dr., New Britain, CT, 06051
Bridgestone Tire Co. of America, Inc., 2160 W. 190 St., Torrence, CA, 90509
Bridgestone Tire Co., Ltd., 1-10-1, Kyobashi, Chuo-ku, Tokyo, Japan, 104
Bridon American Corp., P.O. Box 6000, Wilkes-Barre, PA, 18773
Bristol-Babcock Div., Acco Industries Inc., 40 Bristol St., Waterbury, CT, 06708
Bristol Engineering Co., P.O. Box 696, Yorkville, IL, 60560
Bristol Steel & Iron Works, Inc., P.O. Box 471, Bristol, VA, 24201
British Coal International, Hobart House, Grosvenor Pl., London SW1X 7AE, England
British Jeffrey Diamond, Div. of Dresser Europe S.A. (U.K. Branch), Thornes Works, Wakefield, W. Yorks, England
Broderick & Bascom Rope Co., 10440 Trenton Ave., St. Louis, MO, 63132
Brookville Locomotive Div., Pennbro Corp., P.O. Box 130, Brookville, PA, 15825
Brown, G. A. & Son, Inc., P.O. Box 1589, Fairmont, WV, 26554
Browning Mfg. Div., Emerson Electric Co., Box 687, Maysville, KY, 41056
Bruening Bearings, Inc., 3600 Euclid Ave., Cleveland, OH, 44115
Brunner & Lay, Inc., 9300 King St., Franklin Park, IL, 60131
Brush Transformers Ltd., P.O. Box 70, Loughborough, Leicestershire, England, LE11 1HN
Bucyrus Blades, Inc., 260 E. Beal Ave., Bucyrus, OH, 44820
Bucyrus-Erie Co., P.O. Box 56, S. Milwaukee, WI, 53172
Budd Co., Polychem Div., P.O. Box 257 CA, Phoenixville, PA, 19460
Buffalo Wire Works Co., Inc., P.O. Box 129, Buffalo, NY, 14240
Bullard, E. D. Co., 2680 Bridgeway, Sausalito, CA, 94965
Burndy Corp., Richards Ave., Norwalk, CT, 06856
Burnside Steel Foundry Co., 1300 E. 92nd St., Chicago, IL, 60619
Burrell Construction & Supply Co., One Fifth St., New Kensington, PA, 16058
Bussman Div., McGraw-Edison Co., P.O. Box 14460, St. Louis, MO, 63178
Butler Mfg. Co., Bolted Tank Group, 7400 E. 13th St., Kansas City, MO, 64126
Byron Jackson Pump Div., Borg Warner Corp., P.O. Box 2017-Terminal Annex, Los Angeles, CA, 90051

C

CCS Hatfield Mining Products, 12 Commerce Dr., Cranford, NJ, 07016

C & D Batteries, Div. of ELTRA Corp., 3043 Walton Rd., Plymouth Meeting, PA, 19462
C-E Power Systems, Combustion Eng., Inc., 1000 Prospect Hill Rd., Windsor, CT, 06095
C-E Raymond, 300 N. Cedar, Abilene, KS, 67410
C-E Raymond, Combustion Engineering, Inc., 200 W. Monroe St., Chicago, IL, 60606
C-E Tyler, W.S. Tyler, Inc., sub. of Combustion Engineering Inc., 8200 Tyler Blvd., Mentor, OH, 44060
CEA Carter-Day Co., 500 73rd Ave. N.E., Minneapolis, MN, 55432
C F & I Steel Corp., P.O. Box 1830, Pueblo, CO, 81002
CHR Industries, Inc., Box 1911, New Haven, CT, 06509
CM Chain Div., Columbus McKinnon Corp., One Fremont St., Tonawanda, NY, 14150
CM Hoist Div., Columbus McKinnon Corp., One Fremont St., Tonawanda, NY, 14150
CM Longwall Mining Chain Div., Columbus McKinnon Corp., P.O. Box 74, McKees Rocks, PA, 15136
CMI Corp., P.O. Box 1985, Oklahoma City, OK, 73101
CR Industries (Chicago Rawhide), 2515 Pan Am Blvd., Elk Grove Village, IL, 60007
CRP Industries, Inc., Carteret, NJ, 07008
CSE Corp., 600 Seco Rd., Monroeville, PA, 15146
CWI Distributing Co., 655 Brea Canyon Rd., Walnut, CA, 91789
Cable Belt Conveyors Inc., 5200 N.W. 84th Ave., Ste. 203, Miami, FL, 33166
Cable Belt Ltd., Yorktown House, 8 Frimley Rd., Camberley, Surrey, England, GU15 3 HS
Cabot, Samuel, Inc., One Union St., Boston, MA, 02108
Calgon Corp., P.O. Box 1346, Pittsburgh, PA, 15230
California Perforating Screen Co., P.O. Box 77106X, 655 Bryant St., San Francisco, CA, 94107
Call, Inc., Ray C., P.O. Box 8245, So. Charleston, WV, 25303
Calweld, Div. of Smith International, Inc., P.O. Box 2875, 9200 Sorensen Ave., Sante Fe Springs, CA, 90670
Cambelt Intl. Corp., 2420 West 1100 South, Salt Lake City, UT, 84104
Cam-Lok Div., Empire Products, Inc., 10540 Chester Rd., Cincinnati, OH, 45215
Campbell, E. K. Co., 1809 Manchester Trafficway, Kansas City, MO, 64126
Campbell Chain Co., 3990 E. Market St., P.O. 3056, York, PA, 17402
CAM-RAL Chain Co., Inc., 450 Ragland Rd., Beckley, WV, 25801
Can-Tex Industries, P.O. Box 340, Mineral Wells, TX, 76067
Canton Stoker Corp., P.O. Box 6058, Canton, OH, 44706
Capital City Industrial Supply Co., 544 Broad St., Charleston, WV, 25323
Capital Conservation Group, Fifth Ave. E. & 18th St., Hibbing, MN, 55746
Capital Controls Co., 201 Advance Lane, P.O. Box 211, Colmar, PA, 18915
Capital Conveyor of Columbus, P.O. Box 510, Worthington, OH, 43085
Carboloy Mining Products, General Electric Co., P.O. Box 7428, Houston, TX, 77008
Carborundum Co., Refractories Div., Manor Oak #1, Suite 500, Pittsburgh, PA, 15220
Card Corp., P.O. Box 117, Denver, CO, 80201
Cardinal Cap & Jacket Co., Highway 460-22, Box 275, Grundy, VA, 24614

Cardinal Scale Mfg. Co., 203 E. Daugherty, Webb City, MO, 64870
Carman Industries, Inc., 1005 W. Riverside Dr., Jeffersonville, IN, 47130
Carmet Co., Minetool Div., P.O. Box 208, Rte. 161-N, Kings Mountain, NC, 28086
Carol Cable Co., Div. of Avnet, Inc., 249 Roosevelt Ave., Pawtucket, RI, 02862
Carver Pump Co., 1056 Hershey Ave., Muscatine, IA, 52761
Carus Chemical Co., 1500 8th St., LaSalle, IL, 61301
Case, J. I., Co., C. E. Div., 700 State St., Racine, WI, 53404
Case, J. I. Drott Div., P.O. Box 1087, Wausau, WI, 54401
Cashco, Inc., P.O. Box A, Hwy. 140 W., Ellsworth, KS, 67439
Catalytic, Inc., 1500 Market St., Centre Square West, Philadelphia, PA, 19102
Caterpillar Industrial Products, Inc., Peoria, IL, 61629
Caterpillar Tractor Co., 100 N.E. Adams, Peoria, IL, 61629
Caudill Seed Co., 1201 Story Ave., Louisville, KY, 40206
Celanese Chemical Co., Inc., 1250 W. Mockingbird Lane, Dallas, TX, 75247
Celanese Fibers Marketing Co., 1211 Ave. of Americas, New York, NY, 10036
Celtite, Inc., 13670 York Rd., Cleveland, OH, 44133
Cementation Co. of America, Inc., P.O. Box 9, Brampton Ont., Canada, L6V 2K7
Cementation Mining Ltd., Bentley Works, Bentley, Doncaster, England, DN5 OBT
Central Engineering Co., 4429 W. State St., Milwaukee, WI, 53208
Central Mine Equipment Co., 6200 N. Broadway, St. Louis, MO, 63147
Central States Industries Inc., 100 Executive Commons, 29425 Chagrin Blvd., Pepper Pike, OH, 44122
Centrifugal & Mechanical Industries, Inc., 146 President St., St. Louis, MO, 63118
Centurion Industries Inc., 845 Hickory St., Pewaukee, WI, 53072
Century Geophysical Corp., 6650 E. Apache, Tulsa, OK, 74115
Century Hulburt Inc., 2200 E. Castor Ave., Philadelphia, PA, 19134
Century Oils Ltd., P.O. Box 2, Century Works, Hanley, Stoke-on-Trent ST1 5HU, England
Cerro Wire & Cable Co. (Maspeth), 5500 Maspeth Ave., Maspeth, NY, 11378
Certain Teed Corp., Pipe & Plastics Group, Box 860, Valley Forge, PA, 19482
Challenge-Cook Bros., Inc., P.O. Box 1272, Industry, CA, 91749
Chemetron Corp., Railway Products, 111 E. Wacker Dr., Chicago, IL, 60601
Chemical Services, Inc., P.O. Box 6370, 205D Monongalia St., Charleston, WV, 25302
Chessie System, 2 No. Charles St., Baltimore, MD, 21201
Chesterton, A.W., Company, Middlesex Industrial Park, Rt. 93, Stoneham, MA, 02180
Chevron U.S.A., Inc., 575 Market St., San Francisco, CA, 94105
Chicago Pneumatic Equipment Co., 191 Howard St., Franklin, PA, 16323
Chlorinators Inc., 733 NE Dixie Highway, Jensen Beach, FL, 33457
Christensen Diamin Tools, Inc., P.O. Box 387, Salt Lake City, UT, 84110
Chromalloy, Shunk Blade Div., 1460 Auto Ave., P.O. Box 431, Bucyrus, OH, 44820
Ciba-Geigy Corp., Pipe Systems Dept., 9800 Northwest Freeway, Houston, TX, 77092
Cincinnati Gear Co., Wooster Pike & Mariemont Ave., Cincinnati, OH, 45227
Cincinnati Mine Machinery Co., 2980 Spring Grove Ave., Cincinnati, OH, 45225
Cincinatti Rubber Mfg. Co. Inc., 4900 Franklin Ave., Cincinnati, OH, 45212
Cisco Fabricating Div., Central Illinois Steel Co., P.O. Box 75, Carlinville, IL, 62626
CIT Corp., 650 Madison Ave., New York, NY, 10022
Citizens Fidelity Energy Co., Citizens Plaza, Louisville, KY, 40296
Clark Equipment Co., 1030 E. Jefferson Blvd., South Bend, IN, 46617
Clark Equipment Co., Austin-Western Div., 601 N. Farnsworth Ave., Aurora, IL, 60507
Clark Equipment Co., Construction Machinery Div., P.O. Box 547, Benton Harbor, MI, 49022
Clark Equipment Co., Crane Div., 1046 S. Main St., Lima, OH, 45802
Clark Equipment Co. Melroe Div., 112 N. University Dr., Fargo, ND, 58102
Clarkson Co., 3430 W. Bayshore Rd., Palo Alto, CA, 94303
Clayton Mfg. Co., 4213 No. Temple City Blvd., El Monte, CA, 91731
Cleveland Armstrong, 1108 S. Kilbourn St., Chicago, IL, 60624
Cleveland Wire Cloth & Mfg. Co., 3573 E. 78th St., Cleveland, OH, 44105
Clouth Gummiwerke Aktiengesellschaft, Niehler Strass 92-116, 5000 Koln 60, Germany
Coal Processing Equipment Inc., Box 877, Uniontown, PA, 15401
Coal Washer Rental Corp., Masonic Bldg., Rm. 301, Ebsenburg, PA, 15931
Coe & Clerici, via Martin Piagglo, 17, 16122 Genova, Italy
Collyer Insulated Wire, 100 Higginson Ave., Lincoln, RI, 02865
Colt Industries, Crucible, P.O. Box 226, Midland, PA, 15059
Columbia Electric Mfg. Co., 4508 Hamilton Ave., Cleveland, OH, 44114
Columbia Steel Casting Co., Inc., 10425 N. Bloss Ave., Portland, OR, 97203
Combustion Service & Equipment Co., 2022 Babcock Blvd., Pittsburgh, PA, 15209
Commercial Shearing, Inc., 1775 Logan Ave., Youngstown, OH, 44501
Commercial Testing & Engineering Co., 228 N. La Salle St., Chicago, IL, 60601
Communication & Control Eng. Co. Ltd., Park Rd., Calverton, Nottingham, England
CompAir Construction & Mining Ltd., Camborne, Cornwall, England, TR14 8DS
Compton Electrical Equipment Corp., 720 15th St. W. Box 7398, Huntington, WV, 25776
Computer Sharing Services, Inc., 2498 West Second Ave., Denver, CO, 80223
Comtrol Corp., 500 Pennsylvania Ave., Irwin, PA, 15642

MINING EQUIPMENT MANUFACTURERS

Conac Corp., P.O. Box 529, St. Charles, IL, 60174
Concrete Equipment Co., Inc., P.O. Box 430, Blair, NE, 68008
Cone Drive Div., Excello Corp., P.O. Box 272, Traverse City, MI, 49684
Connellsville Corp., 120 S. Third, Connellsville, PA, 15425
Connors Steel Co., P.O. Box 577, Birmingham, AL, 35201
Conoco Inc., P.O. Box 2197, Houston, TX, 77001
Conrac Corp., Three Landmark Sq., Stamford, CT, 06901
Consolidated Rail Corp., 1542 Six Penn Center, Philadelphia, PA, 19103
Construction & Mining Services, Inc., P.O. Box 2086, Fairview Heights, IL, 62208
ContiCarriers and Terminals, Inc., 2700 River Rd., Des Plaines, IL, 60018
Continental Conveyor & Equipment Co., Inc., P.O. Box 400, Winfield, AL, 35594
Continental Hydrodyne Systems Inc., P.O. Box F, Rte. 28, Milford, OH, 45150
Continental Rubber Works, Sub. of Continental Copper & Steel Industries, Inc., 2000 Liberty St., Erie, PA, 16512
Contractors Warehouse Inc., 3937 Wheeler Ave., Alexandria, VA, 22304
Control Concepts, Inc., Terry Dr., Newtown, PA, 18940
Control Data Corp., P.O. Box O, Minneapolis, MN, 55440
Control Products, Inc., P.O. Drawer 1087, Beckley, WV, 25801
Controlled Systems Inc., 1106 Chamberlain Ave., Fairmont, WV, 26554
Conveyor Components Co., 130 Seltzer Road, Croswell, MI, 48422
Conwed Corp., P.O. Box 43237, St. Paul, MN, 55164
Co-Ordinated Industries, Rd. #2 Flaugherty Run Rd., Coraopolis, PA, 15108
Coors Porcelain Co., 600 Ninth St., Golden, CO, 80401
Coppinger Machinery Service, P.O. Box 89, Bluefield, WV, 24701
Core Laboratories, Inc., 10703 E. Bethany Dr., Aurora, CO, 80014
Corhart Refractories Co., Div. of Corning Glass Works, 1600 W. Lee St., Louisville, KY, 40210
Corporate Training Systems, Inc., 255 Gateway Towers, Pittsburgh, PA, 15222
Costain Mining Ltd., 111 Westminster Bridge Rd., London, SE1 8EW, England
Crane Co., 300 Park Ave., New York, NY, 10022
Crisafulli Pump Co., Inc., Box 1051, Glendive, MT, 59330
Crosby Group, P.O. Box 3128, Tulsa, OK, 74101
Cross Electronics Inc., 100 Commonwealth Dr., Warrendale, PA, 15086
Crouse-Hinds Co., P.O. Box 4999, Syracuse, NY, 13221
Crouse-Hinds Electro, 15146 Downey Ave., Paramount, CA, 90723
Crown Iron Works Co., P.O. Box 1364, Minneapolis, MN, 55440
Cummins Engine Co., Inc., 1000 5th St., Columbus, IN, 47201
Curry Manufacturing Corp., P.O. Box 618, Glade Spring, VA, 24340
Cushman/OMC Lincoln, P.O. Box 82409, Lincoln, NE, 68501
Custodis Construction Co., 222 S. Riverside Plaza, Chicago, IL, 60606
Custom Hoists Inc., P.O. Box 98, Rt. 30-A West, Hayesville, OH, 44838
Cutler-Hammer Products, Eaton Corp., 4201 N. 27th St., Milwaukee, WI, 53216
Cyclone Machine Corp., P.O. Box 187, Hurricane, WV, 25526
Cypher Co., The, 1201 Washington Blvd., Pittsburgh, PA, 15206

D

D-A Lubricant Co., Inc., 1331 W. 29th St., Indianapolis, IN, 46208
DAP Inc., 855 N. 3rd St., Tipp City, OH, 45431
D & E Tool Co., Inc., P.O. Box 8176, Huntington, WV, 25705
DJB Sales Inc., 9139F Red Branch Rd., Columbia, MD, 21045
D P Way Corp., P.O. Box 09336, Milwaukee, WI, 53209
Dana Corp., Spicer-U-Joint Div., P.O. Box 5367, Detroit, MI, 48211
Dana Industrial, Dana Corp., 23577 Hoover, P.O. Box 40, Warren, MI, 48090
Daniels, C. R., Inc., 3451 Ellicott Center Dr., Ellicott City, MD, 21043
Daniels Company, The, Route 5, Box 203, Bluefield, WV, 24701
Dart Truck Company, P.O. Box 321, Kansas City, MO, 64141
Darworth Co., Tower Lane, Avon, CT, 06001
Davco Corp., P.O. Box 17221, Memphis, TN, 38117
Davey Compressor Co., 11060 Kenwood Rd., Cincinnati, OH, 45242
David Round & Sons, Inc., P.O. Box 39456, Cleveland, OH, 44139
Davis Instrument Mfg. Co., 513 E. 36th St., Baltimore, MD, 21218
Davis, John & Son (Derby) Ltd., 20 Alfreton Rd., Derby, DE2 4AB, England
Davy McKee Corp., 2700 Campus Dr., San Mateo, CA, 94403
Davy McKee (Minerals & Metals) Ltd., Ashmore House, Stockton-on-Tees, England, TS18 3LT
Dayco Corp., 333 W. 1st St., Dayton, OH, 45402
Dayton Automatic Stoker Co., 17 Dell St., P.O. Box 255, N. Dayton Station, Dayton, OH, 45404
Dean Brothers Pumps, Inc., P.O. Box 68172, Indianapolis, IN, 46268
Dearborn Chemical (U.S.) Chemed Corp., 300 Genesee St., Lake Zurich, IL, 60047
Deere & Co., John Deere Rd., Moline, IL, 61265
Deister Concentrator, Co., Inc., The, 901 Glasgow Ave., Ft. Wayne, IN, 46801
Deister Machine Co., Inc., P.O. Box 5188, Ft. Wayne, IN, 46895
Delavan Electronics, Inc., 14605 North 73rd St., Scottsdale, AZ, 85260
Delavan Corp., P.O. Box 100, West Des Moines, IA, 50265
Demco, 845 S.E. 29th St., P.O. Box 94700, Oklahoma City, OK, 73143
Derby Automation Consultants, Ltd., 811 Parkway View Dr., Pittsburgh, PA, 15205
Deron Corp., P.O. Box 603, Morgantown, WV, 26505
Derrick Mfg. Co., 588 Duke Rd., Buffalo, NY, 14225
DESA Industries, A Unit of AMCA Intl. Corp., 25000 S. Western Ave., Park Forest, IL, 60466

Design Space Intl., Bala Cynwyd Plaza, Bala Cynwyd, PA, 19004
Design Structures, 328 W. Central Ave., Lombard, IL, 60148
Detrick, M. H., Co., 20 N. Wacker Dr., Chicago, IL, 60606
Detroit Diesel Allison Div., General Motors Corp., 13400 W. Outer Dr., Detroit, MI, 48228
Detroit Stoker Co., 1510 E. Frist St., P.O. Box 732, Monroe, MI, 48161
Deutz Corp., 7585 Ponce de Leon Circle, Atlanta, GA, 30340
DeZurik, a Unit of General Signal, Sartell, MN, 56377
Diamond Chain Co., 402 Kentucky Ave., P.O. Box 7045, Indianapolis, IN, 46207
Diamond Crystal Salt Co., 916 S. Riverside Ave., St. Clair, MI, 48079
Diamond Shamrock Corp., 1100 Superior Ave., Cleveland, OH, 44114
Diamond Tool Research Co., Inc., 345 Hudson St., New York, NY, 10014
Dick Inc., R. J., P.O. Box 306, King of Prussia, PA, 19406
Difco, Inc., Box 238, Findlay, OH, 45840
Dings Co., Dynamics Group, 4742 W. Electric Ave., Milwaukee, WI, 53219
Dings Co., Magnetic Group, 4742 W. Electric Ave., Milwaukee, WI, 53219
Diversified Electronics, Inc., 119 N. Morton Ave., Evansville, IN, 47711
Dixie Bearings, Inc., 3600 Euclid Ave., Cleveland, OH, 44115
Dixon Valve & Coupling Co., KRM Bldg., 800 High St., Chestertown, MD, 21620
Dodge Div., Reliance Electric Co., 500 So. Union St., Mishawaka, IN, 46544
Dominion Engineering Works Ltd., P.O. Box 220, Station A, Montreal, Que., Canada, H3C 2S5
Donaldson Co., Inc., P.O. Box 1299 (1400 W. 94 St.), Minneapolis, MN, 55440
Donaldson Co., Majac Div., P.O. Box 43217, St. Paul, MN, 55164
Dorr-Oliver Inc., 77 Havemeyer La., Stamford, CT, 06904
Dorr Oliver Long, Ltd., 174 West St. South, Orillia, Ontario, Canada, L3V 6L4
Dosco Corp., 1020 N. Eisenhower Dr., Beckley, WV, 25801
Douglas Battery Mfg. Co., 500 Battery Dr., Winston-Salem, NC, 27107
Dover Conveyor & Equipment Co., Inc., Box 300, Midvale, OH, 44653
Dow Chemical Co., 2020 Abbott Rd. Center, Midland, MI, 48640
Dow Corning Corp., P.O. Box 1767, Midland, MI, 48640
Dowell Division, Box 21, Tulsa, OK, 74102
Downard Hydraulics, Inc., Box 122, Princeton, WV, 24740
Dowty Corp., Progress St., Cranberry Industrial Park, Zelienople, PA, 16063
Dravo Corp., One Oliver Plaza, Pittsburgh, PA, 15222
Dravo Mechling, One Oliver Plaza, Pittsburgh, PA, 15222
Dravo Wellman Co., 113 St. Clair Ave. N.E., Cleveland, OH, 44114
Dresser Industries, Inc., 1505 Elm St., Dallas, TX, 75201
Dresser Industries, Inc., Crane & Hoist Operations, W. Broadway, Muskegon, MI, 49443
Dresser Industries, Inc., Drilling Equipment, P.O. Box 1879, Columbus, OH, 43216
Dresser Industries, Inc., Industrial Products Div., 900 W. Mount St., Connersville, IN, 47331
Dresser Manufacturing, Div. Dresser Industries, Inc., 41 Fisher Ave., Bradford, PA, 16701
Dresser Mining Services & Equipment Div., P.O. Box 24647, Dallas, TX, 75224
Drilco Industrial, Div. Smith Intl. Inc., Drawer 3135, Midland, TX, 79702
Drill Systems Inc., P.O. Box 5140, Station "A" Calgary, Alberta, Canada, T2H 1X3
DuBois Chemicals Div. of Chemed Corp., DuBois Tower, 10th floor, Cincinnati, OH, 45202
Ducon Co., Inc., The, 147 E. Second St., Mineola, NY, 11501
Duff-Norton Co., P.O. Box 32605, Charlotte, NC, 28232
du Pont de Nemours, E. I. & Co. Inc., 1007 Market St., Wilmington, DE, 19898
Duquesne Mine Supply Co., 2 Cross St., Pittsburgh, PA, 15209
Durakool Inc., 1010 North Main St., Elkhart, IN, 46514
Durex Products, Inc., 10 Durex Parkway, Luck, WI, 54853
Duriron Co., Inc., The, 425 N. Findlay, Dayton, OH, 45401
Dyna Med Inc., 6200 Yarrow Dr., Carlsbad, CA, 92008
Dynex/Rivett Inc., 770 Capitol Dr., Pewaukee, WI, 53072
Dyson, Jos. & Sons Inc., 53 Freedom Rd., Painesville, OH, 44077

E

E.A.R. Corp., 7911 Zionsville Rd., Indianapolis, IN, 46268
EBSP/Envirotech Corp., 5755 Granger Rd., Independence, OH, 44131
EG & G Geometrics, 395 Java Dr., Sunnyvale, CA, 94086
ERL Inc., P.O. Box 631, New Albany, IN, 47150
Eagle Crusher Co., Inc., 4250 S.R. 309, Galion, OH, 44833
Eagle Iron Works, 129 Holcomb Ave., Des Moines, IA, 50313
East Penn Mfg. Co., Deka Rd., Lyon Station, PA, 19536
Eaton Corp., World Headquarters, 100 Erieview Plaza, Cleveland, OH, 44114
Eaton Corp., Axle Div., 739 E. 140 St., Cleveland, OH, 44110
Eaton Corp., Industrial Drives Operations, 9921 Clinton Rd., Cleveland, OH, 44144
Eaton Corp., Samuel Moore Oper., Synflex Div., Mantua, OH, 44255
Eaton Corp., Transmission Div., 222 Mosel Ave., Kalamazoo, MI, 49007
Ecology and Environment, Inc., 195 Sugg Rd., P.O. Box D, Buffalo, NY, 14225
Edgar Allen Mining Products Ltd., P.O. Box 93, Sheffield, England, S9 1RX
Edmont-Wilson, Div. of Becton, Dickinson & Co., 3172 Walnut St., Coshocton, OH, 43812
Edwards & Jones, Inc., 563 Eagle Rock Ave., Roseland, NJ, 07068
Eickhoff America Corp., 480 Manor Oak Two, Cochran Rd., Pittsburgh, PA, 15220
Eimco Elkhorn, P.O. Box 2068, Monroe, NC, 28110
Eimco Mining Machinery, Envirotech Corp., P.O. Box 1211, Salt Lake City, UT, 84110
Electric Apparatus Co., 409 N. Roosevelt St., Howell, MI, 48843

MINING EQUIPMENT MANUFACTURERS

Electric Machinery Mfg. Co., 800 Central Ave., Minneapolis, MN, 55413

Electric Products Div., Portec Inc., 1725 Clarkstone Rd., Cleveland, OH, 44003

Electric Wheel Co., Div. of Firestone Tire & Rubber Co., Dept.144, Quincy, IL, 62301

Electrical Automation, 9 Horseshoe Rd., Leola, PA, 17540

Electrofact, 2300 Berkshire Lane N., Plymouth, MN, 55441

Electro Lite Battery Co., 1225 East 40th St., Chattanooga, TN, 37407

Electro Switch Corp., King Ave., Weymouth, MA, 02188

Electronized Chemicals Corp., S. Bedford St., Burlington, MA, 01803

El-Jay (a Div. of Iowa Mfg. Co.), 916 16th St., N.E., Cedar Rapids, IA, 52402

ELMAC Corp., P.O. Box 1692, Huntington, WV, 25701

Eireco Corp., The, 2000 Central Ave., P.O. Box 14178, Cincinnati, OH, 45214

E/M Lubricants Inc., Box 2200, West Lafayette, IN, 47906

Emaco Inc., 111 Van Riper Ave., Elmwood Park, NJ, 07407

Endress & Hauser, Inc., 2350 Endress Pl., Greenwood, IN, 46142

Energy Packaging, Inc., P.O. Box 22, Virginia, MN, 55792

Energy Sciences & Consultants, Inc., Box B, Biwabik, MN, 554708

ENERPAC, Div. of Applied Power Inc., Butler, WI, 53007

English Drilling Equipment Co. Ltd., Lindley Moor Rd., Huddersfield HD3 3RW, Yorkshire, England

Ensign-Bickford Co., The, P.O. Box 7, Simsbury, CT, 06070

Ensign Electric Div., Harvey Hubbell Inc., 914 Adams Ave., P.O. Box 7758, Huntington, WV, 25778

Enterprise Fabricators, Inc., Box 151, Bristol, VA, 24201

Entoleter Inc., P.O. Box 1919, New Haven, CT, 06509

Environeering, Inc., 363 Third Ave., Des Plaines, IL, 60016

Envirex Inc., 1901 So. Prairie Ave., Waukesha, WI, 53186

Enviro-Clear, a Div. of Amstar Corp., Readington Rd. & Industrial Pkwy, Somerville, NJ, 08876

Environmental Control Systems, Inc., P.O. Box 167, Gallaway, TN, 38036

Environmental Equip. Div., FMC Corp., 1800 FMC Dr. West, Itasca, IL, 60143

Environmental Research & Technology, Inc. (ERT), 1716 Heath Pkwy., P.O. Box 2105, Ft. Collins, CO, 80521

Envirosphere Co., 19 Rector St., New York, NY, 10006

Envirotech Coal Services Corp., P.O. Box 1046, Beckley, WV, 25801

Envirotech Corp., Eimco PMD Div., 669 W. 2nd South, Salt Lake City, UT, 84110

Envirotechnics, Inc., P.O. Box 610, Roosevelt, UT, 84066

E-Power Industries Co., 406 B. Jalich Dr., Wichita Falls, TX, 76307

Equipment Corp. of America, Box 306, Coraopolis, PA, 15108

Equipment Mfg. Services, Inc., P.O. Box 942, Harmony, PA, 16037

Erico Products, Inc., 34600 Solon Rd., Solon, OH, 44139

Eriez Magnetics, 381 Magnet Dr., Erie, PA, 16512

ESCO Corp., 2141 N.W. 25th St., Portland, OR, 97210

Euclid, Inc., 22221 St. Clair Ave., Cleveland, OH, 44117

Europower Hydraulics Ltd., High St., Market Weighton, York YO4 3AD, England

Eutectic Corp., 40-40 172nd St., Flushing, NY, 11358

Evergreen Weigh Inc., 15125 Highway 99, Lynnwood, WA, 98036

Everson Electric Co., Lehigh Valley Industrial Park #1, 2000 City Line Rd., P.O. Box 2688, Bethlehem, PA, 18001

Exide Corp., 101 Gibraltar Rd., Horsham, PA, 19044

Exxon Co., U.S.A., P.O. Box 2180, Houston, TX, 77001

F

FAG Bearings Corp., Hamilton Ave., Stamford, CT, 06904

FMC Corp., Agricultural Machinery Div., 5601 E. Highland Dr., Jonesboro, AR, 72401

FMC Corp., Bearing Div., 7601 Rockville Rd., Box 85, Indianapolis, IN, 46206

FMC Corp., Chain Div., 220 S. Belmont, Box 346B, Indianapolis, IN, 46206

FMC Corp., Cable Crane & Excavator Div., 1201 Sixth St., S.W., Cedar Rapids, IA, 52406

FMC Corp., Drive Div., 2045 W. Hunting Park Ave., Philadelphia, PA, 19140

FMC Corp., Material Handling Equipment Div., Homer City, PA, 15748

FMC Corp., Material Handling Systems Div., 3400 Walnut, St., Colmar, PA, 18915

FMC Corp., Mining Equipment Div., Drawer 992, Fairmont, WV, 26554

FMC Corp., Steel Products Div., Box 1030, Anniston, AL, 36202

Fabreeka Products Co., P.O. Box F/1190 Adams St., Boston, MA, 02124

Fabri-Valve, Div. of ITT Grinnel Valve Co., Inc., P.O. Box 2713, Portland, OR, 97208

Fafnir Bearing, Div. of Textron, 37 Booth St., New Britain, CT, 06051

Fagersta, Inc., 2 Henderson Dr., P.O. Box 598, W. Caldwell, NJ, 07006

Failing, George E., Co., A Div. of Azcon Corp., 2215 S. Van Buren, P.O. Box 872, Enid, OK, 73701

Fairbanks Co., The, 2 Glenwood Ave., Binghamton, NY, 13905

Fairbanks Morse Engine Div., Colt Industries, 701 Lawton Ave., Beloit, WI, 53511

Fairbanks Weighing Div., Colt Industries, 711 E. St. Johnsbury Rd., St. Johnsbury, VT, 05819

Fairchild Inc., P.O. Box 1184, Beckley, WV, 25801

Fairey Filtration Ltd., Cranford Lane, Heston, Middlesex, TW5 9 NQ, England

Fairfield Engineering Co., 324 Barnhart St., Marion, OH, 43302

Fairmont Supply Co., Box 501, Washington, PA, 15301

Falk Corp., The, Sub. of Sundstrand Corp., Box 492, Milwaukee, WI, 53201

Family Lines System, P.O. Box 32290, Louisville, KY, 40232

Farr Co., 2301 Rosecrans, El Segundo, CA, 90245

Fate Intl. Ceramic & Processing Equip., Div. of Plymouth Locomotive Works, Inc., Bell & High Sts., Plymouth, OH, 44865

Federal Hose Mfg. Corp., P.O. Box 548, Painesville, OH, 44077

Federal-Mogul Corp., P.O. Box 1966, Detroit, MI 48235

Federal Supply & Equipment Co., Inc., Box 127, 4000 Parkway Lane, Hilliard, OH, 43026

Federated Metals Corp., Sub. of Asarco Inc., 120 Broadway, New York, NY, 10005

Feeco International, Inc., 3913 Algoma Rd., Green Bay, WI, 54301
Fenner America Ltd., 400 East Main St., Middletown, CT, 06457
Fenner International Ltd., Marfleet Hull, Yorkshire, England, HU9 5RA
Ferguson Gear, P.O. Box 160, 3021 Northwest Blvd., Gastonia, NC, 28052
Ferguson, H. K., Co., One Erieview Plaza, Cleveland, OH, 44114
Fermont Div. DCA, 141 North Ave., Bridgeport, CT, 06606
Ferro Corp., Composites Div., 34 Smith St., Norwalk, CT, 06852
Ferro Corp., Electro Div., 661 Willet Rd., Buffalo, NY, 14218
Ferro Corp., Temtek-Allied Div., 150 E. Dartmoor Dr., Crystal Lake, IL, 60014
Ferro-Tech, Inc., 467 Eureka Rd., Wyandotte, MI, 48192
Fiat-Allis Construction Machinery, Inc., Box F, 106 Wilmot Rd., Deerfield, IL, 60015
Fiberglass Resources Corp., Motor Ave., Farmingdale, NY, 11735
Fibre-Metal Products Co., Box 248, Concordville, PA, 19331
Fidelity Electric Co. Inc., 332 North Arch St., Lancaster, PA, 17604
Fil-T-Vac Corp., P.O. Box 27451, Tempe, AZ, 85282
Finn Equipment Co., 2525 Duck Creek Rd., Cincinnati, OH, 45208
Fire Protection Supplies Inc., 501 Mercer St., Princeton, WV, 24740
Firestone Industrial Products, 1700 Firestone Blvd., Noblesville, IN, 46060
Firestone Tire & Rubber Co., 1200 Firestone Pkwy., Akron, OH, 44317
First Colony Corp., P.O. Box 296, Greene & Acme Sts., Marietta, OH, 45750
Firstmark Morrison Div., 107 Delaware Ave., Buffalo, NY, 14202
First National Bank of Maryland, Energy Resources Div., 25 S. Charles St., Baltimore, MD, 21202
Fisher Controls Co., P.O. Box 190, Marshalltown, IA, 50158
Fisher Scientific Co., 711 Forbes Ave., Pittsburgh, PA, 15219
Flat Top Insurance Agency, 320 Federal St., Bluefield, WV, 24701
Fleetguard, Inc., 8204 Elmbrook, Suite 250, Dallas, TX, 75247
Flender Corp., 105 W. Fay Ave., P.O. Box 405, Addison, IL, 60101
Fletcher, J. H. & Co., P.O. Box 2143, Huntington, WV, 25722
Fletcher Sutcliffe Wild, Ltd., Horbury, Wakefield, England
Flexaust Co., The, 11 Chestnut St., Amesbury, MA, 01913
Flexible Steel Lacing Co., 2525 Wisconsin Ave., Downers Grove, IL, 60515
Flexible Valve Corp., 9 Empire Blvd., South Hackensack, NJ, 07606
Flex-Kleen Corp., 222 S. Riverside Plaza, Chicago, IL, 60606
Flexo Products, Inc., 24864 Detroit Rd., Westlake, OH, 44145
Flexowall Corp., One Heritage Park, Clinton, CT, 06413
Flight Systems, Inc., P.O. Box 25, Hempt Rd., Mechanicsburg, PA, 17055

Flowers Transportation, Inc., P.O. Box 1588, Greenville, MS, 38701
Fluid Controls Inc., 8341 Tyler Blvd., Mentor, OH, 44060
Fluidrive Engineering Co. Ltd., Broad Lane, Bracknell, Berkshire, Great Britain, RG12 3BH
Fluor Mining & Metals, Inc., 177 Bovet Rd., San Mateo, CA, 94402
Flygt AB, P.O. Box 1309, S-171 25 Solna, Sweden
Flygt Corp., 129 Glover Ave., Norwalk, CT, 06856
Foamcraft, Inc., 30450 Little Mack, Roseville, MI, 48066
Foote-Jones Operations, Dresser Industries, Inc., 603 Rogers St., Downers Grove, IL, 60515
Foote Mineral Co., Route 100, Exton, PA, 19341
Foraky Ltd., Colwick, Nottingham, NG 4 2BB, England
Ford, Bacon & Davis Utah, Inc., 375 Chipeta Way, Salt Lake City, UT, 84108
Ford Motor Co., P.O. Box 2053, Bldg. 1, Dearborn, MI, 48121
Ford Motor Credit Co., P.O. Box 1729, Dearborn, MI, 48121
Ford Tractor Operations, 2500 E. Maple Rd., Troy, MI 48084
Fort Pitt Steel Casting, 200 25th St., McKeesport, PA, 15134
Fostoria Industries, Inc., Box E. 1200 N. Main St., Fostoria, OH, 44830
Foxboro Co., The, 38 Neponset Ave., Foxboro, MA, 02035
Frazer & Jones Co., Div. of The Eastern Co., Box 4955, Syracuse, NY, 13221
Fredrik Mogensen AB, Box 78, S-544 00 HJ0, Sweden
Frick-Gallagher Mfg. Co, The, 201 S. Michigan Ave., Wellston, OH, 45692
Friemann & Wolf GmbH, Meidericher Str-6-8, 4100 Duisburg 1, Germany
Frog Switch Mfg. Co., East Louther St., Carlisle, PA, 17013
Fruco Engineers, Inc., 9666 Olive Blvd., St. Louis, MO, 63132
Fruehauf Div., Fruehauf Corp., 10900 Harper, Detroit, MI, 48232
Fuller Co., A Gatx Co., 2040 Avenue C, P.O. Box 2040, Bethlehem, PA, 18001
Funk Mfg. Div. of Cooper Industries Inc., 1211 W. 12th, Coffeyville, KS, 67333

G

GAF Corp., 140 W. 51st St., New York, NY, 10020
GCA Environmental Instruments, Burlington Rd., Bedford, MA, 01730
GEC Mechanical Handling Ltd., Birch Walk, Erith, Kent DA8 1QH, England
GS Industries, Inc., 60 Kansas Ave., Kansas City, KS, 66105
GTE Products Corp., Electrical Equipment, One Stamford Forum, Stamford, CT, 06904
G & W Electric Specialty Co., 3500 W. 127th St., Blue Island, IL, 60406
G & W Energy Products Group, 2222 Kensington Ct., Oak Brook, IL, 60521
Gai-Tronics Corp., 400 E. Wyomissing Ave., Monhnton, PA, 19540
Galigher Co., The, 440 W. 8th S., P.O. Box 209, Salt Lake City, UT, 84110

MINING EQUIPMENT MANUFACTURERS

Gammeter, W. F., Co., P.O. Box 307, Cadiz, OH, 43907
Gardner-Denver/Cooper Ind., Mining & Construction Group, 8585 Stemmons Frwy., Dallas, TX, 75247
Gardner-Denver/Cooper Ind., Petroleum & Exploration Equip. Group, P.O. Box 47647, Dallas, TX, 75247
Gardner-Denver/Cooper Ind., Pneutronics Div., 1333 Fulton St., Grand Haven, MI, 49417
Gardner-Denver, Industrial Machinery Div., 1800 Gardner Expy., Quincy, IL, 62301
Garland Mfg. Co., Box 26, Ironton, MN, 56455
Gates Engr. Co., 201 N. Kanawha St., Beckley, WV, 25801
Gates Rubber Co., The, P.O. Box 5887, Denver, CO, 80217
Gauley Sales Co., P.O. Drawer C, Hico, WV, 25854
Gebruder Kulenkampff, P.O. Box 10-38-69, Bremen, W. Germany
General Aviation Div., Rockwell International, 5001 N. Rockwell Ave., Bethany, OK, 73008
General Battery Corp., Box 1262, Reading, PA, 19603
General Battery Corp., Hertner Div., 12690 Elmwood Ave., Cleveland, OH, 44111
General Cable Co., Div. GK Technologies, 500 W. Putnam Ave., Greenwich, CT, 06830
General Electric Co., 1 River Rd., Bldg. 23, Schenectady, NY, 12345
General Electric Co., Carboloy Mining Products, P.O. Box 919, State Rte. 1717, Bristol, VA, 24201
General Electric Co., DC Motor & Generator Dept., 3001 E. Lake Rd., Erie, PA, 16531
General Electric Co., Gen. Purpose Control Dept., P.O. Box 2913, Bloomington, IL 61701
General Electric Co., Instrument Products Operation, 40 Federal St., Lynn, MA, 01910
General Electric Co., Insulating Materials, 1 Campbell Road, Schenectady, NY, 12345
General Electric Co., Lamp Div., Nela Park, Cleveland, OH, 44112
General Electric Co., Lighting Systems Dept., Hendersonville, NC, 28739
General Electric Co., Mobile Communications Div., P.O. Box 4197, Lynchburg, VA, 24502
General Electric Co., Power Circuit Breaker Dept., Section 1, 6901 Elmwood Ave., Philadelphia, PA, 19142
General Electric Co., Transportation Systems Business Div., 2901 E. Lake Rd., Erie, PA, 16501
General Electric Co., Wire and Cable Dept., 1285 Boston Ave. Bldg. 26BEL, Bridgeport, CT, 06602
General Electric Co., Wiring Device Product Dept., 95 Hathaway St., Providence, RI, 02907
General Electric Credit Corp., 260 Long Ridge Rd., Stamford,CT, 06904
General Energy Development Corp., 661 Highland Ave., Needham Heights, MA, 02194
General Equipment & Mfg. Co., Inc., P.O. Box 13226, Louisville, KY, 40213
General Kinematics Corp., 777 Lake Zurich Rd., Barrington, IL, 60010
General Plastics Inc., 502 Schrock Rd., Columbus, OH, 43229
General Refractories Co., U.S. Refractories Div., 600 Grant St., Pittsburgh, PA, 15219
General Resource Corp., 201 S. 3rd St., Hopkins, MN, 55343
General Scientific Equipment Co., Limekiln Pike & Williams Ave., Philadelphia, PA, 19150
General Splice Corp., P.O. Box 158, Rt. 129, Croton-on-Hudson, NY, 10510
General Tire & Rubber Co., The, One General St., Akron, OH, 44329
GenRad, 300 Baker Ave., Concord, MA, 01742
Geo Space Corp., 5803 Glenmont Dr., Houston, TX, 77081
GEOMIN, Calea Victoriei 109, Bucharest, Romania
George Evans Corp., The, 121 37th St., Moline, IL, 61265
Giant, 3156 Bellevue Rd., Toledo, OH, 43606
Gilbert Associates, Inc., P.O. Box 1498, Reading, PA, 19603
Gilson Company Inc., P.O. Box 677, Worthington, OH, 43085
Glastech Engineering, P.O. Box 4007, Green Bay, WI, 54303
Globe Safety Products, Inc., 125 Sunrise Pl., Dayton, OH, 45407
Glover Plastics Ltd., Grove Rd., Millbrook, Stalybridge, Cheshire, England
Golder Associates, Inc., 10628 N.E. 38th Pl., Kirkland, WA, 98033
Goodall Rubber Co., P.O. Box 8237, Trenton, NJ, 08650
Goodman Equipment Corp., 4834 South Halsted St., Chicago, IL, 60609
Goodrich, B.F., Chemical Group, 6100 Oak Tree Boulevard, Cleveland, OH, 44131
Goodrich, B.F., Co., Engineered Products Group, 500 S. Main St., Akron, OH, 44318
Goodrich, B.F., Company, 500 So. Main St., Akron, OH, 44318
Goodrich, B.F., Transport Marketing, Box 340, Troy, OH, 45373
Goodyear Industrial Products, Box 52, Akron, OH, 44309
Goodyear Tire & Rubber Co., 1144 E. Market, Akron, OH, 44316
Gorman-Rupp Co., The, P.O. Box 1217, Mansfield, OH, 44902
Gould Inc., Distribution & Controls Div., R. B. Denison, 103 Broadway, Bedford, OH, 44146
Gould Inc., Electric Motor Div., 1831 Chestnut St., St. Louis, MO, 63166
Gould Inc., Electrical Products Group, 10 Gould Center, Rolling Meadows, IL, 60008
Gould Inc., Fluid Components Div., 6565 W. Howard, Chicago, IL, 60648
Gould Inc., Hose and Couplings Div., 1440 N. 24th St., Manitowoc, WI, 54220
Gould Inc., Industrial Battery Div., 2050 Cabot Blvd. W., Langhorne, PA, 19047
Goulds Pumps, Inc., 240 Fall St., Seneca Falls, NY, 13148
Goyne Pumps, Sub. of Goulds Pumps Inc., East Centre St., Ashland, PA, 17921
Grace, W. R. & Co., Construction Products Div., 62 Whittemore Ave., Cambridge, MA, 02140
Grad-Line, Inc., P.O. Box 245, Woodinville, WA, 98072
Graemross Plant & Equipment Ltd., Automation House, Rosebery Rd., Anstey, Leicester, England
Grass Growers, Inc., 424 Cottage Pl., Plainfield, NJ, 07060
Great Lakes Instruments, Inc., 8855 N. 55th St., Milwaukee, WI, 53223
Greenbank Cast Basalt Eng. Co. Ltd., Gate St., Blackburn,Lancs., England
Greengate Industrial Polymers Ltd., Irwell Works, Ordsall Lane, Salford M5 4TD, England

Greening Donald Co. Ltd., P.O. Box 430, Hamilton, Ont., Canada, L8N 3J3
Greenings/N. Greening Ltd., P.O. Box 22, Warrington, U.K., WA5 5JX
Greenville Steel Car Co., Union St.,, Greenville, PA, 16125
Griffolyn Co., Inc., P.O. Box 33248, Houston, TX, 77033
Griphoist, Inc., 392 University Ave., P.O. Box 707, Westwood, MA, 02090
Grove Mfg. Co., Div. of Walter Kidde & Co. Inc., Shady Grove, PA, 17256
Gruendler Crusher & Pulverizer Co., 2917 N. Market St., St. Louis, MO, 63106
Guidler Co., The, P.O. Box 96, Roslyn Heights, NY, 11577
Gulf Oil Chemicals Co., Explosives Dept., P.O. Box 2900, Shawnee Mission, KS, 66201
Gulf Oil Corp., Dept. DM, P.O. Box 1563, Houston, TX, 77001
Gullicie Dobson Intl. Ltd., P.O. Box 12, Wigan, Lancashire, England, WN1 3DD
Gundlach, T. J., Machine Div., Rexnord Inc., P.O. Box 385, Belleville, IL, 62222
Gunson's Sortex (Mineral & Automation) Ltd., 40 Warton Rd., Stratford, London E15 2JU, U.K.
Gustin-Bacon Div., Aeroquip Corp., P.O. Box 927, Lawrence, KS, 66044
Guyan Machinery Co., P.O. Box 150, Logan, WV, 25601

H

HPI-Nichols, a W. H. Nichols Co., P.O. Box 458, Sturtevant, WI, 53177
HRB-Singer, Inc., Energy & Natural Resource Systems, P.O. Box 60, State College, PA, 16801
Hach Chemical Co., P.O. Box 389, 57th St. & Lindbergh Parkway, Loveland, CO, 80537
Hacker Instruments Inc., P.O. Box 657, Fairfield, NJ, 07006
Hackman-Skeehan Inc., 200 Keystone Dr., Carnegie, PA, 15106
Hahn Industries, Mine & Mill Specialties, 50 Broadway, New York, NY, 10004
Halliburton Services-Research Center, P.O. Box 1431, Duncan, OK, 73533
Hallite Seals Inc., 1929 Lakeview Dr., Fort Wayne, IN, 46808
Hammermills, Inc., Sub. of Pettibone Corp., 625 C Ave., N.W. Cedar Rapids, IA, 52405
Hammond, J. V. Co., N. 1st St., Spangler, PA, 15775
Hanco International Div. of Hannon Electric Co., 1605 Waynesburg Rd., Canton, OH, 44707
Hansen Transmissions Inc., P.O. Box 710, Branford, CT, 06405
Hardman Inc., Belleville, NJ, 07109
Hardy Plants, 587 Harmony Rd., New Brighton, PA, 15066
Harnischfeger Corp., P.O. Box 554, Milwaukee, WI, 53201
Harrington & King Perforating, 5655 Fillmore St., Chicago, IL, 60644
Harrison R. Cooper Systems, Inc., AMF Box 22014, Salt Lake City, UT, 84122
Hartman-Fabco Inc., 1415 Lake Lansing Rd., Lansing, MI, 48912
Hartzell Propeller Fan Co., P.O. Box 919, Piqua, OH, 45356
Hauck Mfg. Co., P.O. Box 90, Lebanon, PA, 17042
Hauhinco Maschinenfabrik, P.O. Box 639, 4300 Essen 1, W. Germany

Haulmasters, Inc., 100 Mission Woods Rd., Industrial Airport, KS, 66031
Hausherr, Rudolf und Sohne, P.O.B. 1240, D-4322 Sprockhovel, West Germany
Hawker Siddeley Dynamics Engineering Limited, Manor Road, Hatfield, Herts., England
Hawker Siddeley Electric Export Ltd., P.O. Box 20, Loughborough, Leics., LE11 1HN, England
Haws Drinking Faucet Co., 4th & Page Sts., P.O. Box 1999, Berkeley, CA, 94701
Hayden-Nilos Conflow Ltd., Darnall Rd., Sheffield, England, NG7 2GF
Hayward Tyler, Inc., 25 Harbor Ave., Norwalk, CT, 06850
Hazemag, Dr. E. Andreas GmbH & Co., Postfach 3447, Rosnerstr. 6/8, 4400 Munster, Germany
Hazemag USA, Inc., P.O. Box 15515, Pittsburgh, PA, 15244
Hazen Research, Inc., 4601 Indiana St., Golden, CO, 80401
HB Electrical Mfg. Co., Inc., P.O. Box 1466, Mansfield, OH, 44906
Heede Intl., Inc., Slipform Div., Hydraulic & Pneumatic Service, 43 Lindstrom Rd., Stamford, CT, 06902
Heil Process Eqipment, Fiberglass Equip. Div., Dart Environment & Services Co., 34250 Mills Rd., Avon, OH, 44011
Heintzmann Corp., 2662 Cedarvue Dr., Pittsburgh, PA, 15241
Helwig Carbon Products, Inc., 2550 N. 30th St., Milwaukee, WI, 53210
Hemscheidt America Corp., Manor Oak No. 2, Suite 536, Pittsburgh, PA, 15220
Hendrick Mfg. Co., 7th Ave. & Clidco Dr., Carbondale, PA, 18407
Hendrick Screen Co., P.O. Box 369, 2942 Medley Rd., Owensboro, KY, 42301
Hendrix Mfg. Co., Inc., P.O. Box 919, Mansfield, LA, 71052
Hensley Industries Inc., 2108 Joe Field Rd., Dallas, TX, 75229
Hercules Inc., 910 Market St., Wilmington, DE, 19801
Herold Mfg. Co., Inc., 215 Hickory St., Scranton, PA, 18505
Hewitt-Robins Conveyor Equipment Div. Litton Systems, Inc., 270 Passaic Ave., Passaic, NJ, 07055
Hewitt-Robins Div., Litton Systems, Inc., P.O. Box 1481, Columbia, SC, 29202
Hewlett-Packard, 815 14th St., S.W., P.O. Box 301, Loveland, CO, 80537
Heyl & Patterson, Inc., P.O. Box 36, Pittsburgh, PA, 15230
HIAB Cranes & Loaders, Inc., HIAB Circle, Newark, DE, 19711
Hi-Flex Intl. Inc., 2001 Lowell Ave., Erie, PA, 16506
Hitachi Construction Machinery Co./Marubeni America Corp., 200 Park Ave., New York, NY, 10017
Hobart Bros. Co., 600 W. Main St., Troy, OH, 45373
Holmes Bros. Inc., 510 Junction Ave., Danville, IL, 61832
Holz Rubber Co., Inc., a Randtron Sub., P.O. Box 109, 1129 So. Sacramento St., Lodi, CA, 95240
Homelite Div., Textron Inc., P.O. Box 7047, Charlotte, NC, 28217
Homer Magnetics Div., 915 Shawnee Rd., Lima, OH, 45805
Homestead Industries, Jenny Div., P.O. Box 348, Coraopolis, PA, 15108
Honeywell Inc., Process Control Div., 1100 Virginia Dr., Fort Washington, PA, 19034

MINING EQUIPMENT MANUFACTURERS

Horsburgh & Scott Co., 5114 Hamilton Ave., Cleveland, OH, 44114
Hossfeld Mfg. Co., 440 W. Third St., Winona, MN, 55987
Houdaille Hydraulics, 537 E. Delavan Ave., Buffalo, NY, 14211
Houghton & Co., E. F., Madison & Van Buren Aves., Valley Forge, PA, 19482
Howe Richardson Scale Co., 680 Van Houten Ave., Clifton, NJ, 07015
Hoyt Wire Cloth Co., 10 Abraso St., Box 1577, Lancaster, PA, 17604
Hughes, L. J. & Sons, Inc., 320 Turnpike Rd., Summersville, WV, 26651
Hughes (Robert) Associates, 7839 Churchill Way, Suite 130, Dallas, TX, 75251
Hughes Tool Co., P.O. Box 2539, Houston, TX, 77001
Humphreys Engineering Co., 2219 Market St., Denver, CO, 80205
Hunslet Holdings Ltd., Hunslet Engine Works, Leeds LS10 1BT, England
Huntec (70) Ltd., 25 Howden Rd., Scarborough, Ont., Canada, M1R 5A6
Huron Mfg. Corp., P.O. Box 1398, Huron, SD, 57350
Huwood-Irwin Co., Box 409, Irwin, PA, 15642
Huwood Limited, Gateshead, Tyne & Wear, NE11 OLP, England
Hycaloader Co., P.O. Box 749, Lake Providence, LA, 71254
HYCO, Dana Industrial, Fluid Drives & Controls Div., 1401 Jacobson Ave., Ashland, OH, 44805
Hydra-Mac, Inc., Box N, Thief River Falls, MN, 56701
Hydreco, A Unit of General Signal, 9000 E. Michigan Ave., Kalamazoo, MI 49003
Hydrophilic Industries, Inc., 5815 N. Meridian, Puyallup, WA, 98371
Hy Test Safety Shoes Div. International Shoe Co., 1509 Washington Ave., St. Louis, MO, 63166

I

I & M Mfg. & Sales, Inc., R. #1, Box 28M, Bourbon, IN, 46504
ITT Grinnell Corp., 260 W. Exchange St., Providence, RI, 02901
ITT Harper, 8200 Lehigh Ave., Morton Grove, IL, 60053
ITT Holub Industries, 1701 W. Bethany Rd., Sycamore, IL, 60178
ITT Marlow Pumps, Pumps and Compressors Div., P.O. Box 200, Midland Park, NJ, 07432
ITT Royal Electric, 95 Grand Ave., Pawtucket, RI, 02862
Illinois Gear/Wallace Murray Corp., 2138 N. Natchez Ave., Chicago, IL, 60635
Imperial Oil & Grease Co., 10960 Wilshire Blvd., Los Angeles, CA, 90024
INCO Safety Products Co., P.O. Box 1733, Reading, PA, 19603
Independent Explosives Co., 20950 Center Ridge Rd., Cleveland, OH, 44114
Indiana Steel & Fabricating Co., Box 767, Indiana, PA, 15701
Industrial Electric Reels, 10102 F St., P.O. Box 3129, Omaha, NE, 68103
Industrial Resources, Inc., Barry Addition, Box 352, Fairmont, WV, 26554
Industrial Resources, A. L. Lee Corp., P.O. Box 295, Beckley (Lester), WV, 25865

Industrial Steel Co., P.O. Box 504, Carnegie, PA, 15106
Ingersoll-Rand Co., 200 Chestnut Ridge Rd., Woodcliff Lake, NJ, 07675
Ingersoll-Rand Cyclone Drill, East Chestnut St., Orrville, OH, 44691
Inland Motor/Specialty Products Div., 501 First St., Radford, VA, 24141
Inland Steel Co., 30 W. Monroe St., Chicago, IL, 60603
Inryco Inc., P.O. Box 393, Milwaukee, WI, 53201
Insley Mfg., A Unit of AMCA Int'l Corp., 801 N. Olney, P.O. Box 11308, Indianapolis, IN, 46201
International Alloy Steel Div., 30403 Bruce Industrial Pkwy., Solon, OH, 44139
International Harvester, Pay Line Group, 600 Woodfield, Schaumburg, IL, 60196
International Salt Co., Clarks Summit, PA 18411
Interroll Corp., 60 Hoffman Ave., Hauppauge, NY, 11787
Interstate Equipment Corp., 300 Mt. Lebanon Blvd., Pittsburgh, PA, 15234
Interstate Fabricators & Constructors, P.O. Box 1427, Pleasant Valley, Fairmont, WV, 26554
Iowa Industrial Hydraulics, Inc., Industrial Park Rd., Pocahontas, IA, 50574
Iowa Manufacturing Co., 916 16th St., N.E., Cedar Rapids, IA, 52402
Iowa Mold Tooling Co., Inc., 500 Highway 18 West, Garner, IA, 50438
Irad Gage, Inc., Etna Rd., Lebanon, NH, 03766
Irathane Systems, Inc., Industrial Park, Hibbing, MN, 55746
IRECO Chemicals, Kennecott Bldg., Suite 726, Salt Lake City, UT, 84133
Irvin-McKelvy Co., The, P.O. Box 767, Indiana, PA, 15701
ISCO, 3621 N.W. 36th St., Lincoln, NE, 68524
Izumi Chain Co., 398 Wrightwood Ave., Elmhurst, IL, 60126

J

J & S Sieger Ltd., 31, Nuffield Estate, Poole, Dorset, BH17 7RZ, England
J-Tec Associates, Inc., 317 7th Ave., S.E., Cedar Rapids, IA, 52401
Jabco, Inc., 526 Ogle St., Ebensburg, PA, 15931
Jaeger Machine Co., 550 W. Spring St., Columbus, OH, 43216
James, D. O. Gear Mfg. Co., Unit of Ex-Cell-O Corp., 1140 W. Monroe St., Chicago, IL, 60607
Janes Manufacturing Inc., 7625 S. Howell Ave., Oak Creek, WI, 53154
Jarva, Inc., 29125 Hall St., Solon, OH, 44139
Jaswell Drill Corp., Sanderson Rd., Greenville, RI, 02828
Jeffrey Mfg. Div., Dresser Industries Inc., P.O. Box 3080, Greenville, SC, 29602
Jeffrey Mining Machinery Div., Dresser Industries Inc., 274 E. First Ave., P.O. Box 1879, Columbus, OH, 43216
Jeffrey Mining Machinery Div., Drilling Equipment, 274 E. First Ave., Columbus, OH, 43216
Jenkins of Retford Ltd., Retford, Notts DN22 7AN, England
Jennmar Corp., P.O. Box 187, Cresson, PA 16630
Jersey Chain & Metal Co., Inc., 198 Rte. 206 So., Somerville, NJ, 08876
Jet Lube Inc., P.O. Box 21258, 4849 Homestead Rd., Houston, TX, 77026
Jim Pyle Co., Junction City, KY, 40440

Johnson Blocks Div., Hinderliter Energy Equipment Corp., 1240 N. Harvard, P.O. Box 4699, Tulsa, OK, 74104
Johnson Div. UOP Inc., P.O. Box 43118, St. Paul, MN, 55164
Johnson-March Corp., The, 3018 Market St., Philadelphia, PA, 19104
Johnston-Morehouse-Dickey Co., 5401 Progress Blvd., P.O. Box 173, Bethel Park, PA, 15102
Johnston Pump Co., 1775 E. Allen Ave., Glendora, CA, 91740
Jold Mfg. Co., Inc., Box 341, Oakwood, VA, 24631
Jones & Laughlin Steel Corp., 3 Gateway Center, Pittsburgh, PA, 15230
Jones & Laughlin Steel Corp., Conduit Products, McKees Lane, Niles, OH, 44446
Joy Mfg. Co., Henry W. Oliver Bldg., Pittsburgh, PA, 15222
Joy Mfg. Co., Air Moving Products, 338 S. Broadway, New Philadelphia, OH, 44663
Joy Mfg. Co., Denver Equipment Div., 621 So. Sierra Madre, Colorado Springs, CO, 80903
Joy Mfg. Co., Electrical Products, Rt. 4, Box 156, La Grange, NC, 28551
Joy Service Center, Div. Joy Mfg. Co., P.O. Box 687, Bluefield, WV, 24701
Judsen Rubber Works, Inc., 4107 W. Kinzie St., Chicago, IL, 60624

K

KHD Industrieaniagen AG, Humboldt Wedag, Wiersbergstrasse, D 5 Koeln 91, Fed. Rep. of Germany
KW Battery Co., 3555 Howard St., Skokie, IL, 60076
Kaelble, Carl GmbH, P.O. Box 1320, 7150 Backnang, West Germany
Kaiser Aluminum & Chemical Corp., 300 Lakeside Dr., Rm. 1123 KB, Oakland, CA, 94643
Kaiser Engineers, Inc., 300 Lakeside Dr., P.O. Box 23210, Oakland, CA, 94623
Kalenborn, Dr. Ing. Mauritz KG, D-5461 Kalenborn near Linz on Rhine, Germany
Kanawha Mfg. Co., P.O. Box 1786, Charleston, WV, 25326
Kanawha Valley Bank, NA, P.O. Box 1793, Charleston, WV, 25326
Kay-Ray Inc., 516 W. Campus Dr., Arlington Heights, IL, 60005
Keenan Oil Co., 2350 Seymour Ave., Cincinnati, OH, 45212
Keene Corp., 200 Park Ave., S., New York, NY, 10022
Kelley Corp., 1135 S. Beech St., Casper, WY, 82601
Kennametal Inc., P.O. Box 346, Latrobe, PA, 15650
Kennedy-McMaster Inc., P.O. Box 304, Taylorville, IL, 62568
Kennedy Metal Products & Buildings, Inc., Jack, Box 38, 200 S. Jayne St., Taylorville, IL, 62568
Kennedy Van Saun Corp., Railroad St., Danville, PA, 17821
Kent Air Tool Co., 711 Lake St., Kent, OH, 44240
Kentucky Road Equipment, Inc., 13124 Aiken Rd., Anchorage, KY, 40223
Kenworth Truck Co., P.O. Box 1000, Kirkland, WA, 98033
Kern Instruments Inc., Geneva Road, Brewster, NY, 10509
Kersey Mfg. Co., Div. ATO-Inc., P.O. Box 151, Bluefield, VA, 24605
Key Bellevilles, Inc., R.D. #2, Leechburg, PA, 15656
Keystone Bolt Co., Sub. of Jenmar Corp., 600 Arch St., Cresson, PA, 16630

Keystone Div., Pennwalt Corp., 21 & Lippincott Sts., Philadelphia, PA, 19132
Kidde Belleville, Div. of Walter Kidde & Co., 675 Main St., Belleville, NJ, 07109
Kilborn/NUS Inc., 720 S. Colorado Blvd. #930, Denver, CO, 80222
Kilo-Wate Inc., Box 798, Georgetown, TX, 78626
Klein Tools, Inc., 7200 McCormick, Chicago, IL, 60645
Koch Engineering Co., Inc., 161 E. 42nd St., New York, NY, 10017
Koch Transporttechnik GmbH, Am Bahnhof, Wadgassen, W. Germany, 6622
Kockum Landsverk AB, Box 512, S-261 24 Landskrona, Sweden
Koehring Crane & Excavator Group, P.O. Box 2060, Milwaukee, WI, 53201
Kolberg Mfg. Corp., Box 20, W. 21st St., Yankton, SD, 57078
Komatsu America Corp., 555 California St., San Francisco, CA, 94104
Koppers Co., Inc., 1900 Koppers Bldg., Pittsburgh, PA, 15219
Koppers Co. Inc., Engineered Metal Products Group, P.O. Box 298, Baltimore, MD, 21203
Koppers Co. Inc., Mineral Processing Systems Div., Box 312, York, PA, 17405
Kothari Eng. Co., 111 Hill St., Roswell, GA, 30075
Krauss Maffei Corp., Process Equipment Div., P.O. Box 9104, Wichita, KS, 67277
Krebs Engineers, 1205 Chrysler Dr., Menlo Park, CA, 94025
Kress Corp., 400 Illinois St., Brimfield, IL, 61517
Krupp Industrie-und Stahlbau, Franz-Schubert-Strasse 1-3, 4100 Duisburg-Rheinhausen, Germany
K-Tron Corp., P.O. Box 548, Glassboro, NJ, 08028
Kue-Ken Div., Process Technology Corp., 8383 Baldwin St., Oakland, CA, 94621
Kugler Ltd., Case Postale 240, La Jonction, CH-1211 Geneve 8, Switzerland

L

L & M Radiator, Inc., 1414 E. 37th St., Hibbing, MN, 55746
LaBour Pump Co., P.O. Box 1187, Elkhart, IN, 46515
Ladish Co., 5401 S. Packard Ave., Box F, Cudahy, WI, 53110
La Font Corp., 1319 Town St., Prentice, WI, 54556
Lake Shore, Inc., P.O. Box 809, Iron Mountain, MI, 49801
LaMarche Manufacturing Co., 106 Bradrock Dr., Des Plaines, IL, 60018
Lambert Industries, Inc., P.O. Box 1127, Virginia, MN, 55792
Lancaster Steel Products, 9 Horseshoe Rd., Leola, PA, 17540
Lane Metal Products, Inc., 3705 Trindle Rd., Camp Hill, PA, 17011
Laser Alignment, Inc., 6330 28th St., S.E., Grand Rapids, MI, 49506
Laubenstein Mfg. Co., 418 S. Hoffman Blvd., Ashland, PA, 17921
Lawnel Corp., P.O. Box 206, Bluefield, WV, 24605
Lawrence Pumps, Inc., 371 Market St., Lawrence, MA, 01843
Lebco, Inc., Illinois Div., Hiway 14E., P.O. Box 656, Benton, IL, 62812

MINING EQUIPMENT MANUFACTURERS

Lebus International Inc., Box 2352, Longview, TX, 75606
Leco Corp., 3000 Lakeview Ave., St. Joseph, MI, 49085
Lee, A. L., Corp., 1166 Cleveland Ave., P.O. Box 8218, Columbus, OH, 43201
Leeco Steel Products Inc., 1600 S. Kostner, Chicago, IL, 60623
Leeds & Northrup, a unit of General Signal, Sumneytown Pike, North Wales, PA, 19454
Lee-Norse Co., P.O. Box 2863, Pittsburgh, PA, 15230
Lee Steel Corp., Box 98, Baxter, KY, 40806
Lehigh Safety Shoe Co., 1100 E. Main St., Endicott, NY, 13760
Le Roi Div., Dresser Industries, Inc., Main & Russell Rd., Sidney, OH, 45365
Leman Machine Co., 1049 So. Railroad Ave., Portage, PA, 15946
Leschen Wire Rope Co., Box 407, St. Joseph, MO, 64502
Liebherr America, Inc., 4100 Chestnut Ave., Drawer O, Newport News, VA, 23605
Lightning Industries, Inc., 801 Woodswether Rd., Kansas City, MO, 64105
Lima Electric Co., Inc., 200 E. Chapman Rd., Lima, OH, 45802
Linatex Corp. of America, P.O. Box 65, Stafford Springs, CT, 06076
Lincoln Electric Co., The, 22801 St. Clair Ave., Cleveland, OH, 44117
Lincoln St. Louis Div. of McNeil Corp., 4010 Goodfellow Blvd., St. Louis, MO, 63120
Linden-Alimak, Inc., 6295 E. 56th Ave., Commerce City, CO, 80022
Line Power Manufacturing Corp., 329 Williams St., Bristol VA, 24201
Lippmann-Milwaukee, Inc., 4603 W. Mitchell, Milwaukee, WI, 53214
Lively Mfg. & Equipment Co., P.O. Box 339, Glen White, WV, 25849
Loftus, Peter F., Corp., Chamber of Commerce Bldg., Pittsburgh, PA, 15219
Logan Corp., 555 7th Ave., P.O. Box 1895, Huntington, WV, 25719
Logan Mine Supply Co., Inc., P.O. Box 831, W. Logan, WV, 25601
Long-Airdox Co. A Div. of the Marmon Group, Inc., P.O. Box 331, Oak Hill, WV, 25901
Long-Airdox Construction Co., a Member of the Marmon Group, P.O. Box 56, Boswell, PA, 15531
Longwall Mining Div., Columbus McKinnon Corp., One Fremont St., Tonawanda, NY, 14150
Longyear Co., 925 Delaware St. S.E., Minneapolis, MN, 55414
Louis Allis Div., Litton Industrial Products, Inc., 427 E. Stewart St., P.O. Box 2020, Milwaukee, WI, 53201
Lubrication Engineers, Inc., P.O. Box 7128, Ft. Worth, TX, 76111
Lubriplate Div., Fiske Brothers Refining Co., 129 Lockwood St., Newark, NJ, 07105
Lubriquip Houdaille, 18901 Cranwood Parkway, Cleveland, OH, 44128
Lucas Fluid Power, 23645 Mercantile Rd., Cleveland, OH, 44122
Ludlow-Saylor Div. G.S.I., 8474 Delport Dr., St. Louis, MO, 63114
Lukens Steel Co., W. Lincoln Highway, Coatesville, PA, 19320

Lummus Co., 1515 Broad St., Bloomfield, NJ, 07003
Lukenheimer Co., Div. of Conval Corp., Sub. of Condec Corp., Beckman at Waverly Ave., Cincinnati, OH, 45214
Lupo, G. J. Company, Inc., 2482 So. 3270 West, Salt Lake City, UT, 84119
Lurgi Kohle und Mineralolitechnik GmbH, Gervinusstr, 17/19, D-6 Frankfurt/M. Fed. Rep. Germany
Lyon Metal Prods. Inc., P.O. Box 671, Aurora, IL, 60507

M

3 M Co., 3M Center, St. Paul, MN, 55144
MCC MARPAC, a unit of Mark Controls Corp., 1900 W. Dempster, Evanston, IL, 60204
MMP Co., 400 So. Main St., Box 398, Fountain Inn, SC, 29644
Mac Products, Inc., 60 Pennsylvania Ave., Kearny, NJ, 07032
Macauley, H. C., Foundry Co., 811 Carleton St., Berkeley, CA, 94710
Macawber Engineering Ltd., Ogden Rd., Doncaster DN2 4SQ, England
MacDonald Engineering Co., 22 W. Madison St., Chicago, IL, 60602
Machinery Center, Inc., 1201 S. 7th West P.O. Box 964, Salt Lake City, UT, 84110
Mack Trucks, Inc., Box M, Allentown, PA, 18105
Macwhyte Wire Rope Co., 2931 14th Ave., Kenosha, WI, 53141
Mag-Con, Inc., 1626 Terrace Dr., St. Paul, MN, 55113
Magco Ltd., Lake Works, Portchester Fareham, Hampshire, England, PO16 9DS
M.A.N. Maschinenfabrik Augsburg-Nurnberg, Div. GHH Sterkrade, Bahnhofstrasse 66, 42 Oberhausen 11, West Germany
Manheim Mfg. & Belting, 311 W. Stiegel St., Manheim, PA, 17545
Manitowoc Engineering Co., Div. Manitowoc Co., 500 S. 16th St., Manitowoc, WI, 54220
Mannesmann Demag Corp., 1100 Jorie Blvd., Oakbrook, IL, 60521
Mannesmann Demag Baumaschinen Bereich Lauchhammer, Forststrasse 16, Postfach 180 230 D-4000, Dusseldorf 13, FRG
Manson Services, Inc., R.D. #1, Box 307-A, Greensboro, PA, 15338
Manufacturers Equipment Co., The, 35 Enterprise Dr., Middletown, OH, 45042
Manufacturers Hanover Leasing Corp., 30 Rockefeller Plaza, New York, NY, 10020
Marathon Coal Bit Co., Inc., Box 391, Montgomery, WV, 25136
Marathon Letourneau Co., Longview Div., P.O. Box 2307, Longview, TX, 75606
Marathon Mfg. Co., 600 Jefferson, 1900 Marathon Bldg., Houston, TX, 77002
Marathon Metallic Bldg. Co., P.O. Box 14240, Houston, TX, 77021
Marathon Steel Co., P.O. Box 6598, Phoenix, AZ, 85005
Marietta Concrete Co., P.O. Box 254, Marietta, OH, 45750
Marion Manufacturing Co., 6501 Barberton Ave., Cleveland, OH, 44102
Marion Power Shovel Div., Dresser Industries, Inc., P.O. Box 505, 617 W. Center St., Marion, OH, 43302

Mark Controls Corp., 1900 Dempster St., Evanston, IL, 60204

Mark Equipment Co., 6033 Manchester Ave., St. Louis, MO, 63110

Marland One-Way Clutch Div., Zurn Industries, Inc., P.O. Box 308, La Grange, IL, 60525

Marmon Transmotive Div., Sanford Day Products, P.O. Box 1511, Knoxville, TN, 37901

Marquis Delta, Inc., 6060 Northwest Hwy., Chicago, IL, 60631

Mars Mineral Corp., Box 128, Valencia, PA, 16059

Marsh, E. F., Engineering Co., 1400 Hanley Industrial Dr., St. Louis, MO, 63144

Martec International, One World Trade Center, Ste. 2845, New York, NY, 10048

Martin-Decker, 1928 So. Grand Ave., Santa Ana, CA, 92705

Martin Engrg. Co., U.S. Rte. 34, Neponset, IL 61345

Martindale Electric Co., 1307 Hird Ave., Cleveland, OH, 44107

Maschinenfabrik Buckau R. Wolf Aktiengesellschaft, Postfach 100 460, D-4048, Grevenbroich 1, West Germany

Massey-Ferguson, Inc., 1901 Bell Ave., Des Moines, IA, 50315

Material Control, Inc., 719 Morton Ave., Aurora, IL, 60506

Mathews, Abe W., Engineering Co., 555 West 27th St., Hibbing, MN, 55746

MAT Industries, Inc., P.O. Box 454, W. Frankfort, IL, 62896

MATO, P.O. Box 70, D-6050 Offenbach (Main) 1., W. Germany

Matthew Hall Ortech Ltd., Marsland House, Hope Rd., Sale, Cheshire, M33 3AQ, England

Matthias Spencer and Sons Ltd., P.O. Box 7, Arley St., Sheffield, England, S2 4QQ

McDonnell Douglas Electronics Co., P.O. Box 426, St. Charles, MO, 63301

McDonough-Caperton-Shepherd Group, P.O. Box 1691, Beckley, WV 25801

McGraw-Edison Co., Power Systems Div., P.O. Box 2850, Pittsburgh, PA, 15230

McJunkin Corp., 1400 Hansford St., Charleston, WV, 25312

McKey Perforating Co., Inc., 3036 So. 166th St., New Berlin, WI, 53151

McLanahan Corp., 200 Wall St., Hollidaysburg, PA, 16648

McLaughlin Mfg. Co., P.O. Box 303, Plainfield, IL, 60544

McNally Pittsburg Mfg. Corp., P.O. Box 15, Pittsburg, KS, 66762

McNichols Co., 7723 Holiday Dr., Sarasota, FL, 33581

McQuay-Perfex, Perfex Group, 500 W. Oklahoma Ave., Milwaukee, WI, 53207

Megator Corp., 136 Gamma Dr., Pittsburgh, PA, 15238

Merkel Forsheda Corp., 5375 Naiman Parkway, Cleveland, OH, 44139

Merrick Scale, 180 Autumn St., Passaic, NJ, 07055

Mescher Mfg. Co. Inc., P.O. Box 789, Grundy, VA, 24614

Metal Carbides Corp., 6001 Southern Blvd., Youngstown, OH, 44512

Metal Craft Inc., P.O. Box 862, Tazewell, VA, 24651

Metcalf & Eddy, Inc., 50 Staniford St., Boston, MA, 02114

Metropolitan Wire Corp., N. Washington St. & George Ave., Wilkes-Barre, PA, 18705

M/G Transport Service, Inc., 111 E. 4th St., Cincinnati, OH, 45202

Michael Baker Jr. Inc., 4301 Dutch Ridge Rd., Beaver, PA, 15009

Michael Walters Ind., 6th & Pine St., Kenova, WV, 25530

Michelin Tire Corp., Earthmover Tire Dept., 1 Marcus Ave., Lake Success, NY, 11040

Microdot, Inc., 475 Steamboat Rd., Greenwich, CT, 06830

Micro Switch, A Div. of Honeywell, 11 W. Spring St., Freeport, IL, 61032

Midland Affiliated Inc., 580 Walnut St., Cincinnati, OH, 45202

Midland Pipe & Supply Co., 6111 W. 28th St., Cicero, IL, 60650

Midland Pump, LFE Fluids Control Div., 100 Skiff St., Hamden, CT, 06514

Midland Ross Corp., National Castings Div., 700 So. Dock St., Sharon, PA, 16146

Midway Equipment Inc., 2380 Cassens Dr., Fenton, MO, 63026

Midwestern Industries, Inc., Screen Heating Transformers Div., 915 Oberlin Rd., SW, Massillon, OH, 44646

Mid-West Conveyor Co., 450-B E. Donovan Rd., Kansas City, KS, 66115

Midwest Industrial Supply Inc., P.O. Box 8431, Canton, OH, 44711

Midwest Steel Div., Midwest Corp., P.O. Box 271, Charleston, WV, 25321

Midwest Telecommunications, Div. of Unarco Industries, 300 First Ave., Nitro, WV, 25143

MikroPul Corp., 102 Chatham Rd., Summit, NJ, 07901

Miller Electric Mfg. Co., 718 S. Bounds St., Appleton, WI, 54912

Milltronics Inc., 2409 Ave. J, Arlington, TX, 76011

Milwaukee Electric Tool Corp., 13135 W. Lisbon Rd., Brookfield, WI, 53005

Milwaukee Tool & Equipment, 2773 S. 29th St., Milwaukee, WI, 53215

Mine Equipment Co., 2304 Industrial Dr., Mt. Vernon, IL, 62864

Mine Management Systems, 623 Hawley Bldg., Wheeling, WV, 26003

Mine Safety Appliances Co., 600 Penn Center Blvd., Pittsburgh, PA, 15235

Mine & Smelter, P.O. Box 16067, Denver, CO, 80216

Mine Systems Inc., P.O. Box 7, Cloverdale, VA, 24077

Mine Ventilation Systems, Inc., Box 385, Madison, WV, 25130

Minemet Inc., 450 Park Ave., New York, NY, 10022

Minerals Processing Co., Div. of Trojan Steel Co., 6318 MacCorkle Ave., S.W., St. Albans, WV, 25177

Mining & Transport Engineering B.V., P.O. Box 3084, 1003 AB Amsterdam, Holland

Mining Controls, Inc., P.O. Box 1141, Beckley, WV, 25801

Mining Developments Ltd., Crown Lane, Horwich, Bolton, BL6 5HN, England

Mining Equipment Mfg. Corp., 3319 Four Mile Rd., Racine, WI, 53404

Mining Machine Parts, 6345 Norwalk Rd., Medina, OH, 44256

Mining Progress, Inc., 605 Boulevard Tower, Charleston, WV, 25301

Mining Supplies, Ltd., Hillcrest Works, Carr Hill, Balby, Doncaster, S. Yorks, England, DN4 8DH

Mining Tools, Inc., 7700 St. Clair Ave., Mentor, OH, 44060

Minnesota Automotive Inc., Box 2074, North Mankato, MN, 56001

MINING EQUIPMENT MANUFACTURERS

Mintec/International, Div. of Barber-Greene, 400 N. Highland Ave., Aurora, IL, 60507
Mitchell Industrial Tire Co. (MITCO), 1400 E. 40th St., Chattanooga, TN, 37407
Mixing Equipment Co., A Unit of General Signal, 135 Mt. Read Blvd., Rochester, NY, 14603
Mobile Drilling Co., Inc., 3807 Madison Ave., Indianapolis, IN, 46227
Mobile Oil Corp., 150 E. 42nd St., New York, NY, 10017
Modern Engineering Co., P.O. Box 14858, St. Louis, MO, 63178
Modern Welding Co., 2880 New Hartford Rd., Owensboro, KY, 42301
Modular Technology Corp., P.O. Box 6, Plato Center, IL, 60170
Molded Dimensions Inc., 701 Sunset Rd., Pt. Washington, WI, 53074
Monitor Manufacturing, Drawer AL, Elburn, IL, 60119
Monitrol Mfg. Co. Inc., P.O. Box 6296, Tyler, TX, 75711
Mono Group Inc., 847 Industrial Dr., Bensenville, IL, 60106
Monogram Industries, Inc., Jet-O-Matic Div., 1945 E. 223rd St., Long Beach, CA, 90810
Monsanto Co., 800 N. Lindbergh Blvd., St. Louis, MO, 63166
Montgomery Elevator Co., 30 20th St., Moline, IL, 61265
Montreal Engineering Co. Ltd., P.O. Box 6088, Station A, Montreal Que., H3C 3Z8 Ca
Moog Inc., Industrial Div., P.O. Box 8, E. Aurora, NY, 14052
Moore Co., Inc., The, P.O. Box 753, Charleston, WV, 25323
Moore Industrial Battery Co., 4312-20 Spring Grove Ave., Cincinnati, OH, 45223
Morgantown Machine & Hydraulics, Inc., Div. Natl. Mine Service Co., P.O. Box 986, Morgantown, WV, 26505
Morris Pumps, Inc., 31 E. Genesee St., Baldwinsville, NY, 13027
Morrison-Knudsen Co., Inc., One Morrison-Knudsen Plaza, Box 7808, Boise, ID, 83729
Morse Bros. Machinery Co., 1290 Harlan St., Denver, CO, 80214
Morse Chain, Div. of Borg-Warner Corp., So. Aurora St., Ithaca, NY, 14850
Morton Salt, Div. Morton-Norwich, 110 N. Wacker Dr., Chicago, IL, 60606
Motoren-Werke Mannheim AG, Postbox 1563, D-6800 Manheim 1, Fed. Rep. Germany
Motorola Inc., Communications Group, 1301 E. Algonquin Rd., Schaumburg, IL, 60196
Multi-Amp Corp., 4271 Bronze Way, Dallas, TX, 75237
Muncie Power Products, P.O. Box 548, Muncie, IN, 47305
Myers, F.E., Company, 400 Orange St., Ashland, OH, 44805
Myers-Whaley Co., P.O. Box 4265, Knoxville, TN, 37921

N

NL Bearings/NL Industries, P.O. Box 934, Toledo, OH, 43694
NL Industries, Inc., 1230 Ave. of the Americas, New York, NY, 10020
Nagle Pumps, Inc., 1249 Center Ave., Chicago Heights, IL, 60411
Nalco Chemical Co., 2901 Butterfield Rd., Oak Brook, IL, 60521
Nash Engineering Co., 310 Wilson Ave., Norwalk, CT, 06856
National Air Vibrator Co., 6880 Wynnwood Lane, Houston, TX, 77008
National Car Rental Systems Inc., Mudcat Div., P.O. Box 16247, St. Louis Park, MN, 55416
National Electric Cable, Div. National Electric Control Co., 1730 Elmhurst Rd., Elk Grove Village, IL, 60007
National Electric Coil Div. of McGraw-Edison Co., 941 Chatham Lane, Suite 301, Columbus, OH, 43221
National Engineering Co., 20 North Wacker Dr., Suite 2060, Chicago, IL, 60606
National Environmental Instruments, Inc., P.O. Box 590, Pilgrim Station, Warwick, RI, 02888
National Filter Media Corp., 1717 Dixwell Ave., Hamden, CT, 06514
National Foam System Inc., 150 Gordon Dr., Lionville, PA, 19353
National Iron Co., 50 Ave. W. & Ramsey St., Duluth, MN, 55807
National Mine Service Co., 4900/600 Grant St., Pittsburgh. PA, 15219
National Photographic Laboratories, Inc., 1926 W. Gray, Houston, TX, 77019
National-Standard Co., Perf. Metals Div., 166 Dundaff St., Carbondale, PA, 18407
National Supply Co., Div. of Armco Inc. 1455 W. Loop South, Houston, TX, 77027
Native-Plants, 360 Wakara Way, Salt Lake City, UT, 84108
Nature Seeds, Inc., P.O. Drawer 438, McBee, SC, 29101
Naylor Pipe Co., 1265 E. 92 St., Chicago, IL, 60619
Neese Industries, Inc., P.O. Box 628, Gonzales, LA, 70737
New Concepts Inc., 2470 N. Jackrabbit, Tucson, AZ, 85703
Newport News Industrial, Sub. of Newport News Shipbuilding, 230 41st St., Newport News, VA, 23607
Nicholson Engineered Systems, P.O. Box 11336, Ft. Worth, TX, 76109
Niles Expanded Metals, 403 No. Pleasant Ave., Niles, OH, 44446
Nissan Industrial Equipment Co., P.O. Box 16104, Memphis, TN, 38116
Nolan Co., The, Box 201, Bowerston, OH, 44695
Non-Fluid Oil Corp., 298 Delancy St., Newark, NJ, 07105
North American Galis Co., Rte. 7 East, P.O. Box 3158, Morgantown, WV, 26505
North American Hydraulics, Inc., P.O. Box 15431, Baton Rouge, LA, 70895
North American Mfg. Co., 4455 E. 71st St., Cleveland, OH, 44105
North American Mining Consultants, Inc., One Penn Plaza, 250 W. 34 St., New York, NY, 10001
North American Roller, Inc., P.O. Box 215, Stapleton, AL, 36578
Northwest Engineering Co., 201 West Walnut St., Green Bay, WI, 54305
Norton Co., 1 New Bond St., Worcester, MA, 01606
Norton Co., Safety Products Div., 2000 Plainfield Pike, Cranston, RI, 02920
Norwood, Inc., 8507 Perry Highway, Pittsburgh, PA, 15237
Numonics Corp., 418 Pierce St., Lansdale, PA, 19446
NUS Corp., 4 Research Pl., Rockville, MD, 20850

O

O&K Orenstein & Koppel, Karl-Funke-Str. 30, D-4600 Dortmund 1, W. Germany
O&K Orenstein & Koppel Inc., 700 Route 46, P.O. Box 479, Clifton, NJ, 07015
O&K Orenstein & Koppel (Canada) Ltd., 21 Hatt St., Dundas, Ont., Canada, L9H 5P9
OTC Power Team Div. of Owatonna Tool Co., North St., Owatonna, MN, 55060
OCENCO, Inc., P.O. Box 8, Rt. 22 East, Blairsville, PA, 15717
Ohio Brass Co., a Sub. of Harvey Hubbell Inc., 380 N. Main St., Mansfield, OH, 44902
Ohio Carbon Co., 12508 Berea Rd., Cleveland, OH, 44111
Ohio Gear, Box 238, Liberty, SC, 29657
Ohio River Co. The, P.O. Box 1460, Cincinnati, OH, 45201
Ohio Thermal, Inc., 7030-A Huntley Rd., Columbus, OH, 43229
Ohio Transformer Corp., P.O. Box 191, Louisville, OH, 44641
Ohmart Corp., 4241 Allendorf Dr., Cincinnati, OH, 45209
Oil Center Research, 320 Heymann Boulevard, Lafayette, LA, 70503
Okonite Co., P.O. Box 340, Ramsey, NJ, 07446
Oldham Batteries Ltd., Nelson Street, Denton, Manchester M34 3AT, England
Olin Chemicals Group, Olin Corp., 120 Long Ridge Rd., Stamford, CT, 06904
Onan Corp., 1400 73rd Ave., N.E., Minneapolis, MN, 55432
Onox, Inc., 240 Hamilton Ave., Palo Alto, CA, 94301
Ore Reclamation Co., 301 N. Connell Ave., Picher, OK, 74360
Ortner Freight Car Co., 2652 Erie Ave., Cincinnati, OH, 45208
Ortran, P.O. Box 705, Superior, WI, 54880
Oshkosh Truck Corp., P.O. Box 2566, Oshkosh, WI, 54903
Osmose Wood Preserving Co. of America Inc., 980 Ellicott St., Buffalo, NY, 14209
Outokumpu Oy, Technical Export Div., P.O.B. 27, 02201 Espoo 20, Finland
Over-Lowe Co., Inc., 2767 S. Tejon, Englewood, CO, 80110
Owens Mfg., Inc., P.O. Box 1490, Bristol, VA, 24201

P

P&R Systems Corp., 3460 Lexington Ave. No., St. Paul, MN, 55112
PHB Material Handling Corp., 7 Pearl Court, Allendale, NJ, 07401
PLM, Inc., 50 California St. San Francisco, CA, 94111
PPG Industries, Inc., Chemicals Group, One Gateway Center, Pittsburgh, PA, 15222
Pace Transducer Co., Div. of C.J. Enterprises, P.O. Box 834, Tarzana, CA, 91356
Paceco, Inc., a Sub. of Fruehauf Corp., 2320 Blanding Ave., Alameda, CA, 94501
Padley & Venables Ltd., Callywhite Lane, Dronfield, Sheffield S18 6XT, England
Page Engineering Co., Clearing P.O., Chicago, IL, 60638
Pall Corp., 30 Sea Cliff Ave., Glen Cove, NY, 11542
Palm Industries, Box 562, Litchfield, MN, 55355
Parker-Hannifin Corp., Hose Products Div., 30240 Lakeland Blvd., Wickliffe, OH, 44092
Parker-Hannifin Corp., Power Units Div., 930 Penn Ave., Orrville, OH, 44667
Parker-Hannifin Corp., Tube Fittings Div., 17325 Euclid Ave., Cleveland, OH, 44112
Parkson Corp., P.O. Box 24407, Ft. Lauderdale, FL, 33307
Parsch Inc., Union City, PA, 16438
Patent Scaffolding Co., 2125 Center Ave., Fort Lee, NJ, 07024
Patterson-Kelley Co., Div. of Harsco Corp., 132 Burson St., East Stroudsburg, PA, 18301
Pattin Manufacturing Co., Div. The Eastern Co., P.O. Box 659, Marietta, OH, 45750
Paulsen Wire Rope Corp., 440 Josephine St., New Orleans, LA, 70130
Paurat GmbH, Nordstrasse, 4223 Voerde 2, W. Germany
Peabody Barnes, 615 N. Main St., Mansfield, OH, 44902
Peabody Galion, P.O. Box 607, Galion, OH, 44833
Peabody International, 722 Post Rd., Darien, CT, 06820
Peabody Intl. Corp., Peabody ABC Group, P.O. Box 77, Warsaw, IN, 46580
Peerless Conveyor & Mfg. Co., Inc., 3341 Harvester Rd., Kansas City, KS, 66115
Peerless Div. Lear Siegler, Inc., P.O. Box 447, Tualatin, OR, 97062
Peerless Hardware Mfg. Co., 210 Chestnut St., Columbia, PA, 17512
Peerless Pump, 1200 Sycamore St., Montebello, CA, 90640
Pekor Iron Works, Inc., P.O. Box 909, Columbus, GA, 31902
Pemco Corp., Box 1338, Bluefield, WV, 24701
Penetone Corp., 74 Hudson Ave., Tenafly, NJ 07670
Penh Machine Co., 106 Station St., Johnstown, PA, 15905
Pennekamp & Hueseker KG, Kunstsoffwerk, P.O. Box 126, 4426 Vreden, West Germany
Pennsylvania Crusher Corp., P.O. Box 100, Broomall, PA, 19008
Pennwalt Corp., Keystone Div., 21st & Lippincott St., Philadelphia, PA, 19132
Pennwalt Corp., Sharples Div., Three Parkway, Philadelphia, PA, 19102
Penreco, Div. Pennzoil Co., 106 S. Main St., Butler, PA, 16001
Pente Industries Inc., P.O. Box 92, Worthington, OH, 43085
Perard Engineering Ltd., Brittain Dr., Codnor Gate Ind. Estate, Ripley, Derbyshire DE5 3QB England
Perfection Electronic Products Corp., 1530 Rochester Rd., Box 280A, Royal Oak, MI, 48068
Perkin-Elmer Corp., Main Ave., Norwalk, CT, 06856
Permco Inc., 1500 Frost Rd., Streetsboro, OH, 44240
Persingers Inc., P.O. Box 1886, 520 Elizabeth St., Charleston, WV, 25327
Peterson Filters Corp., P.O. Box 606, Salt Lake City, UT, 84110
Petrolite Corp., Tretolite Div., 369 Marshall Ave., St. Louis, MO, 63119
Petron Corp., 16700 Glendale Dr., New Berlin, WI, 53151
Pettibone Corp., 4710 W. Div. St., Chicago, IL, 60651
Pettibone Corp., Pettibone New York Div., 1212 E. Dominick St., Rome, NY, 13440
Phelps Dodge Industries, Inc., P.O. Box 1126, Wall St. Station, New York, NY, 10005

Philadelphia Gear Corp., 181 S. Gulph Rd., King of Prussia, PA, 19406
Philadelphia Resins Corp., 20 Commerce Dr., Montgomeryville, PA, 18936
Philippi-Hagenbuch, Inc., 7500 West Plank Rd., Peoria, IL, 61604
Phillips Driscopipe, Inc., 12200 Ford Rd., Suite 400, Dallas, TX, 75234
Phillips Mine & Mill, Inc., P.O. Box 70, Bridgeville, PA, 15017
Phillips Stamping Co., Inc., 18th St., Bellaire, OH, 43906
Phillipsburg Brake Lining Warehouse, P.O. Box 669, Phillipsburg, PA, 16866
Phoenix Products Co., Inc., 4715 North 27th St., Milwaukee, WI, 53209
Pipe Benders, Inc., P.O. Box 396, Duluth, MN, 55801
Pipe Systems, Inc., P.O. Box 137, Fenton, MO, 63026
Pitcraft Summit Ltd., Platts Common Industrial Estate, Hoyland Nether, Barnsley, S. Yorkshire, U.K.
Pitman Mfg., Co., Div. A.B. Chance Co., P.O. Box 120, Grandview, MO, 64030
Pittsburgh Corning Corp., 800 Presque Isle Dr., Pittsburgh, PA, 15239
Pittsburgh Forgings Co., Coraopolis, PA, 15108
Plastic Techniques of Pennsylvania, R.D. #3, Box 91, Clarks Summit, PA, 18411
Plibrico Company, 1800 Kingsbury St., Chicago, IL, 60614
Plymouth Locomotive Works, Inc., Plymouth Locomotive Div., Bell & High Sts., Plymouth, OH, 44865
Plymouth Rubber Co., Inc., 51 Revere St., Canton, MA, 02021
Pneumafil Corp., P.O. Box 16348, Charlotte, NC, 28216
Poclain, 60330 LePlessis, Belleville, France
Poly-Hi/Menasha Corp., P.O. Box 9086, Ft. Wayne, IN, 46899
Poly Pipe Industries, Inc., Box 1219, West Highway 82, Gainesville, TX, 76240
Portadrill, Inc., 3811 Joliet St., P.O. Box 39-P, Denver, CO, 80239
Porta Space, Inc., P.O. Box 515, Cockeysville, MD, 21030
Portec, Inc., Cast Products Div., P.O. Box 75, Kingsbury, IN, 46345
Portec, Inc., Pioneer Div., 3200 Como Ave., S.E. Minneapolis, MN, 55414
Porter, H. K. Co., Inc., Porter Bldg., Pittsburgh, PA, 15219
Porto Pump, Inc., 19735 Ralston, Detroit, MI, 48203
Post Glover, Inc., Box 709, Covington, KY, 41012
Potter & Brumfield Div., AMF Inc., 200 Richland Creek Dr., Princeton, IN, 47671
Power Transmission Div., Dresser Industries, 400 W. Wilson Bridge Rd., Worthington, OH, 43085
Power Transmission Equipment Co., 3839 S. Normal Ave., Chicago, IL, 60609
Precision National Corp., P.O. Box 789, Mt. Vernon, IL, 62864
Preco, Inc., P.O. Box 449, Boise, ID, 83701
Preiser/Mineco Div., Preiser Scientific Inc., Jones & Oliver St., St. Albans, WV, 25177
Premier Pneumatics, 606 North Front, Salina, KS, 67401
Prestolite Battery Div. of Eltra Corp., 511 Hamilton St., Toledo, OH, 43694
Prestolite Electronics Div. of Eltra Corp., P.O. Box 931, Toledo, OH, 43694
Prestolite Motor Div. of Eltra Corp., P.O. Box 931, Toledo, OH, 43694

Prestolite Wire Div. of Eltra Corp., 3529 24th St., Port Huron, MI, 48060
Prince Mfg. Co., 700 Lehigh St., Bowmanstown, PA, 18030
Process Equipment Builders, Inc., P.O. Box 1479, Paducah, KY, 42001
Prosser Industries, Div. of Purex Corp., P.O. Box 3818, Anaheim, CA, 92803
Prox Company, Inc., P.O. Box 1484 1201 S. 1st St., Terre Haute, IN, 47808
Pullman Torkelson Co., 5077 W. Wiley Post Way, Salt Lake City, UT, 84116
Pulmosan Safety Equip. Co., 30-48 Linden Pl., Flushing, NY, 11354
Pulverizing Machinery, Div. of MikroPul Corp., 102 Chatham Rd., Summit, NJ, 07901
Pure Carbon Co., Inc., 441 Hall Ave., St. Marys, PA, 15857
Pyle National Co., 1334 North Kostner, Chicago, IL, 60651
Pyott-Boone Electronics, P.O. Box 809, Tazewell, VA, 24651
Pyott-Boone Manufacturing, Drawer U, Saltville, VA, 24370

Q

Quest Electronics, 510 Worthington St., Oconomowoc, WI, 53066
Qunicy Compressor Div., Colt Industries, 217 Maine St., Quincy, IL, 62301

R

RCA Corp., 30 Rockefeller Plaza, New York, NY, 10020
RCA Mobile Communications Systems, Product Info Center, Meadow Lands, PA, 15347
RKL Controls Inc., Ark Road, Lumberton, NJ, 08048
Racal Airstream, Inc., 1151 Seven Locks Rd., Rockville, MD, 20854
Raco International, Inc., 3350 Industrial Blvd., Bethel Park, PA, 15102
Racor Industries, Inc., Filtration Div., P.O. Box 3208, Modesto, CA, 95353
RAHCO, R. A. Hanson Co., W. 601 Main, Suite 305, Spokane, WA, 99201
Railweight, Inc., 1701 Nicholas Blvd., Elk Grove Village, IL, 60007
Ramsey Engineering, Co., 1853 W. County Rd. C., St. Paul, MN, 55113
Ransomes & Rapier Ltd., P.O. Box 1, Waterside Works, Ipswich 1P2 8HL, England
Rapid Electric Co., Inc., Grays Bridge Rd., Brookfield, CT, 06804
Raybestos Friction Materials Co. (a Raybestos Manhattan Co.), 100 Oakview Dr., Trumbull, CT, 06611
Raybestos Manhattan Industrial Products Co., Box 5205, No. Charleston, SC, 29406
Raychem Corp., 300 Constitution Dr., Menlo Park, CA, 94025
RayGo, Inc., P.O. Box 1362, Minneapolis, MN, 55440
Red Valve Co., Inc., 500 Bell Ave., Carnegie, PA, 15106
Red Wing Shoe Co., Inc., 419 Bush St., Red Wing, MN, 55066
Redding Co., James A., 615 Washington Rd., Pittsburgh, PA, 15228

Reed Mining Tools, Inc., P.O. Box 90750, Houston, TX, 77090
Reedrill, Inc., Box 998, Sherman, TX, 75090
Reggie Industries, Inc., 122 W. 22nd St., Ste. 332, Oak Brook, IL, 60521
Reinco, Inc., P.O. Box 584, Plainfield, NJ, 07061
Reiss Viking Corp., P.O. Box 3336, 1300 Georgia Ave., Bristol, TN, 37620
Reliance Electric, 25001 Tungsten Rd., Cleveland, OH, 44117
Rema-Tech, 200 Paris Ave., Northvale, NJ, 07647
Renold Inc., P.O. Box A, Westfield, NY, 14787
Republic National Bank of Dallas, Dallas, TX, 75200
R. Republic Steel Corp., P.O. Box 6778, 1441 Cleveland, Cleveland, OH
Research-Cottrell, Inc., P.O. Box 1500, Somerville, NJ, 08876
Research Energy of Ohio, 237 Charleston St., P.O. Box 312, Cadiz, OH, 43907
Resisto-Loy Co., 1251 Phillips Ave., S.W., Grand Rapids, MI, 49507
Revere Corp. of America, North Colony Rd., Wallingford, CT, 06492
Rexarc, Inc., Rexarc Place, West Alexandria, OH, 45381
Rexnord Inc., P.O. Box 2022, Milwaukee, WI, 53201
Rexnord Inc., Bearing Div., P.O. Box F, 2400 Curtiss St., Downers Grove, IL, 60515
Rexnord Inc., Corp. Relations Dept., P.O. Box 2022, Milwaukee, WI, 53201
Rexnord Inc., Process Machinery Div., Box 383, Milwaukee, WI, 53201
Rexnord Inc., Resin Systems Div., 6120 E. 58th Ave., Commerce City, CO, 80022
Rexnord Inc., Vibrating Equipment Div., 3400 Fern Valley Rd., Louisville, KY, 40213
Reynolds Metals Co., P.O. Box 27003, Richmond, VA, 23261
Rheinbraun-Consulting GmbH P.O.B. 41 07 62 Stuttgenweg 2, 5000 Colonge 41, Fed. Rep. of Germany
Rhino Sales Corp., 620 Andrews Ave., Kewanee, IL, 61443
Richards Industries, 4 Fairfield Crescent, W. Caldwell, NJ, 07006
Richmond Machine Co., Amer. Reducer Div., Richmond & Ontario Sts., Philadelphia, PA, 19134
Richmond Mfg. Co., P.O. Box 188, Ashland, OH, 44805
Ridge Tool Co., Sub. of Emerson Electric Co., 400 Clark St., Elyria, OH, 44036
Riley Stoker Corp., P.O. Box 547, Worcester, MA, 01613
Rimpull Corp., Box 748, U.S. 169 South, Olathe, KS, 66061
Ripco, Inc., 251 S. 3rd St., Oxford, PA, 19363
Rish Equipment Co., 100 Bluefield Ave., Bluefield, WV, 24701
Robbins Co., The, 650 S. Orcas St., Seattle, WA, 98108
Robbins Div., Joy Mfg. Co., 300 Fleming Rd. (P.O. Box 6505), Birmingham, AL, 35217
Robbins & Myers, Inc., 1345 Lagonda Ave., Springfield, OH, 45501
Roberts & Schaefer Co., 120 S. Riverside Plaza, Chicago, IL, 60606
Robicon Corp., 100 Sagamore Hill Rd., Plum Ind. Park, Pittsburgh, PA, 15239
Robins Engineers & Constructors (Hewitt-Robins), 711 Union Blvd., Totowa, NJ, 07511

Robinson Industries, Inc., P.O. Box 100, Zelienople, PA, 16063
Robinson & Robinson Div., NUS Corp., 1517 Charleston National Plaza, Charleston, WV, 25321
Rochester Corp., P.O. Box 312, Culpeper, VA, 22701
Rock Master Inc., 1022 Santerre, Grand Prairie, TX, 75050
Rock Tools, Inc., 8 Columbine Lane, Littleton, CO, 80123
Rockbestos Co., The, Nicoll and Canner Sts., New Haven, CT, 60504
Rockwell International Automotive Operations, 2135 W. Maple Rd., Troy, MI, 48084
Rockwell International Flow Control Div., 400 N. Lexington Ave., Pittsburgh, PA, 15208
Rockwell International, Power Tool Div., 400 N. Lexington Ave., Pittsburgh, PA, 15208
Rohm and Haas Co., Independence Mall West, Philadelphia, PA, 19105
Roller Corp., P.O. Box 12606, Pittsburgh, PA, 15241
Rome Cable Corp., 421 Ridge St., Rome, NY, 13440
Roof Control Systems, 1161 Murfreesboro Rd., Suite 320, Nashville, TN, 37217
Rose Mfg. Co., 2250 S. Tejon St., Englewood, CO, 80110
Rost, H. & Co., P.O. Box 90 11 68, D-2100 Hamburgh 90, W. Germany
Rousselle Corp., Davey/Rousselle Drill Rig Div., 2310 W. 78th St., Chicago, IL 60620
Rung, D. G., Industries, Inc., 780 S. Michigan St., Seattle, WA, 98108
Rust Engineering Co., Sub. of Wheelabrator-Frye, P.O. Box 101, 1130 South 22nd St., Birmingham, AL, 35201
Rust-Oleum Corp., 11 Hawthorn Parkway, Vernon Hills, IL, 60061
Ruttmann Companies, 425 W. Walker St., P.O. Box 120, Upper Sandusky, OH, 43351
Ryerson, Joseph T., & Son, Inc., P.O. Box 8000A, Chicago, IL, 60680

S

S & S Corp., Route 3, Box 70, Cedar Bluff, VA, 24609
SGS Control Service, Inc., Minerals & Chemicals Div., 17 Battery Pl., No., New York, NY, 10004
SKF Industries, Inc., 1100 First Ave., King of Prussia, PA, 19406
Safetran Systems Corp., Mining & Urban Transit, Signal & Control Div., 4780 Crittenden Dr., Louisville, KY, 40221
Sala International, S-733 00 Sala, Sweden
Sala Machine Works Ltd., 3136 Mavis Rd., Mississauga, Ont., Canada, L5C 1T8
Salem Tool Co., The, 767 S. Ellsworth Ave., Salem, OH, 44460
Salisbury & Dietz, Inc., S. 1815 Lewis St., Spokane, WA, 99204
Sanford-Day/Marmon Transmotive, Div. of the Marmon Group, Inc., P.O. Box 1511, Gov. John Sevier Hwy., Knoxville, TN, 37901
Sangamo Weston, Inc., P.O. Box 48400, Atlanta, GA, 30362
Sauerman Bros., Inc., 620 S. 28th Ave., Bellwood, IL, 60104
Savage, W. J. Co., 912 Clinch Ave., S.W., Knoxville, TN, 37901

MINING EQUIPMENT MANUFACTURERS

Scandura Inc., 1801 N. Tryon St., Charlotte, NC, 28230
Schaefer Brush Mfg. Co., 117 W. Walker St., Milwaukee, WI, 53204
Scharf Co., 1010 Ohio River Blvd., Pittsburgh, PA, 15202
Schauenburg Flexadux Corp., 12 A Buncher Ind. Dist., Leetsdale, Pittsburgh, PA, 15056
Scholten, T. H. & Co., Postfach 204, 4020 Mettmann, Germany, 02104-2727
Schopf Maschinenbau GmbH, P.O. Box 750-360, D7 Stuttgart 75, W. Germany
Schramm Inc., 800 E. Virginia Ave., Dept. F-1, West Chester, PA, 19380
Schroeder Bros. Corp., Nichol Ave., Box 72, McKees Rocks, PA, 15136
Scott Aviation, A Div. of A-T-O, Inc., 225 Erie St., Lancaster, NY, 14086
Screen Equipment Co., Div. Hobam Inc., 40 Anderson Rd., Buffalo, NY, 14225
SEI Ltd., Peel Works, Barton Lane, Eccles, Manchester, England
Seiberling Tire & Rubber Co., 345 15th St., N.W., P.O. Box 189, Barberton, OH, 44203
Seminole Products Co., Inc., Box 123, Glendora, NJ, 08029
Semperit of America, Inc., 156 Ludlow Ave., Northvale, NJ, 07647
Senior Conflow, P.O. Box 265, Clinton, PA, 15026
Sepor, Inc., 718 N. Fries, Wilmington, CA, 90748
Serpentix Conveyor Corp., 1550 S. Pearl St., Denver, CO, 80210
Service Supply Co., Inc., 603 E. Washington St., Indianapolis, IN, 46206
Servus Rubber Co., 1136 Second St., Rock Island, IL, 61201
Seton Name Plate Corp., 1654 Boulevard, New Haven, CT, 06505
Sevcon, Div. Technical Operations, Inc., 40 North Avenue, Burlington, MA, 01803
Shamy Coal Systems Inc., P.O. Box 09574, Bexley, OH, 43209
Shaw-Almex Industries Ltd., P.O. Box 430, Parry Sound, Ont., Canada, P2A 2X4
Shell Oil Co., One Shell Plaza, Houston, TX, 77002
Shingle, L. H., Co., 500 Gravers Rd., Plymouth Meeting, PA, 19462
Shwayder Co., 2335 E. Lincoln, Birmingham, MI, 48008
Siemens-Allis Inc., P.O. Box 2168, Milwaukee, WI, 53201
Siemens Corp., 186 Wood Ave., South, Iselin, NJ, 08830
Sigmaform Corp., Mine Products Div., 1233 Burdette Ave., Evansville, IN, 47715
Sii Smith-Gruner, Mining & Industrial Products, Div. of Smith Intl. Inc., Drawer 911, Ponca City, OK, 74601
Simmons Machinery Co., P.O. Drawer B, Birmingham, AL, 35073
Simonacco Ltd., Durranhil, Carlisle CA1 3ND, England
Simon-Carves of Canada Ltd., Two Lansing Square, Willowdale, Ont., Canada, M2J 1W2
Simplicity Engineering, 212 S. Oak St., Durand, MI, 48429
Sioux Steam Cleaner Corp., Route 46 & I-29, Beresford, SD, 57004
Skelly and Loy, 2601 N. Front St., Harrisburg, PA, 17110
Sly, W. W., Mfg. Co., P.O. Box 5939, Cleveland, OH, 44101
Smico Corp., 500 N. MacArthur Blvd., Oklahoma City, OK, 73127
Smit, J. K., & Sons, Inc., 571 Central Ave., Murray Hill, NJ, 07974
Smith, A. O. Inland, Inc. Reinforced Plastics Div., 2700 West 65th St., Little Rock, AR, 72209
Smith, Elwin G. Div., Cyclops Corp., 100 Walls St., Pittsburgh, PA, 15202
Smith Tool, 17871 Von Karman Ave., Irvine CA, 92714
Snap-On Tools Corp., 2801 80th St., Kenosha, WI, 53140
Snap-tite, Inc., 201 Titusville Rd., Union City, PA, 16438
Snorkel Div. of ATO Inc., Box 65 Stock Yards Station, St. Joseph, MO, 64504
Solitest, Inc., 2205 Lee St., Evanston, IL, 60202
Solarflo Corp., 22901 Aurora Rd., Bedford Hts., OH, 44146
Solids Flow Control Corp., 4 Fairfield Crescent, West Caldwell, NJ, 07006
Solidur Plastics Co., 200 Plum Industrial Court, Pittsburgh, PA, 15239
Sollami Co., P.O. Box 627, Herrin, IL, 62948
Sonic Development Corp., 305 Island Rd., Mahwah, NJ, 07430
Sortex Co. of North America, Inc., P.O. Box 160, Lowell, MI, 49331
Southern Tire Co., 1414 Broadway, Sheffield, AL
Spang & Co., P.O. Box 751, Butler, PA, 16001
Spatz Paint Industries, Inc., 1601 N. Broadway, St. Louis, MO, 63102
Speakman Co., P.O. Box 191, Wilmington, DE, 19899
Specialty Chemicals & Plastics, Union Carbide Corp., 270 Park Ave., New York, NY, 10017
Specialty Services, Inc., 6152 Steeplechase Dr., S.W., Salem, VA, 24153
Spectrum Infrared Inc., 246 E. 131st St., Cleveland, OH, 44108
Spendrup Fan Co., 746 Ouray Ave., Grand Junction, CO, 81501
Sperry Vickers Div., Sperry Corp., 1401 Crooks Rd., Troy, MI, 48084
Sperry Vickers, Tulsa Div., P.O. Box G, Tulsa, OK, 74112
Spider Inc., 4001 Gratiot, St. Louis, MO, 63110
Sprague & Henwood, Inc., 221 W. Olive St., Scranton, PA, 18501
Spraying Systems Co., North Ave. at Schmale Rd., Wheaton, IL, 60187
Sprengnether Instruments, Inc., 4567 Swan Ave., St. Louis, MO, 63110
Sprout-Waldron Div., Koppers Co., Inc., Muncy, PA, 17756
Square D Co., Executive Plaza, Palatine, IL, 60067
St. Regis Paper Co., 150 E. 42nd St., New York, NY, 10017
Stamler, W. R., Corp., The, 600 Main St., Millersburg, KY, 40348
Stanadyne/Hartford Div., Box 1440, Hartford, CT, 06102
Stanco Mfg. & Sales Inc., 310 E. Alton Ave., Santa Ana, CA, 92707
Standard Laboratories, Inc., 3322 Pennsylvania Ave., Charleston, WV, 25302
Standard Metal Mfg. Co., 110 Main St., P.O. Box 57, Malinta, OH, 43535
Stanford Seed Co., 809 N. Bethlehem Pike, Spring House, PA, 19477
Stauffer Chemical Co., Specialty Chemical Div., Westport, CT, 06880
Stearns Magnetics Inc., Div. of Magnetics Intl., 6001 So. General Ave., Cudahy, WI, 53110

Stearns-Roger, 4500 Cherry Creek Dr., P.O. Box 5888, Denver, CO, 80217
Stedman Fdy. & Mach. Co., P.O. Box 209, Aurora, IN, 47001
Steel Heddle Mfg. Co., Industrial Div., 1801 Rutherford St. (P.O. Box 1867), Greenville, SC, 29602
Steele & Associates, P.O. Box 1797, Huntsville, TX, 77340
Steelplank Corp., 671 Grove St., Wyandotte, MI, 48192
Stephens-Adamson Inc., Ridgeway Ave., Aurora, IL, 60507
Sterling Power Systems, Inc., A Sub. of The Lionel Corp., 16752 Armstrong Ave., Irvine, CA, 92714
Stevens International Inc., P.O. Box 619, Kennett Sq., PA, 19348
Stockham Valve & Fittings, Box 10326, Birmingham, AL, 35202
Stonhard, Inc., Park Ave. & Rte. 73, Maple Shade, NJ, 08052
Stoody Co., Box 1901 CA, Industry, CA, 91749
Stoody Co. WRAP Div., 11804 Wakeman St., Whittier, CA, 90607
Strachan & Henshaw, Inc., P.O. Box 153, 12059 S. Western Ave., Blue Island, IL, 60406
Straight Line Filters, Inc., P.O. Box 1911, Wilmington, DE, 19899
Stratoflex, Inc., P.O. Box 10398, Ft. Worth, TX, 76114
Streeteramet, Div. of Mangood Corp., 155 Wicks St., Grayslake, IL, 60030
Strojexport, pzo, Vaclavske Nam 56, Prag 1, Czechoslovakia, 11326
Stromberg-Carlson Corp., P.O. Box 7266, Charlottesville, VA, 22906
Stryker Machine Products Co., 2560 E. State St., P.O. Box 8067, Trenton, NJ, 08650
Sturtevant Mill Co., 22 Sturtevant St., Dorchester, Boston, MA, 02122
Sullair Corp., 3700 E. Michigan Blvd., Michigan City, IN, 46360
Sun Petroleum Products Co., P.O. Box 7438, Philadelphia, PA, 19101
Sundstrand Fluid Handling, P.O. Box FH, Arvada, CO, 80004
Super Products Corp., P.O. Box 27225, Milwaukee, WI, 53227
Sverdrup Corp., 801 N. Eleventh, St. Louis, MO, 63101
Swan Hose Div., P.O. Box 509, Worthington, OH, 43085
SWECO, Inc., 6033 E. Bandini Blvd., P.O. Box 4151, Los Angeles, CA, 90051

T

TBA Industrial Products Ltd., Spotland Rochdale, Lancs., England
TJB Inc., 19940 Ingersoll Dr., Rocky River, OH, 44116
TRC, 125 Silas Deane Highway, Wethersfield, CT, 06109
T & J Industries Inc., P.O. Box 8620, Kansas City, MO, 64114
T & T Machine Co., Inc., Rte. 8 Box 343, Fairmont, WV, 26554
Taber Pump Co., Inc., P.O. Box 1071, Elkhart, IN, 46514
Tabor Machine Co., P.O. Box 3037, Bluefield Station, Bluefield, WV, 24701
Tamrock Drills, 33310 Tampere 31, Finland
Tanner Systems, Inc., P.O. Box 87, Sauk Rapids, MN, 56379

Taylor Instrument Co., 95 Ames St., Rochester, NY, 14601
Taylor-Wharton, Div. of Harsco Corp., 2900 William Penn Highway, Easton, PA, 18042
Tecnetics Industries, Inc., 4599 N. Chatsworth, St. Paul, MN, 55112
Teledyne McKay, 850 Grantley Rd., York, PA, 17405
Teledyne Wisconsin Motor, 1910 S. 53rd St., Milwaukee, WI, 53219
Telsmith Div., Barber-Greene Co., 532 E. Capitol Dr., Milwaukee, WI, 53212
Tema Inc., a Siebtechnik Co., 4015 Executive Park Dr., Cincinnati, OH, 45241
Templeton, Kenly & Co., 2525 Gardner Rd., Broadview, IL, 60153
Terex Div., GMC, Hudson, OH, 44236
Terrell Industries, P.O. Box 5467, Huntington, WV, 25703
Terrell Machine Co., Industrial Products Div., P.O. Box 240868, Charlotte, NC, 28224
Teton Big Hole Drillers, Inc., P.O. Drawer A-1, Casper, WY, 82602
Texaco Inc., 2000 Westchester Ave., White Plains, NY, 10650
Texas Instruments Inc., P.O. Box 144, M/S 6705, Houston, TX, 77001
Texas Nuclear, 9101 Research Rd. (P.O. Box 9267), Austin, TX, 78757
Thayer Scale Hyer Industries, Rt. 139, Pembroke, MA, 02359
Thiele, Inc., Route 56 at Spruce St., Windber, PA, 15963
Thomas Foundries Inc., P.O. Box 96, Birmingham, AL, 35201
Thor Power Tool Co., 175 N. State St., Aurora, IL, 60507
Throwaway Bit Corp., 624 N.E. Everett, Portland, OR, 97232
Thurman Scale Co. Div. Thurman Mfg. Co., 1939 Refugee Rd., Columbus, OH, 43215
Thyssen (Great Britain) Group of Co.'s Bynea, Llanelli, Dyfed. SA14 9SU, U.K.
Thyssen Mining Equipment, 928 Washington Trust Bldg., Washington, PA, 15301
Thyssen Mining Equipment Div., 400 E. DeYoung St., Marion, IL, 62959
Tidewater Supply Co., 330 Third Ave., Huntington, WV, 25706
Tiger Equipment & Services, 111 W. Jackson Blvd., Chicago, IL, 60604
Timberland Equipment Ltd., Box 490, Woodstock, Ont., N4S 7Z2 Canada
Timken Co., The, 1835 Dueber Ave., S.W., Canton, OH, 44706
Titan Mfg. Co. Pty. Ltd., Woodstock St., Mayfield, NSW 2304, Australia
Toledo Scale, P.O. Box 1705, Columbus, OH, 43216
Tol-O-Matic, 1028 S. 3rd St., Minneapolis, MN, 55415
Tonemaster Mfg. Co., 8727 N. Pioneer Rd., Peoria, IL, 61614
Topcon Instruments Corp. of America, 9 Keystone Pl., Paramus, NJ, 07652
Torit Div. Donaldson Co. Inc., P.O. Box 1299, Minneapolis, MN, 55116
Torque Tension Ltd., Claylands Ave., Workshop, Notts., England, S81 7BQ
Torrington Co., The Bearings Div., 59 Field St., Torrington, CT, 06790
TOTCO, 600 Rock Creek Rd. E., Norman, OK, 73069

Townley Engineering & Mfg. Co., Inc., P.O. Box 221, Candler, FL, 32624
Toyo Tire (USA) Corp., 3136 E. Victoria St., Compton, CA, 90221
Tracy, Bertrand P. Co., 919 Fulton St., Pittsburgh, PA, 15233
Transall Div., Dick-Precismeca, Inc., 200 Industrial Rd., Alabaster, AL, 35007
Transamerica Delaval Inc., Wiggins Connectors Div., 5000 Triggs St., Los Angeles, CA, 90022
Tread Corp., P.O. Box 13207, Roanoke, VA, 24032
Treadwell Corp., 1700 Broadway, New York, NY, 10019
Trelleborg AB, Box 501, S-231 01 Trelleborg, Sweden
Trelleborg, Inc., 30700 Solon Ind. Pkw. Solon, OH, 44139
Triangle/PWC, Inc., A Sub. of Triangle Industries, Inc., Box 711,Triangle & Jersey Aves., New Brunswick, NJ, 08903
Trico Mfg. Corp., 1235 Hickory St., Pewaukee, WI, 53072
Tricon Metals & Services, Inc., P.O. Box 6634, Birmingham, AL, 35210
Trojan Div. I.M.C. Chemical Group, Inc., 666 Garland Pl., Des Plaines, IL, 60016
Trojan Industries, Inc., Trojan Circle, Batavia, NY, 14020
Trowelon, Inc., 973 Haven Dr., P.O. Box 3126, Green Bay, WI, 54303
TRW Bearings Div., 402 Chandler St., Jamestown, NY, 14701
TRW/J. H. Williams Div., 400 Vulcan St., Buffalo, NY, 14207
TRW Mission Mfg. Co., Div. of TRW Inc., P.O. Box 40402, Houston, TX, 77040
Tube-Lok Products Div. of Portland Wire & Iron, 4644 S.E. 17th Ave., Portland, OR, 97202
Tube Turns Div. of Allegheny Ludlum Industries, 2900 W. Broadway, Louisville, KY, 40211
Turmag (G.B.) Ltd., Whalley Rd., Barnsley, S. Yorks, S75 1HT England
TWECO Products, Inc., P.O. Box 12250, Wichita, KS, 67277
Twin Disc, Inc., 1328 Racine St., Racine WI, 53403
Twisto-Wire Fire Systems, Inc., 1010 Norumbega Dr., Monrovia, CA, 91016

U

U.I.P. Engineered Products Corp., 2020 Estes Ave., Elk Grove Village, IL, 60007
Underground Mining Machinery Ltd., P.O. Box 19, Aycliffe Industrial Estate, Newton Aycliffe, Co. Durham, England, DL5 6DS
Unilok Belting Co., Div. of Georgia Duck and Cordage Mill, Scottdale, GA, 30079
Union Carbide Corp., 270 Park Ave., New York, NY, 10017
Union Chain Div., P.O. Box 651, 1010 Edgewater Dr., Sandusky, OH, 44870
Union Iron Works, P.O. Box 248, Warrensburg, IL, 62573
Union Oil of California, 1650 E. Golf Rd., Schaumburg, IL, 60196
Union Switch & Signal Div., Amer. Standard Inc., 1789 Braddock Ave., Pittsburgh, PA, 15218
Uniroyal, Inc., P.O. Box 1126, Wall St. Station, New York, NY, 10005

Uniroyal Industrial Protective Products, Benson Rd., Middlebury, CT, 06751
Unit Crane & Shovel Corp., 1915 South Moorland Rd., New Berlin, WI, 53151
Unit Rig & Equipment Co., P.O. Box 3107, Tulsa, OK, 74101
United Bank of Denver, Energy & Minerals Group, 1740 Broadway, Denver, CO, 80217
United Detector Technology, 2644 30th St., Santa Monica, CA, 90405
United McGill Corp., 2 Mission Park, P.O. Box 85, Groveport, OH, 43125
U.S. Electrical Motors Div. Emerson Electric Co., 125 Old Gate Lane, Milford, CT, 06460
U.S. Filter Corp., 12442 E. Putnam St., Whittier, CA, 90602
U.S. Gypsum Co., 101 S. Wacker Dr., Chicago, IL, 60606
U.S. Steel Corp., 600 Grant St., Pittsburgh, PA, 15230
U.S. Thermit Inc., Ridgeway Blvd., Lakehurst, NJ, 08733
United Tire & Rubber Co. Ltd., 275 Belfield Rd., Rexdale, Ont., Canada, M9W5C6
Uni-Tool Attachments, Inc., 1607 Woodland Ave., Columbus, OH, 43219
Universal Atlas Cement Co., 600 Grant St., 12th Fl., Pittsburgh, PA, 15230
Universal Engineering Corp., 625 C. Ave., N.W., Cedar Rapids, IA, 52405
Universal Industries, 1575 Big Rock Rd. West, Waterloo, IA, 50701
Universal Road Machinery Co., 27 Emerick St., Kingston, NY, 12401
Universal Vibrating Screen Co., P.O. Box 1097, 1745 Deane Blvd., Racine, WI, 53405
Utex Industries, Inc., 5200 Clinton Dr., Houston, TX, 77020

V

VME-Nitro Consult Inc., 8707 Skokie Blvd., Skokie, IL, 60077
Valenite, Mining Products Div., P.O. Box D, Tri-County Industrial Park, Piney Flats, TN, 37686
Valley Steel Products Co., P.O. Box 503, St. Louis, MO, 63166
Van Gorp Corp., Sub. Emerson Elec. Co., Box 123, Pella, IA, 50219
Varel Mfg. Co., 9230 Denton, Dallas, TX, 75220
Varian Associates, 611 Hansen Way, Palo Alto, CA, 94303
Vereinigte Edelstahlwerke AG, Schreyvogelgasse 2, A-1010 Vienna, Austria
Vibco Inc., P.O. Box 8, Stilson Rd., Wyoming, RI, 02898
Vibranetics, Inc., 7310 Grade Lane, Louisville, KY, 40219
Vibra Screw Inc., 755 Union Bldg., Totowa, NJ, 07511
Victaulic Co. of America, P.O. Box 31, Easton, PA 18042
Victor Products (Wallsend) Ltd., P.O. Box Wallsend, Tyne and Wear NE28 6PP, England
Viking Machinery Co., Rt. 8, Orebank Rd., Kingsport, TN, 37664
V/O Machinoexport, Mosfilmovskaja, 35, Moscow, USSR, 117330
Voest-Alpine Intl. Corp., 60 E. 42nd St., 46 Fl., New York, NY, 10017
Voith Turbo KG, Postfach 460, D-7180 Crailsheim, W. Germany

Vortex Air Corp., P.O. Box 928, Beckley, WV, 25801
VR/Wesson a Div. of Fanstell, One Tantalum Pl., Waukegan, IL, 60064
Vulcan Material Co., Conveyor Belt Dept., P.O. Box 7324-A, Birmingham, AL, 35253

W

WABCO Construction & Mining Equip., a div. of Amer. Standard Inc., 2300 N.E. Adams St., Peoria, IL, 61639
WABCO Fluid Power Div., an American-Standard Co., 1953 Mercer Rd., Lexington, KY, 40505
Wachs, E. H., Co., 100 Shepard St., Wheeling, IL, 60090
Wagner Mining Equip. Co., P.O. Box 20307, Portland, OR, 97220
Wajax Industries Ltd., 350 Sparks St., Ste. 1105, Ottawa, Ont., Canada, K1R 7S8
Waldon Inc., Fairview, OK, 73737
Walker Parkersburg Textron, 620 Depot St., Parkersburg, WV, 26101
Wall Colmonoy Corp., 19345 John R. St., Detroit, MI, 48203
Wallacetown Engineering Co. Ltd., Heathfield Rd., Ayr KA89 SR, Scotland
Walter Nold Co., 24 Birch Rd., Natick, MA, 01760
Warman International, Inc., 2701 S. Stoughton Rd., Madison, WI, 53716
Warn Industries Inc., 19450 68th Ave., So., Kent, WA, 98031
Warner & Swasey Construction Equipment, Gradall Div., 406 Mill Ave., S.W., New Philadelphia, OH, 44663
Warren Pumps Div., Houdaille Industries, Inc., 783 Bridges Ave., Warren, MA, 01083
Warren Rupp Co., The, P.O. Box 1568, Mansfield, OH, 44901
Watt Car & Wheel Co., Box 71, Barnesville, OH, 43713
Waukesha Engine Div., 1000 St. Paul Ave., Waukesha, WI, 53186
Wear Technology Div., Cabot Corp., 1020 W. Park Ave., Kokomo, IN, 46901
Weatherhead Div., Dana Corp., 300 E. 131st St., Cleveland, OH, 44108
Webb, Jervis B., Co., Webb Dr., Farmington Hills, MI, 48018
Webster Mfg. Co., W. Hall St., Tiffin, OH, 44883
Wedge Wire Corp., P.O. Box 157, Wellington, OH, 44090
Weir, Paul, Co., Inc., 20 N. Wacker Dr., Chicago, IL, 60606
Wellman, S.K., Corp., The, 2000 Egbert Rd., Bedford, OH, 44146
WEMCO Div., Envirotech Corp., P.O. Box 15619, Sacramento, CA, 95813
Wen-Don Corp., P.O. Box 13905, Roanoke, VA, 24034
Wescott Steel Inc., 425 Andrews Rd., Trevose, PA, 19047
Weserhuette Inc., Railway Exchange Bldg., 611 Olive, Ste. 1762, St. Louis, MO, 63101
WESMAR Industrial Systems Div., 905 Dexter Ave. N., Box C19074, Seattle, WA, 98109
West Virginia Armature Co., P.O. Box 1100, Bluefield, WV, 24701
West Virginia Belt Sales & Repairs Inc., P.O. Box 32, Mount Hope, WV, 25880
Western Precipitation Div., Joy Mfg. Co., P.O. Box 2744, Terminal Annex, Los Angeles, CA, 90051

Westfalia Lunen, Industrie Str., D4670 Lunen, P.O. Box, Germany
Westinghouse Credit Corp., Three Gateway Center, Pittsburgh, PA, 15222
Westinghouse Electric Corp., Westinghouse Bldg., Gateway Center, Pittsburgh, PA, 15222
Westlake Plastics Co., Lenni Rd., Lenni, PA, 19052
Wheel Trueing Tool Co., P.O. Box 1317, Columbia, SC, 29202
Wheelabrator-Frye Inc., Air Pollution Control Div., 600 Grant St., Pittsburgh, PA, 15219
Wheelabrator-Frye, Inc., Materials Cleaning Systems, 1476 S. Brykit St., Mishawaka, IN, 46544
Wheeling Corrugating Co., Div. of Wheeling-Pittsburgh Steel Corp., Wheeling, WV, 26003
Wheeling-Pittsburgh Steel, Four Gateway Center, Pittsburgh, PA, 15230
White Engines, Inc., 101-11th St., S.E. Canton, OH, 44707
White Motor Corp., 35129 Curtis Blvd., Eastlake, OH, 44094
Whiting Corp., 15700 Lathrop, Harvey, IL, 60426
Whitmore Mfg. Co., The, P.O. Box 488, Cleveland, OH, 44127
Whittaker Corp., 10880 Wilshire Blvd., Los Angeles, CA, 90024
Wichita Clutch Co., Industrial Group, Dana Corp., 2800 Fisher Rd., P.O. Box 1550, Wichita Falls, TX, 76307
Wiegand, Edwin L. Div. Emerson Elec. Co. #4 Allegheny Center, 10th Fl., Pittsburgh, PA, 15212
Wild Heerbrugg Insts. Inc., 465 Smith St., Farmingdale, NY, 11735
Wilfley, A. R. & Sons, P.O. Box 2330, Denver CO, 80201
Williams Patent Crusher & Pulv. Co., 810 Montgomery St., St. Louis, MO, 63102
Willis & Paul Corp., The, 400 Morris Ave., Denville, NJ, 07834
Wilmont Engineering Co., Berwick St., White Haven, PA, 18661
Wilson, R. M. Co., Box 6274, Wheeling, WV, 26003
Wing Co., Div. of Wing Industries, 125 Moen Ave., Cranford, NJ, 07016
Winslow Scale Co., P.O. Box 1523, Terre Haute, IN, 47808
Wire Rope Corp. of America, Box 288, St. Joseph, MO, 64502
Wisconsin Protective Coating Corp., P.O. Box 3396, Green Bay, WI, 54303
Witco Chemical Corp., 277 Park Ave., New York, New York, 10017
Wood's T. B. Sons Co., 440 N. Fifth Ave., Chambersburg, PA, 17201
Workman Developments, Inc., 1741 Woodvale Rd., Charleston, WV, 25314
Worthington Pump Inc., 270 Sheffield St., Mountainside, NJ, 07092
Wyandotte, Div. of The Diversey Corp., 1532 Biddle Ave., Wyandotte, MI, 48192

X

XTEK, Inc., 211 Township Ave., Cincinnati, OH, 45216
Xtek: Pipe, 11175 Reading, Cincinnati, OH, 45241

Y

Yardney Electric Corp., 82 Mechanic St., Pawcatuck, CT, 02891

Yokohama Rubber Co. Ltd., c/o Marubeni America Corp., 200 Park Ave., New York, NY, 10017

Yokohama Tire Corp., 1530 Church Rd., Montebello, CA, 90640

Young Corp., Box 3522, Seattle, WA, 98124

Youngstown Sheet & Tube Co., The, Post Office Box 900, Youngstown, OH, 44501

Z

Zeni Drilling Co., 324 Eighth St., Morgantown, WV, 26505

Zweigniederlassung der Salzgitter Maschinen und Aniagen AG, Postfach 1220, D-4408 Dulmen Westfalen; W. Germany

Index

A

Active listening, 237
American Mining Congress (AMC), 28, 41
Anchorage in
    belt drive installations, 154–155
    belt line installations, 166
    limestone, 9
    partings or splits, 7
    sandstones, 9
    shale or siltstone, 9

B

Bang-boards
    chutes, 139
    designs, 137–139
    materials, 139
    sideboards, 139
Belt drives, 144–150
Belt line extension
    materials, 173
    splicing the belt, 174
    stringing the structure, 174
    tensioning the ropes, 175
    tools, 171
Belt structure, 166–170
    training, 180
Belt wipers and cleaners
    costs, 142
    designs, 142

C

Capital investment decisions
    alternative choices, 263–264
    decision variables, 264–265
    discounted payback method, 267–268
    net present value method, 265–266
    payback method, 265–267
Cleats. *See* Joints.
Coal Mine Health and Safety Act of 1969
    (CMHSA), 15, 16, 18, 19–20, 32, 63,
    199
Computer design tools
    mainframe software, 241–243
    mini, micro software, 252–253
Conglomerates. *See* Lithology.

Conveyor belt, 132–133
    capacity, 132
    typical mine lengths, 144
Conveyor Equipment Manufacturers Association
    (CEMA), 131–132, 166–167, 143
Cost
    control process, 268
    definitions, 259–260
    estimating
        drainage, 275–276
        general approach, 270–272
        haulage, 276–278
        project systems, 279
        transportation, 272–274
        ventilation, 274–275
    measuring
        direct cost, 261–262
        indirect cost, 262–263
    nature of project costs, 260–261
Critical Path Method (CPM), 230, 234
    fundamentals, 243–247
    networks for support construction
        haulage, 249–250
        pumping, 248
        transportation, 247–248
        ventilation, 248
    techniques via computer, 250–252
Crusher stations, 197
    design alternatives, 216–218

D

Drainage, 1
    definitions, 1
    designing effective systems, 99–113
    gas affecting projects, 12
    mining height affecting projects, 12
    seam pitch affecting projects, 14
    water affecting projects, 13
    weathering of projects, 14

E

Engineering and construction firms, 197, 201,
    220–221

F

Faults, 6–7
    dip, 6
    footwall, 6
    hanging wall, 6
Fire clay. *See* Lithology.
Folds, 3
    anthracite folding, 5
    drag, 3, 5
    monocline, 3
    structural terrace, 3
Foreman office, 197
    design alternatives, 218–220
Full-entry doors
    costs, 94
    materials, 94

G

Gantt chart, 230
Geologic parameters, 1, 3–15
    definition, 3
    effect on construction decisions, 198
    strata characteristics, 9–11
    tectogenic stresses, 3–8

H

Haulage, 1
    belt line crossovers, 182
    definition, 1
    gas affecting projects, 12
    mining height affecting projects, 12
    seam pitch affecting projects, 15
    sensor and deluge systems, 143–144
    spillage, 134–135
    water affecting projects, 13
    weathering of projects, 14
Haulage construction
    belt drive, head roller, and takeup, installing
        definitions, 144–150
        installation, 153–158
        operation and maintenance, 158
        preparation, 150–153
    belt line, installing
        belt line extension, 171–177
        design parameters, 163–171
        operation and maintenance, 179–182
        tailpiece assembly, 177–179
    design factors, 131–133
    efficient system design, 133–144
    equipment manufacturers, 183–196
    transfer points, installing
        building, 160–161
        definitions, 158–159
        design parameters, 159–160
        operation and maintenance, 161–162
Headroller template, 157
Horsebacks. *See* Intrusions.

I

Inherent strata characteristics, 3
    definitions, 3
    engineering properties, 3
    rock lithology, 3
Intrusions
    depositional, 7
    injective, 7

J

Joints, 3, 5–6
    butt cleats, 5–6
    face cleats, 5–6

L

Limestone and dolomite. *See* Lithology.
Line of Balance (LOB), 230, 234
Lithology
    conglomerates, 9, 10
    definition, 9
    limestone and dolomite, 9, 10
    sandstones, 6, 9
    siltstones, shales, and argillites, 6, 9

M

Management, construction, 225
    communications, 227–228
    organization, 225–227
    project control, 230–231
    responsibilities, 228–230
    safety, 231–233
    training, 231, 233
Man doors
    costs, 91–93
    materials, 91
Master Design Simulator, 241
Middleman. *See* Partings and splits.
Mining parameters
    effect on construction decisions, 198
    equipment, 199
    labor, 199
    material, 199
Motivation, worker, 234
    individual needs, 236
    paternalistic, 236
    reward and punishment, 236
    self-actualization, 236
    worker satisfiers, 236
Mudstone. *See* Shale.

O

Operational parameters, 21–23
    cost, 22
    labor, 22
    planning and management, 23
    project material, 22
    safety, 21–22
    supply, 22
Overcasts
    components, 71
    ramping operation, 73
    rib hitching, 78
    types, 65–68
    weathering and degradation, 81–82

P

Partings and splits, 7
Physical parameters
    definitions, 11
    effects on construction decisions, 198
    gas, 12–13
    humidity and weathering, 13–14
    mining height, 11–12
    seam pitch, 14–15
    water, 13
Piping
    joints, 109–110
    materials, 108
Plane of weakness. *See* Unconformity.
Program Evaluation and Review Technique (PERT), 230, 243
Progress Analysis Planning (PAP), 268–269
Pumps
    centrifugal, 107–108
    characteristics, 102–107
    turbine, 102–104
Pump station construction
    definitions, 113–114
    installing the main pump, 118–121
    operation and maintenance, 121–122
    portable pump station, 122–124
    preparing the site, 114–118
    pumping equipment manufacturers, 124–129

R

Railroad construction
    design factors
        clearance, 25–26
        locomotion, 27
        maintenance scheduling, 27
        maximum transported loads, 25
        mining entry, 26
        precautions for transporting equipment, 31–32
        railcar and traffic, 26
        rail weight and tie spacing, 28–31
        roadbed, 26–27
    equipment manufacturers, 59–61
    track, laying
        assembly, 34–39
        maintenance, 39–41
        preparation, 32–34
    trolley wire, installing
        assembly, 57
        cutting a trolley switch, 57–58
        maintenance, 58–59
        preparation, 54–56
    turnouts, installing
        definitions, 41–46
        installation, 48–53
        preparation, 46–48
RAILSIM, 241–242
Regulators
    designs, 83–85
    setting the, 85
Regulatory parameters
    drainage, 19
    effect on construction decisions, 198–199
    haulage, 20
    transportation, 19–20
    ventilation, 16–19
        doors, 17–18
        overcasts, 16–17
        stopping, 18–19
ROM coal
    characteristics of, 131
Rotary dumps, 197
    design alternatives, 210–216
    equipment manufacturers, 220–221

S

Sandstone. *See* Lithology.
Sealants, 68–70
    mixing machines, 69–70
Shale. *See* Lithology.
Shop, underground, 197
    assembly, 208–210
    maintenance, 210
    preparation, 205–208
Siltstone. *See* Lithology.
Skirtboard, 139, 160–161
Splits. *See* Partings.
Stoppings
    costs, 86, 89
    designs, 86
    hitching, 87, 91
    weathering, 89
Sump, 116–118, 121–122, 123
Supervisory leadership
    directing functions, 237–239
    leadership and coordination, 237

Surge bins, 197
   assembly, 204
   cost parameters, 201
   design concepts, 199–201
   equipment manufacturers, 220–221
   operation and maintenance, 204–205
   preparation, 201–204

T

Tailpiece
   aligning, 178
   anchoring, 178
   breaking down, 177
   bunker, 182
   pulling, 177
Takeup units, 135–137
Tectogenic stresses, 3
   faults, 3, 5
   folds, 3
   intrusions, 3
   joints, 3, 5–6
   partings, 3
   unconformities, 3
Track
   ballast, 27, 39
   bolts, 31
   bonding, 39, 52–53
   butterfly, or rail bender, 42, 52
   clearances, 26
   joints, 31
   laying, *see* Transportation construction
   rail, 28–29
   roadbed
      grading, 33–34
      sites, 32–33
   scheduling the work, 35
   splices, 31
   supply delivery, 34
   ties
      steel, 28–29
      wooden, 29–30
   tools, 35
Transportation, 1
   definition, 1
   gas affecting projects, 12
   mining height affecting projects, 12
   seam pitch affecting projects, 15
   water affecting projects, 13
   weathering of projects, 14
Trolley wire
   accessories, 56, 57, 58
   applicator, 59
   sites, 54
Turnouts
   parts, 41–43
   sites, 46–48

U

Unconformities, 7
US Bureau of Mines (USBM), 9, 10

V

Ventilation, 1
   definition of projects, 1
   gas affecting projects, 12
   mining height affecting projects, 12
   quantity, 64–65
   requirements, 12
   seam pitch affecting projects, 14
   water affecting projects, 13
   weathering of projects, 14
Ventilation construction
   door, constructing
      full-entry, 94–96
      man doors, 91–94
   equipment manufacturers, 96–97
   overcasts, building
      alternative designs, 64–72
      definitions, 63
      floor construction, 79
      foundation construction, 76–78
      maintenance and inspection, 81–82
      sealing, 79–81
      site preparation, 72–76
      wall construction, 78–79
   regulator, constructing
      building, 84–85
      definitions, 82–83
      designs, 83–84
      maintaining, 85–86
   stopping, constructing
      definitions, 86
      permanent, 86–89
      temporary, 89–91
VENTSIM, 242

W

Washouts. *See* Intrusions.